Insect Collection and Identification

Arthropod Collection and Identification, Second Edition

Insect Collection and Identification
Techniques for the Field and Laboratory

Timothy J. Gibb
Department of Entomology, Purdue University, West Lafayette, IN, United States

Christian Oseto
Department of Entomology, Purdue University, West Lafayette, IN, United States

ACADEMIC PRESS
An imprint of Elsevier

Academic Press is an imprint of Elsevier
125 London Wall, London EC2Y 5AS, United Kingdom
525 B Street, Suite 1650, San Diego, CA 92101, United States
50 Hampshire Street, 5th Floor, Cambridge, MA 02139, United States
The Boulevard, Langford Lane, Kidlington, Oxford OX5 1GB, United Kingdom

Notices
Knowledge and best practice in this field are constantly changing. As new research and experience broaden our
understanding, changes in research methods, professional practices, or medical treatment may become necessary.

Practitioners and researchers must always rely on their own experience and knowledge in evaluating and using
any information, methods, compounds, or experiments described herein. In using such information or methods
they should be mindful of their own safety and the safety of others, including parties for whom they have a
professional responsibility.

To the fullest extent of the law, neither the Publisher nor the authors, contributors, or editors, assume any liability
for any injury and/or damage to persons or property as a matter of products liability, negligence or otherwise, or
from any use or operation of any methods, products, instructions, or ideas contained in the material herein.

British Library Cataloguing-in-Publication Data
A catalogue record for this book is available from the British Library

Library of Congress Cataloging-in-Publication Data
A catalog record for this book is available from the Library of Congress

ISBN: 978-0-12-816570-6

For Information on all Academic Press publications
visit our website at https://www.elsevier.com/books-and-journals

Publisher: Charlotte Cockle
Acquisition Editor: Anna Valutkevich
Editorial Project Manager: Ruby Smith
Production Project Manager: Poulouse Joseph
Cover Designer: Christian Bilbow

Cover credit: John Obermeyer, Purdue University

Typeset by MPS Limited, Chennai, India

Working together
to grow libraries in
developing countries

www.elsevier.com • www.bookaid.org

Dedication

To Insects.
For some, dedicating a text to insects may seem absurd but, for us,
insects have shaped the course of our professional lives. We have spent
our entire careers sharing our knowledge about these extraordinary
creatures. Even so, we have only scratched the surface. It is an honor
to dedicate our lives and this book to their study.

Contents

Preface to the second edition xiii

Part 1 Basic tools and general techniques 1

CHAPTER 1 Equipment and collection methods 5
1.1 Equipment 5
    1.1.1 Forceps 5
    1.1.2 Sample vials 6
    1.1.3 Killing bottles 6
    1.1.4 Small containers 6
    1.1.5 Small envelopes 7
    1.1.6 Aspirators 7
    1.1.7 Absorbent tissue 7
    1.1.8 Notebook 7
    1.1.9 Tools for cutting or digging 7
    1.1.10 Brush 8
    1.1.11 Bags 8
    1.1.12 Hand lens 8
    1.1.13 Summary 9
1.2 Collecting nets 9
1.3 Killing containers and agents 13
    1.3.1 Freezing insects 13
    1.3.2 Injecting insects with alcohol 13
    1.3.3 Killing jars for field collecting 14
    1.3.4 Liquid killing agents 15
    1.3.5 Solid killing agents 16
1.4 Aspirators and suction devices 18
1.5 Other collection devices 20
    1.5.1 Beating sheets 20
    1.5.2 Drag cloth 20
    1.5.3 Sifters 20
    1.5.4 Separators and extractors 21
1.6 Traps 24
    1.6.1 Windowpane trap 25
    1.6.2 Interception nets and barriers 26
    1.6.3 Malaise traps 26
    1.6.4 Pitfall and dish traps 27

1.6.5 Emergence traps and rearing cages 29
1.6.6 Traps using the lobster or eel trap principle 30
1.6.7 Light traps 30
1.6.8 Light sheets 33
1.6.9 Color traps 34
1.6.10 Sticky traps 35
1.6.11 Snap traps 35
1.6.12 Artificial refuge traps 36
1.6.13 Electrical grid traps 36
1.6.14 Combination traps for student studies 36
1.7 Baits, lures, and other attractants 36
1.7.1 Sugaring for insects 37
1.7.2 Feces 38
1.7.3 Oatmeal 39
1.8 Pheromones and other attractants 39
1.8.1 Carbon dioxide 40
1.8.2 Sounds 40
1.9 Collecting aquatic insects 40
1.10 Collecting soil insects 42
1.11 Collecting ectoparasites 42
1.12 Collecting regulated insects 42
1.13 Collecting insects for pest management audits 43
1.14 Collecting insects for forensic or medico-criminal investigations 44
1.15 Rearing 45
1.15.1 Containers for rearing 45
1.15.2 Rearing conditions and problems 48
1.16 Collecting insects for molecular research 51
1.17 Preparation of insects for molecular research 52
1.17.1 Killing agents 52
1.17.2 Preservation of specimens 62
1.17.3 Preservatives in traps 65
1.17.4 Pinned specimens (Natural History Museum Collections) 66
1.17.5 Nondestructive methods 68
1.18 Vouchuring specimens 69

CHAPTER 2   Agents for killing and preserving 71

CHAPTER 3   Storage of specimens 73
3.1 Temporary storage 73
3.1.1 Refrigeration 73

3.1.2 Dry preservation...73
3.1.3 Papering..74
3.2 Mounting specimens 75
3.2.1 Preparing dry specimens for mounting 76
3.2.2 Degreasing...78
3.2.3 Preparing liquid-preserved specimens.............................79
3.2.4 Direct pinning 80
3.2.5 Double mounts 83
3.2.6 Pinning blocks..87
3.2.7 Mounting and spreading boards and blocks......................88
3.2.8 Mounting specimens for microscopic examination 96
3.3 Labeling..109
3.3.1 Paper...110
3.3.2 Pens 110
3.3.3 Ink...110
3.3.4 Lettered and printed labels 110
3.3.5 Size of labels...111
3.3.6 Label data 111
3.3.7 Placing the labels 113
3.3.8 Labeling vials 114
3.3.9 Labeling microscope slides 114
3.3.10 Identification labels 115
3.4 Care of the collection...115
3.4.1 Housing the collection 115
3.4.2 Protecting specimens from pests and mold.......................118
3.5 Packaging and shipping specimens.......................................119
3.5.1 Packing materials 119
3.5.2 Pinned specimens...119
3.5.3 Specimens in vials 120
3.5.4 Loading cartons..121
3.5.5 Shipping microscope slides 121
3.5.6 Shipping live specimens 122

Part 2 Classification of insects and mites 125
CHAPTER 4   Classification of insects and mites.....................................**129**
4.1 Key to classes of Arthropoda..132
4.2 Class Arachnida...135
4.2.1 Key to orders of Arachnida 135

4.3 Subclass Acari 139

    4.3.1 Key to some primary groups of the subclass Acari 140

4.4 Classes Diplopoda, Chilopoda, Pauropoda, and Symphyla 142

4.5 Class Crustacea 142

4.6 Class Hexapoda (Insecta) 143

CHAPTER 5 Synopsis of insect orders **147**

5.1 Subclass Entognatha: primitive wingless hexapods 147

5.2 Subclass Ectognatha: primitive wingless hexapods 147

5.3 Subclass Pterygota (Insecta): winged and secondarily wingless
insects 148

    5.3.1 Exopterygota 148

    5.3.2 Endopterygota 150

5.4 Key to orders of hexapoda (Insecta) 151

CHAPTER 6 Descriptions of hexapod orders **187**

6.1 Protura 187

6.2 Diplura 187

6.3 Collembola 188

6.4 Microcoryphia 189

6.5 Thysanura 190

6.6 Ephemeroptera 190

6.7 Odonata 192

6.8 Orthoptera 193

6.9 Blattodea (Blattaria) 195

6.10 Mantodea 196

6.11 Phasmatodea (Phasmida) 197

6.12 Grylloblattodea (Grylloblattaria) 198

6.13 Dermaptera 199

6.14 Isoptera 200

6.15 Embiidina (Embioptera) 202

6.16 Plecoptera 203

6.17 Psocoptera 205

6.18 Zoraptera 206

6.19 Phthiraptera 207

6.20 Thysanoptera 208

6.21 Hemiptera 209

6.22 Coleoptera 212

6.23 Strepsiptera 215

6.24 Mecoptera 216

6.25 Neuroptera 217

6.26 Trichoptera 219

6.27 Lepidoptera 220

6.28 Diptera 224

6.29 Siphonaptera 228

6.30 Hymenoptera 229

Summary 235

Appendix I: Liquid preservation formulas...237

Appendix II: Guidelines for mounting small and soft-bodied specimens
(Systematic Entomology Laboratory) 239

Appendix III: Directory state extension service directors and administrators,
March 2019 243

Appendix IV: Submitting specimens for identification to Systematic Entomology
Laboratory Communications & Taxonomic Services Unit 263

Glossary 271

Bibliography 303

Index 331

Preface to the second edition

More than 100 years have passed since C.V. Riley (1892) published *Directions for Collecting and Preserving Insects*. That work marked the starting point in North America of publications on the collection, preservation, and curation of insects and mites. Many of these works are now unavailable, and some of the concepts contained in them involve obsolete technology.

In 1986, *Insects and Mites: Techniques for Collection and Preservation* (Steyskal et al., 1986) was published by the U.S. Department of Agriculture. The book was written by George Steyskal, a research entomologist specializing in the taxonomy of Diptera; William Murphy, information specialist; and Edna Hoover, a materials processor for the Systematic Entomology Laboratory. Their avid interest in collecting and curating insects is contagious and much of their original work is reflected throughout this book. Michael E. Schauff published an electronic version of the original text together with several useful updates. We recognize his valuable contributions in the printed publication of this text.

In *Arthropod Collection and Identification Techniques*, we have updated information and techniques as well as added new material to make the information more useable and complete. This book describes effective methods and equipment for collecting, identifying, rearing, examining, and preserving insects and mites and for storing and caring for specimens in collection. It also provides instructions for the construction of many kinds of collecting equipment, traps, rearing cages, and storage units, as well as updated and illustrated keys for identification of the classes of arthropods and the orders of insects. Such information not only aids hobbyists and professionals in preparing insect collections, but it has become essential in documenting and standardizing the collection of entomological evidence in forensic as well as pest management sciences.

Insect Collection and Identification: Techniques for the Field and Laboratory second edition provides detailed and up-to-date taxonomic keys to insects and related arthropods given recent classification changes to various insect taxa. Revised preservation materials and techniques for forensic, molecular and genomic studies are also included.

We hope that this work will be a useful reference and resource for anyone fascinated by the world of "bugs."

Timothy J. Gibb and Christian Oseto
Entomologists, Purdue University, West Lafayette, IN, United States

Basic tools and general techniques

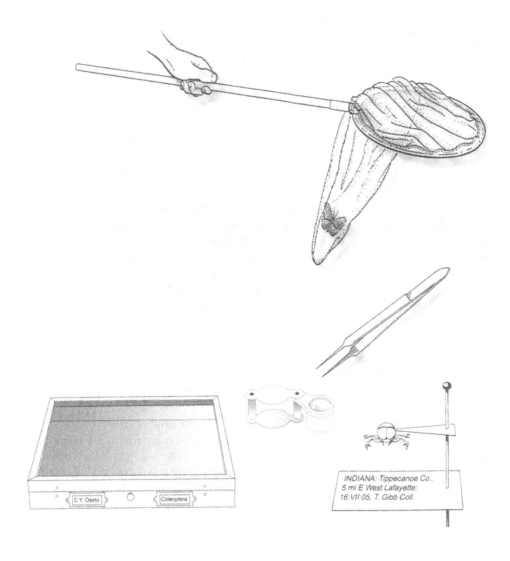

INDIANA: Tippecanoe Co.,
5 mi E West Lafayette;
16:VII:05, T. Gibb Coll.

C.Y. Oseto   Coleoptera

Introduction 2
Chapter 1: Equipment and collection methods 5
Chapter 2: Agents for killing and preserving 71
Chapter 3: Storage of specimens 73

Introduction

Many reasons exist to collect and preserve insects. Hobbyists, nature enthusiasts, amateur collectors, 4-H or high school students, scientists, and criminal investigators each have different purposes to collect insects. Regardless of the purpose, however, insects collected must be preserved and processed according to established protocols. For example, a box of collected insects without accompanying collection information is of little scientific value. On the other hand, properly collected and preserved insects accompanied by collecting data can be invaluable.

Insects have long been considered indicators of environmental disturbance or pollution. Types and numbers of insects found in various habitats over time provide a measure of change within a system if documented correctly. For example, aquatic insects are sensitive indicators of water pollution. Collecting and preserving aquatic samples requires different collection techniques than other types of insect collecting. Proper equipment and techniques must be understood and used if results are to be of value. Over time, a historical baseline of insects and arthropod species composition and population numbers can be established. Environmentalists look for significant changes from established records as evidence of changes (both positive and negative) to the environment in which they live.

In urban pest management, insect audits have become an extremely important tool in establishing the relative health of buildings. Pharmaceutical and food manufacturing plants and warehouses are some of the places where the presence of insect must be carefully monitored. Insect-infested products, where a single insect or even as much as a part of an insect is detected, can mean the loss of millions of dollars in potential revenues. Collecting and identifying insects in these sensitive situations requires great care and precision and a deep and working understanding of the life history and behavior of the insects in question.

New pests continue to threaten invasion. Counties, states, and countries spend many billions of dollars each year to prevent insects or other potentially harmful pests from invading and establishing themselves. Pests can be unintentionally introduced into new areas via commerce or tourism or by natural dispersion. Most often, when pests arrive in a new habitat, they cause extreme damage because natural controls, which may keep the pests in check in the original areas, do not exist in new habitats, thus allowing the pests to flourish unchecked. Intensive eradication efforts are sometimes implemented in an effort to exterminate the pest once detected. Regulatory officials have been more successful by either thoroughly regulating the movement of infested materials or by monitoring all commodities originating from infested regions. In either case, collecting and preserving insects and related pests at ports of entry can help educate people and provide valuable leads in determining how and where to deny pest entry.

Insects found in food products eaten by unsuspecting people are often the subject of serious litigation, and the outcome of a lawsuit often rests on the collection and preservation of the insect sample as evidence. Likewise, insects that damage structures, agriculture or food, or other commodities can result in the loss of millions of dollars. Those who control these pests often come under considerable pressure when insect inspections show infestations. Proper collection and preservation of such insects is essential.

Insects found in hospitals, where sterile environments are critical or where fully dependent patients are housed, can literally mean the difference between life and death in some instances. Insects that bite or sting cause annoyance, human sickness, and/or death. Insect-transmitted diseases continue to be a leading cause of death in the world. Accurate identification, monitoring, and controlling the spread of these insects are assisted by collecting and maintaining arthropods in a proper manner.

Criminal investigations also can hinge on insects as evidence if they are collected and preserved appropriately. Death scene investigators are very aware that natural succession of insects infesting a corpse can provide extremely valuable information in a death investigation. The species and stages of insects found, together with the temperatures in the surrounding areas, can be used to estimate the location as well as the time of death. Variations from the norm can indicate unusual sequences of postmortem events, but they stand up in a court of law only if the evidence is collected and preserved according to established protocol. Opportunities to collect insects at a death scene occur only once, so entomologists must be trained, prepared, and equipped adequately. In these cases, chain of custody and proper labeling of the evidence are especially critical.

Collecting insects from any habitat or for whatever purpose requires an understanding of the insects' specific behavior and ecological needs. A collector must understand why and where an insect lives in order to find and collect it. Various traps, flushing agents, and other technical tools may help in this effort. Rearing adult insects from immature life stages is sometimes appropriate, depending on the purpose of the information. Labeling the collected materials with date, precise location, and collector's name is a minimum. Often, identifying the host or describing the behavior of the insects at the time of collection is also valuable. Sometimes collecting associated materials, such as damaged leaves, wood, cast skins, or fecal material, in addition to the insect is needed. Signs or symptoms of infestations and specific behaviors of the insects just prior to collection are often required. Detailed written accounts or photographs can provide very important information, especially when some time has elapsed. Remember that providing too much information is seldom a problem. *Having only partial information to document a situation that occurred in the past is a much more common frustration.*

Part 1 is devoted to describing the tools and techniques used to collect arthropods and the proper handling of specimens after they have been obtained. Suggested reading *features references that provide further detail for selected topics.*

Equipment and collection methods

Insect collection methods may be divided into two broad categories. In the first, a collector actively finds and collects the insects with the aid of nets, aspirators, beating sheets, or any other apparatus that suits the particular needs. In the second, a collector participates passively and permits traps to do the work. Both approaches may be used simultaneously, and are discussed in the following pages. Using as many different collection methods as possible will permit a collector to obtain the greatest number of specimens in the shortest period of time.

Catching specimens by hand may be the simplest method of collecting; however, this method is not always productive because of the evasive behavior of many insects. Some insects are not active at times and places that the collector finds convenient. Some insects cause injury or discomfort to the collector through bites, stings, repulsive chemicals, or urticating setae. Often, special equipment and methods of catching are needed. Equipment and methods described here have general applications. Advanced studies of specific insect or mite groups have developed unique procedures for collecting and surveying. For example, Agosti (2001) outlines procedures for surveying ground-nesting ant biodiversity. Clever collectors will make adaptations to fit their specific purposes and resources.

The equipment used to assemble an insect or mite collection is not necessarily elaborate or expensive. In many instances, a collecting net and several killing bottles will suffice. However, additional items will permit more effective sampling of a particular fauna. Many collectors carry a bag or wear a vest in which they store equipment. The following items are usually included in the general collectors' bag.

1.1 Equipment

1.1.1 Forceps

Fine, lightweight forceps are strongly recommended for any collector. Specialized forceps may be selected depending upon individual needs (Fig. 1.1). Lightweight spring-steel forceps are designed to prevent crushing of fragile and small insects. Extrafine precision may be achieved with sharp-pointed "watchmaker" forceps; however, care must be taken not to puncture specimens. When possible, grasp specimens with the part of the forceps slightly behind the points. Curved forceps often make this easier. When the forceps are not in use, their tips should be protected. This can be accomplished by thrusting the tips into a small piece of Styrofoam or cork, or by using a small section of flexible tubing as a collar.

Insect Collection and Identification. DOI: https://doi.org/10.1016/B978-0-12-816570-6.00001-0

1.1.2 Sample vials

Sample vials of various sizes containing alcohol or other preservatives are necessary for collecting many species and life stages of insects and mites. Leak-proof caps are recommended for both field and permanent storage (Fig. 1.2).

1.1.3 Killing bottles

Killing bottles or killing jars in various sizes are important to preserve specimens quickly (Fig. 1.3).

1.1.4 Small containers

Small crush-proof containers are necessary for storing and protecting specimens after their removal from killing bottles (Fig. 1.4). These containers may be made of cardboard, plastic, or metal and should be partly filled with soft tissue paper to keep specimens from damage. Some collectors do not recommend the use of cotton in storage containers because specimens become entangled in the fibers and may become virtually impossible to extricate without damage. However, some collectors of minute or fragile insects find that specimens stored in a few wisps of cotton are better protected from damage.

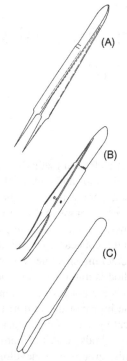

FIGURE 1.1

Forceps for insect collection: (A) fine watchmaker forceps; (B) curved metal collecting forceps; and (C) soft forceps.

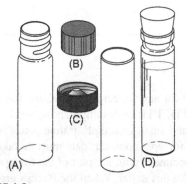

FIGURE 1.2

Sample vials.

FIGURE 1.3

An example of a killing bottle appropriately affixed with a "Poison" label.

1.1.5 Small envelopes

Small envelopes are useful for temporary storage of delicate specimens (Fig. 1.5). Specially designed glassine envelopes, which prevent undue dislodging of butterfly and moth scales, are available from biological supply houses.

1.1.6 Aspirators

Aspirators are necessary for collecting many kinds of small-bodied or agile insects and mites.

1.1.7 Absorbent tissue

Absorbent tissue is highly recommended for use in killing bottles and aspirators.

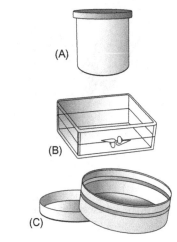

FIGURE 1.4

Examples of small crush-proof collecting containers: (A) empty film canister; (B) plastic box; and (C) tin pill box.

1.1.8 Notebook

A notebook and writing equipment are essential for jotting down notes and label data (Fig. 1.6).

1.1.9 Tools for cutting or digging

A knife, plant clippers, or both are necessary for opening galls, seed pods, twigs, and other kinds of plant material. In addition, a small gardener's trowel (Fig. 1.7) for some kinds of excavation and a heavy knife or small hatchet

FIGURE 1.5

Glassine envelope designed for field storage of moths and butterflies.

FIGURE 1.6

An example of a field journal for recording specimen label data and other notes.

FIGURE 1.7

A gardener's trowel for excavating insects.

FIGURE 1.8

A fine-tip camel's hair brush.

may be helpful for searching under bark or in decaying logs.

1.1.10 Brush

A small, fine brush (camel's hair is best) is needed to aid in collecting minute specimens (Fig. 1.8). Moistening the tip of the brush allows tiny specimens to adhere to it, and they may then be quickly transferred to a killing bottle or vial.

1.1.11 Bags

Using bags for retrieving plant material, rearing material, or Berlese's samples is a good idea (Fig. 1.9). Remember that samples stored in plastic may decompose within a few hours. Samples must be transferred to more permanent containers immediately upon returning from the field. For collecting much plant material, a botanist's vasculum (tin box) is advisable.

1.1.12 Hand lens

A hand lens is helpful and will quickly become an indispensable aid to collectors (Fig. 1.10). A lens worn on a lanyard is convenient and prevents its loss while in the field.

FIGURE 1.9

Bags for collecting miscellaneous plant or soil materials from the field.

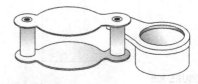

FIGURE 1.10

A collapsible hand lens.

1.1.13 **Summary**

This list may be modified according to the types of insects or mites to be collected. For example, a plant press may be needed to prepare plant specimens for determination or as voucher specimens, especially when leaf-mining insects are studied. When collecting at night, a flashlight or headlamp is essential: the latter is especially useful because it leaves the hands free.

Much of the basic collecting equipment may be obtained from ordinary sources, but equipment especially designed for collecting insects often must be bought from biological supply houses. Their addresses may be found on the Internet or in the yellow pages of telephone directories under "Biological Laboratory Supplies" or "Laboratory Equipment and Supplies." Biological and entomological publications often carry advertisements of equipment suppliers. Because these firms are located in many parts of the country and change names and addresses fairly often, it is not practical to list them here. Biologists at a local university usually can recommend a supplier in their area.

1.2 **Collecting nets**

Collecting nets come in three basic forms: aerial, sweeping, and aquatic (Fig. 1.11). The aerial net is designed especially for collecting butterflies and large-bodied flying insects. Both the bag and the handle are relatively lightweight. The sweeping net is similar to the aerial net, but the handle is stronger and the bag is more durable to withstand dragging through dense vegetation. The aquatic net is used for gathering insects from water and is usually made of metal screening or heavy scrim with a canvas band affixed to a metal rim. A metal handle is advisable because wooden handles may develop slivers after repeated wetting. The choice of net depends on the type of insects or mites intended for collection.

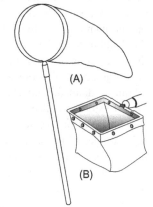

Several types of nets, including collapsible models with interchangeable bags, are available from biological supply houses, but anyone with a little mechanical engineering ability can make a useful net. The advantage of a homemade sweep net is that its size and shape can be adapted to the needs of the user, to the kind of collecting intended, and to the material available. Net-constructing materials include the following.

1. A length of heavy (8-gauge) steel wire for the rim, bent to form a ring 30−38 cm in diameter (Fig. 1.12). Small nets 15 cm or smaller in diameter sometimes are useful, but nets larger than 38 cm are too cumbersome for most collecting.

FIGURE 1.11

Insect-collecting nets: (A) aerial or sweeping net and (B) aquatic net.

2. A strong, light fabric, such as synthetic polyester, through which air can flow freely. Brussels netting

FIGURE 1.12

A common design for the rim of an insect-collecting net made of steel wire.

FIGURE 1.13

General shape of the bag on an aerial or sweep net.

is best, but it may be difficult to obtain; otherwise, nylon netting, marquisette, or good-quality cheesecloth can be used. However, cheesecloth snags easily and is not durable. The material should be double-folded and should be 1.5–1.75 times the rim diameter in length (Fig. 1.13). The edges should be double-stitched (French seams).

3. A strip of muslin, light canvas, or other tightly woven cloth long enough to encircle the rim. The open top of the net bag is sewn between the folded edges of this band to form a tube through which the wire rim is inserted (Fig. 1.14).

4. A straight, hardwood dowel about 19 mm in diameter and 105–140 cm long (to suit the collector). For attachment of the rim to the handle, a pair of holes of the same diameter as the wire are drilled opposite to each other to receive the bent tips of the wire, and a pair of grooves as deep and as wide as the wire are cut from each hole to the end of the dowel to receive the straight part of the wire (Fig. 1.15).

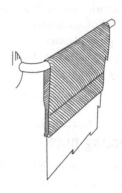

FIGURE 1.14

Attachment of the net to the rim of a collecting net.

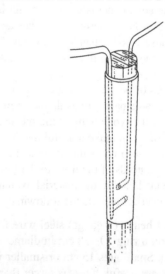

FIGURE 1.15

How to make a handle for an insect net.

5. A tape or wire to lash the ends of the rims tightly into the grooves in the end of the handle. This may be electrician's plastic tape or fiber strapping tape commonly used for packaging. If a wire is chosen, the ends should be bound with tape to secure them and to keep

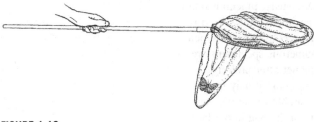

FIGURE 1.16

Proper procedure showing a bag flipped over the rim of a collecting net to prevent escape of the insect.

them from snagging. A close-fitting metal sleeve (ferrule) may be slipped over the rim ends and held in place with a small roundheaded screw instead of tape or wire lashing.

After the net is placed on the rim, the ends of the band should be sewn together and the rim ends should be fastened to the handle. The other end of the handle should be filed to remove sharp edges. The net is then ready for use.

Efficient use of a net is gained only with experience. Collection of specimens in flight calls for the basic stroke: Swing the net rapidly to capture the specimen and then follow through to force the insect into the very bottom of the bag. Twist the wrist as you follow through so that the bottom of the bag hangs over the rim (Fig. 1.16); this will entrap the specimen. If the insect is on the ground or on any other surface, it may be easier to use a downward stroke, quickly swinging down on top of the specimen. With the rim of the net in contact with the ground to prevent the specimen from escaping, hold the tip of the bag up with one hand. Most insects will fly or walk upward into the tip of the bag, which can then be flipped over the rim to entrap the specimen.

Sweeping the net through vegetation, along the sand and seaweed on beaches, or up and down tree trunks will catch many kinds of insects and mites. An aerial net may be used in this way, but a more durable sweeping net is recommended for such rough usage. After sweeping with the net, a strong swing through the air will concentrate anything into the tip of the bag, and then, by immediately grasping the middle of the net with the free hand, the catch will be confined to a small part of the bag. Only the most rugged sweeping net may be used through thistles or brambles. Even some kinds of grasses, such as sawgrass, can quickly ruin a fragile net. Burrs and sticky seeds are also a serious problem.

The catch may be conveyed from the bag to a killing jar in a number of ways. Single specimens are transferred most easily by lightly holding them in a fold of the net with one hand while inserting the open killing jar into the net with the other. While the jar is still in the net, cover the opening until the specimen is overcome; otherwise, it may escape before the jar can be removed from the net and closed. To prevent a butterfly from damaging its wings by fluttering inside the net, squeeze the thorax gently through the netting when the butterfly's wings are closed (Fig. 1.17). This will temporarily paralyze the insect while it is being transferred to the killing jar. Experience will teach you how much pressure to exert.

Obviously, pinching small specimens of any kind is not recommended. When numerous specimens are in the net after prolonged sweeping, it may be desirable to put the entire tip of the bag into a large killing jar for a few minutes to stun the insects. They may then be sorted, and desired specimens placed separately into a killing jar, or the entire mass may be dumped into a killing jar for later sorting. These mass collection methods are especially adapted to obtaining small insects not readily recognizable until the catch is sorted under a microscope.

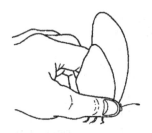

FIGURE 1.17

Technique for paralyzing a butterfly by squeezing its thorax between thumb and index finger.

Removal of stinging insects from a net can be a problem. Wasps and bees often walk toward the rim of the bag and may be made to enter a killing jar held at the point where they walk over the rim. However, many insects will fly as soon as they reach the rim, and a desired

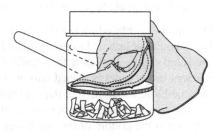

FIGURE 1.18

Technique of stunning insects while in collecting nets using the killing jar.

specimen may be lost. A useful technique involves trapping the insect in a fold of the net, carefully keeping a sufficient amount of netting between fingers and insect to avoid being stung. The fold of the net can then be inserted into the killing jar to stun the insect (Fig. 1.18). After a few moments, the stunned insect may be safely removed from the net and transferred to a killing jar. If the stunned insect clings to the net and does not fall readily into the jar, pry the insect loose with the jar lid or forceps. Do *not* attempt this maneuver with fingers because stunned wasps and bees can sting reflexively.

Several special modifications are necessary to adapt a net for aquatic collection. Aerial nets made of polyester or nylon may be used to sweep insects from water if an aquatic net is not available. The bag will dry quickly if swept strongly through the air a few times. Nets should not be employed for general collection until they are thoroughly dried, or other specimens (especially butterflies) may be damaged.

For more specialized collection, nets can be adapted in many ways. Nets can be attached to the ends of beams that can be rotated about their midlength by a motor drive. Nets also can be adapted to be towed by or mounted on vehicles (Fig. 1.19).

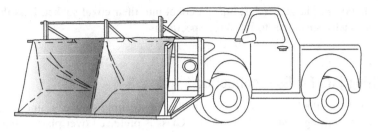

FIGURE 1.19

A specialized vehicle-mounted insect net.

Suggested reading

Collecting nets
- Riley (1957)
- McNutt (1976)
- Martin (1977)
- Rudd and Jensen (1977)
- Dunn and Reeves (1980)
- LeSage (1991)

Vehicle-mounted nets
- Traver (1940)
- Gressitt et al. (1961)
- Sommerman and Simmet (1965)
- Sommerman (1967)
- Hill (1971)

- Almand et al. (1974)
- Rudd and Jensen (1977)
- Holzapfel et al. (1978)
- Kronblad and Lundberg (1978)
- Barnard (1979)

1.3 Killing containers and agents

Killing an insect quickly is paramount to the preservation of a nice specimen. If an insect is allowed to beat its wings or crawl about for extended periods, it will inevitably harm its wings, break its legs and antennae, or lose its color.

1.3.1 Freezing insects

When collecting around the home or school, insects can be killed effectively with minimum damage by placing them immediately into a freezer. This method has two distinct advantages: First, no messy and potentially dangerous chemicals are needed; second, insects may be left in the freezer for long periods of time and need only be thawed before pinning. This convenience alone makes the freezer method attractive to many collectors.

1.3.2 Injecting insects with alcohol

Most large insects and especially large moths are best killed with an injection of alcohol using a hypodermic needle and syringe. For most insects, inject alcohol into their ventral area of the thorax. For large beetles, inject alcohol into their coxal socket. Less than 1 cm^3 of alcohol is usually sufficient to kill large specimens.

1.3.3 Killing jars for field collecting

Any heavy, widemouthed glass jar or bottle with a tight-fitting stopper or metal screw top may be used as a killing container (Fig. 1.20). Glass is preferred over plastic, and jars with relatively thick glass are preferred over thin, fragile glass, for obvious reasons. Olives frequently are sold in bottles that make convenient killing containers. Tops that may be removed with only a quarter turn often are preferred but may not be obtained readily. This type of lid may be quickly removed and returned to the jar with minimum effort. Collectors interested in taking minute or small insects may prefer using small vials that can be carried in a shirt pocket. Parallel-sided vials may be closed with cork stoppers. When collecting small-bodied insects in vials, care must be exercised to ensure that the stopper seats firmly against the wall of the vial. Otherwise, specimens become wedged between the glass and stopper, resulting in damage to some specimens. A crumpled piece of tissue paper placed in the vial helps maintain the specimens clean and disentangled.

(A)

Jars for use with liquid killing agents are prepared in one of the two ways. One way is to pour about 2.5 cm of plaster of paris mixed with water into the bottom of the jar and allow the plaster to dry without replacing the lid. Sufficient amount of killing agent is then added to saturate the plaster; any excess should be removed. The lid is then replaced. This kind of jar can be recharged merely by adding more killing agent. A second method is to place a wad of cotton, paper, or other absorbent material in the bottom of a jar, pour enough liquid killing agent into the jar to nearly saturate the absorbent material, and then press a piece of stiff paper or cardboard, cut to fit the inside of the jar tightly, over it (Fig. 1.20). The paper or cardboard acts as a barrier between the insect and the killing agent, preventing the specimen from contacting the agent directly and keeping the agent from evaporating too rapidly.

(B)

FIGURE 1.20

A homemade killing bottle made of absorbent materials covered by a tight-fitting cardboard disk.

1.3.4 **Liquid killing agents**

Popular liquid killing agents include ethyl acetate ($CH_3CO_2 \cdot C_2H_5Z$), carbon tetrachloride (CCl_4), ether (diethyl ether, $C_2H_5 \cdot O \cdot C_2H_5$), chloroform ($CHCl_3$), and ammonia water ($NH_4OH$ solution). Ethyl acetate is recommended by many entomologists as the most satisfactory liquid killing agent. The fumes of ethyl acetate are less toxic to humans than the fumes of the other substances. Ethyl acetate usually stuns insects quickly but kills them slowly. Specimens, even though they appear dead, may revive if removed from the killing jars too soon. However, a compensating advantage is that most specimens may be left in an ethyl acetate killing jar for several days and remain pliant. (If ethyl acetate is allowed to evaporate from the killing jar, specimens will harden.) For these reasons, a killing jar with ethyl acetate is preferred by many entomologists over a cyanide jar, especially when the jar is used infrequently.

Hobbyists and other collectors have found that fingernail polish remover (acetone) also works as an alternative to ethyl acetate and is more commonly available, especially in an emergency situation.

Carbon tetrachloride was once very popular as a liquid killing agent because it is not flammable and was easily obtained as a spot remover for clothes. Carbon tetrachloride is no longer recommended; however, because it is a carcinogen and a cumulative liver toxin. Specimens killed with carbon tetrachloride often become brittle and difficult to pin.

Ether and chloroform are extremely volatile and flammable and should not be used near an open flame or lighted cigarette. Their high volatility makes them serviceable in a killing jar for only a short period of time. Perhaps the greatest hazard with chloroform is that even when stored in a dark-colored jar, it eventually forms an extremely toxic gas called phosgene (carbonyl chloride, $COCl_2$). Chloroform, however, is useful when other substances cannot be obtained. Chloroform stuns and kills quickly, but it has the disadvantage of stiffening specimens.

Ammonia is irritating to humans, does not kill insects very effectively, and spoils the colors of many specimens. However, ammonia is readily available and will serve in an emergency. Ammonium carbonate, a solid but volatile substance, may also be used as a substitute.

Spray-dispensed insecticides may be used, if not to kill specimens, to at least "knock them down" into a container from which they may be picked up. If they are directed into a container topped with a funnel, they may be allowed to revive and treated further as desired (Clark and Blom, 1979).

Killing agents may be any of the various liquids. Never deliberately inhale the fumes, even momentarily. All killing agents are to some extent hazardous to human health. All killing jars or bottles should be clearly labeled "POISON" and kept away from persons unaware of their danger (Fig. 1.21).

Collectors who travel on airlines to collecting sites, especially on international flights, should be aware that the transportation industry now prohibits poisons on commercial passenger airliners in many parts of the world. Security at airports continues to increase, and the likelihood of being questioned about toxicants within carry-on baggage is high. Customs officials may also object to the importation of killing agents such as cyanides and

FIGURE 1.21

An example of a properly labeled killing jar.

ethyl acetate. When an international collecting trip is anticipated, travel agents and consulate officials should be consulted.

> Never deliberately inhale the fumes, even momentarily. All killing agents can be hazardous to human health. All killing jars or bottles should be clearly labeled "POISON" and kept away from persons unaware of their danger!

1.3.5 Solid killing agents

The solid killing agents used in killing jars are some forms of cyanides: potassium cyanide, sodium cyanide (NaCN), and calcium cyanide (Ca(CN)$_2$). Potassium cyanide is the compound of choice among many collectors. Sodium cyanide is equally effective but it is hygroscopic (i.e., it absorbs water and makes the jar wet). Calcium cyanide is seldom available. Cyanide compounds should be handled with extreme care. They are dangerous, rapid-acting poisons with no known antidote. If even a single grain touches the skin, wash the affected area immediately with water. To avoid handling the cyanide and storing or disposing of surplus crystals, you may be able to find a chemist, pharmacist, or professional entomologist who can make the killing jar for you. If this is not feasible, use utmost care in following the instructions given here.

To make a moderate-sized cyanide killing jar, place cyanide crystals about 15 mm deep in the bottom. A smaller amount of cyanide may be used for smaller jars. Cover the crystals with about 10 mm of sawdust and add about 7 mm of plaster of paris mixed with water to form a thick paste, working quickly before the plaster solidifies. Then add a piece of crumpled absorbent paper to prevent water condensation on the inner surface of the glass. Instead of the plaster of paris, a plug of paper or cardboard may be pressed on top of the sawdust. Be sure that it fits tightly. When the jar is ready for use, place several drops of water on the plaster or paper plug. In an hour or so, enough fumes of hydrocyanic acid will have been produced to make the jar operative. *Do not test this by sniffing the open jar!*

Another substance that has been recommended as a killing agent is dichlorvos (2.2-dichloroethenyl dimethyl phosphate), also called DDVP, Vapona, Nogos, Herkol, and Nuvan. Polyvinyl chloride (PVC) resin impregnated with this chemical and sold commercially as bug strips or No-Pest strips is long-lasting and somewhat less dangerous than other killing agents, but its time-release aspect allows only small quantities of the active agent to be released over time, which may be too small to kill quickly. Therefore PVC-impregnated dichlorvos is more effective as a killing agent in traps than in killing jars.

Every killing jar should be clearly and prominently labeled "POISON." The date the killing jar was fabricated can also be mentioned on the poison label. This will give the collector a rough idea of the life of a killing jar. The bottom outside of the jar must be reinforced with a tape, preferably a cloth, plastic, or clinical adhesive tape. The tape cushions the glass against breakage and keeps its dangerous contents from being scattered if the container breaks.

Laws involving toxic materials are changing, and local officials should be consulted as how best to proceed.

Killing jars or bottles will last longer and give better results if the following simple rules are observed.

1. Place a few narrow strips of absorbent paper in each killing jar to keep it dry and to prevent specimens from mutilating or soiling each other. Replace the strips when they become moist or dirty. This technique is useful for most insects, except those belonging to the order Lepidoptera, which are difficult to disentangle without damage.
2. Do not leave killing jars in direct sunlight because they will "sweat" and rapidly lose their killing power.
3. If moisture condenses in a jar, wipe it dry with absorbent tissue.
4. Keep delicate specimens in separate jars so that larger specimens do not damage them.
5. Do not allow a large number of specimens to accumulate in a jar unless it is to be used specifically for temporary storage.
6. Do not leave insects in a cyanide killing jar for more than a few hours. The fumes will change the colors of some insects, especially yellows to red, and specimens will generally become brittle and difficult to handle.
7. If it is necessary to hold insects for more than several hours, place the specimens in another container and store them in a refrigerator.
8. Keep butterflies and moths in jars by themselves so that the setae and scales that they shed will not contaminate other specimens. Contaminated specimens often make viewing of taxonomic features more difficult. Further, removing scales and setae is time-consuming, often tedious, always annoying, and potentially destructive to the specimen.
9. Never test a killing jar by smelling its contents.
10. Old jars that no longer kill quickly should be recharged or disposed of in a legal and responsible manner. A cyanide jar that has become dry may be reactivated by adding a few drops of water.

Suggested reading

Killing agents
- Lindroth (1957)
- Frost (1964)
- White (1964)
- Pennington (1967)
- Preiss et al. (1970)
- Clark and Blom (1979)
- Banks et al. (1981)

1.4 Aspirators and suction devices

The aspirator (Fig. 1.22), known in England as a "pooter," is a convenient and effective device for collecting small or highly mobile insects and mites. (The device was named in honor of Frederick W. Poos, an American entomologist who employed the device to collect insects belonging to the family Cicadellidae.) The following materials are needed to construct an aspirator:

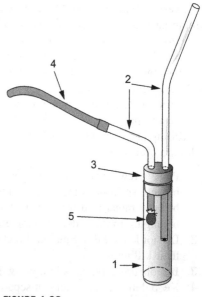

FIGURE 1.22

Components of a vial aspirator:(1) vial; (2) glass or copper tubing; (3) rubber stopper; (4) rubber or plastic tubing; and (5) mesh cloth and rubber band.

1. A vial 2.5–5 cm in diameter and about 12 cm long.
2. Two pieces of glass or copper tubing about 7 mm in diameter, one piece about 8 cm long and the other about 13 cm long. Copper tubing can be obtained from hobby shops and has the obvious advantage of being less fragile than glass.
3. A rubber stopper with two holes in which the tubing will fit snugly.
4. A length of flexible rubber or plastic tubing about 1 m long, with diameter just large enough to fit snugly over one end of the shorter piece of stiff tubing.
5. A small piece of cloth mesh (such as cheesecloth) and a rubber band.

To make an aspirator, bend the glass tubes as in Fig. 1.22. In bending or cutting glass tubes, always protect your fingers by holding the glass between several layers of cloth. Obtain the advice of a chemist or laboratory technician for cutting and bending glass. Moisten one end of the longer tube and insert it through one of the holes in the rubber stopper. Moisten one end of the shorter tube and insert it through the other hole in the stopper. Next, use a rubber band to fasten the cloth mesh over the end that was inserted through the stopper. (This will prevent specimens from being sucked into the collector's mouth when the aspirator is used.) Attach one end of the flexible tubing to the free end of this tube. The length, size, and amount of bend in the tubing will vary according to the user's needs. To complete the assembly, insert the rubber stopper into the vial. To use the aspirator, place the free end of the flexible tubing in the mouth, move the end of the longer glass tube close to a small specimen, and suck sharply. The specimen will be pulled into the vial.

Instead of a vial, some workers prefer a tube. In either method, keep small pieces of absorbent tissue in the vial or tube at all times to prevent moisture from accumulating. Note that there is some danger of inhaling harmful substances or organisms when using a suction-type aspirator (Hurd, 1954).

Either the vial- or tube-type aspirator (Fig. 1.22) may be converted into a blow-type aspirator by removing the 13 cm glass tube and substituting it with a T-shaped attachment (Fig. 1.23). The flexible tubing is attached to one arm of the T. The opposite arm is left open, and the stem of the T is inserted into the vial and covered with a mesh. Upon blowing through the flexible tubing, a current of air passes across the T and creates a partial vacuum in the vial, which produces the suction needed to draw specimens into the vial. This kind of aspirator eliminates the danger of inhaling small particles, fungus spores, or noxious fumes. Aspirators with a squeeze bulb may be purchased or if a valved bulb is available, they may be used with either pressure or suction.

Collection traps have also been devised with the suction feature applied on a much larger scale than with the usual aspirator. These include the handheld converted dustbuster insect vacuums (Fig. 1.24); the 10-vacuum (which employs a backpack motor fan)

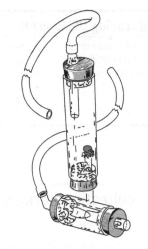

FIGURE 1.23

A blow-type aspirator.

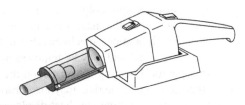

FIGURE 1.24

A motorized suction device (i.e., dustbuster) modified to collect insects.

(Dietrick et al., 1959, 1961); the Insectavac (Ellington et al., 1984a,b); and another device developed for the collection of honey bees (Gary and Marston, 1976). Suction produced by a fan has been employed in traps in conjunction with light or other attractants. Some of these traps are described in the following references and in Section 1.6. Suction is created by a piston in a "slurp-gun" described for aquatic collection of insects (Gulliksen and Deras, 1975). This principle could be adapted for use in air to collect airborne insects and to deposit them in a vial attached to the side of the piston.

Suggested reading

Aspirators and suction devices

- Johnson (1950)
- Hurd (1934)
- Jonasson (1934)

- Johnson and Taylor (1955)
- Woke (1955)
- Johnson et al. (1937)
- Lumsden (1938)
- Dietrick et al. (1959)

- Dietrick (1961)
- Minter (1961)
- Taylor (1962a)
- Singer (1964)
- Wiens and Burgess (1972)

(Continued)

Suggested reading (Continued)

• Coluzzi and Petrarca (1973)	• Gary and Marston (1976)	• Belding et al. (1991)
• Sholdt and Neri (1974)	• Barnard and Mulla (1977)	• DeBarro (1991)
• Bradbury and Morrison (1975)	• Ellington et al. (1984a,b)	• Governatori et al. (1993)
	• Summers et al. (1984)	• Moore et al. (1993)
• Evans (1973)	• Holteamp and Thompson (1985)	• Wilson et al. (1993)
• Azrang (1976)		• Arnold (1994)

1.5 Other collection devices

1.5.1 Beating sheets

A beating sheet should be made of a durable cloth, preferably white, attached to a frame of about 1 m^2, with two pieces of doweling or other light wood crossing each other and fitted into pockets at each corner of the cloth. Variations on the design can include a cloth stretched between two dowels or a triangular cloth that can be operated with one hand (Fig. 1.25). An ordinary light-colored umbrella may also be used as a beating sheet. Place the beating sheet or umbrella under a tree or shrub or limb and sharply beat the branches or foliage with a club or stick. Specimens that fall onto the sheet are clearly visible against the light-colored material and may be removed by hand or using forceps, a moistened brush, or an aspirator. Locating specimens on the sheet is sometimes a problem because leaves or other unwanted materials may also drop onto the sheet. Watching for movement helps locate specimens. Tilting the sheet displaces debris, leaving the insects and mites clinging to the cloth.

Beating sheets are useful for collecting beetles and are particularly effective early and late in the day and when the weather has turned cold. A "ground cloth" also is used in sampling crop fields (Rudd and Jensen, 1977) in much the same manner as a beating sheet.

1.5.2 Drag cloth

A drag cloth or flag (Fig. 1.26) made from durable light-colored cloth attached to a piece of doweling along one edge can be dragged through long grass or small shrubs to collect ticks. A variation on this method is to make a flag on a longer handle and to use it to brush against larger trees and shrubs. Questing ticks easily dislodge from the plants and collect onto the cloth.

1.5.3 Sifters

Sifters serve to collect insects and mites that live in ground litter, leaf mold, rotting wood, mammal and bird nests, fungi, shore detritus, lichens, mosses, and similar materials (Fig. 1.27). Leaf litter sifters can be constructed very simply by collecting leaf litter on a

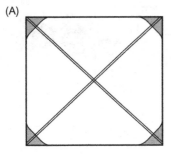

(A)

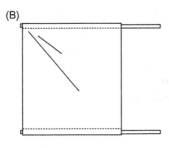

(B)

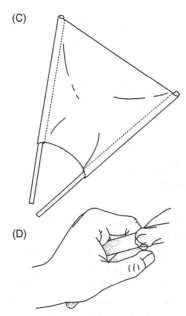

(C)

(D)

container, shaking the contents through a wire screen such as a cooling rack to separate the large leaves and then retaining the reduced sample. Sifters are also useful for collecting insects in diapause especially during winter collection.

Almost any container with a wire-mesh screen bottom will serve as a sifter. The size of the mesh depends on the size of the specimens sought. For general purposes, screening with 2.5−3 meshes per centimeter is satisfactory. To use the sifter, place the material to be sifted into the container and shake it gently over a white pan or a piece of white cloth. As the insects and mites fall onto the cloth, they may be collected with forceps, a brush, or an aspirator.

A similar method is used chiefly to collect mites from foliage. Employing a sifter of 20-mesh screen (about eight meshes per centimeter) with a funnel underneath that leads to a small vial beat pieces of vegetation against the screen to dislodge the mites and cause them to fall through the screen and into the vial below.

1.5.4 **Separators and extractors**

Somewhat similar to the sifter are various devices designed to separate or extract live specimens from substances in which they may be found, such as leaf mold and other kinds of vegetable matter, shore detritus, and dung. A simple extractor can be constructed from a 2 L

FIGURE 1.25

Beating sheets: (A) square beating sheet; (B) two-handled beating sheet; (C) triangular beating sheet, and (D) how to hold a triangular beating sheet.

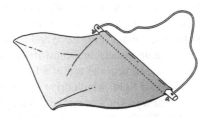

FIGURE 1.26

Drag cloth.

FIGURE 1.27

Sifters.

plastic pop bottle (Fig. 1.28). Separators are also used effectively in net sweepings that include so much foreign matter that makes it difficult to pick out the insects. These devices usually depend on some physical aid, such as light, heat, or dryness, to impel the insects to leave the substrate.

FIGURE 1.28

Pop bottle extractor.

One of the simplest such devices is the sweeping separator (Fig. 1.29), which consists of a carton or wooden box with a tight-fitting lid. A glass jar is inserted near the top of the box on one side. If the jar is made with a screw top, a hole of proper diameter cut on the side of the carton will permit the jar to be screwed onto it. The cover ring, without the lid, from a home-canning jar may be nailed to the periphery of a hole in a wooden box and the jar is then screwed onto the ring.

The sweepings are dumped into the box and the cover is quickly closed. A flashlight, table lamp, or other light source should be placed close to the jar.

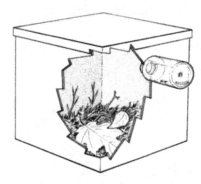

FIGURE 1.29

Sweeping separator box.

The insects in the darkened box soon will be attracted to the lighted glass jar. When all the insects appear to have entered the jar, it can be removed and its contents put into a killing jar. Alternatively, a jar cover containing a piece of blotting paper soaked with xylene may be placed over the jar for a short while to stun the insects, which may then be sorted.

The Berlese funnel (Fig. 1.30) and its modifications are cleaner and more efficient than sifters to separate insects and mites from leaf mold and similar materials. The sample is placed on a screen near the top of a funnel. A light bulb can be placed above the sample to produce heat and light, which drive the insects downward into the funnel. In other designs, heated coils or a jacket around the funnel can be used to dry the sample and make it inhospitable. The insects and mites are directed by the funnel into a container, sometimes containing alcohol, at the bottom of the funnel. Care should be taken not to dry the sample

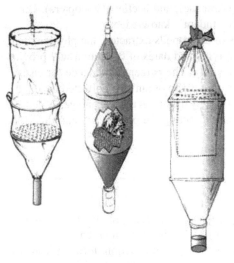

FIGURE 1.30

Berlese funnel and variations.

so rapidly that slow-moving specimens die before they can leave the sample. To prevent large amounts of debris from falling into the container, place the sample on the screen before the container is put in place.

Insects and all other organisms require food, water, and shelter to survive, and each has specific preferences. The habitat beneath our feet contains a myriad of seldom-seen insects and their arthropod relatives. An extractor will allow students to collect specimens not normally obtained by other means and may be used to form the basis of a student study.

A project based on an extractor can compare the types and relative abundance of soil arthropods (insects, mites, and spiders) found in various habitats. Such habitats include sawdust piles, decaying leaf litter and other organic matter, and different soil types found in wooded areas, pastures, and cultivate farm land. When selecting locations to sample, be certain that the materials are moist and contain organic matter. The number of habitat locations is only limited by how much time and effort a student wishes to spend on the project. An advanced student may wish to compare arthropods collected from the same habitat over time, for instance once a month from spring until fall.

For the procedure, refer to Fig. 1.28. To make a plastic pop bottle extractor, cut the top one-third off a plastic pop bottle and place a window screen circle, roughly the same diameter as the bottle, into the inverted top. The screen will prevent the habitat material from plugging the neck of the bottle. Next, place a small volume of rubbing alcohol into the bottle base to preserve extracted specimens and insert the inverted top of the bottle to fit into the base. Place a small volume of the habitat to be sampled into the top of the bottle and an incandescent lamp directly overhead to help dry it out and force living specimens down through the screen and into the preservative.

Use a garden trowel to take samples of moss, sawdust from old sawdust piles, decaying leaves, or other decaying vegetation. Approximately, the same amount of material should be taken for each sample. If different soil types are to be sampled, an inexpensive bulb planter will ensure that the same volume of material is extracted at each site. An enhancement to the project is to purchase an inexpensive probe that measures soil temperature, pH, and soil moisture. These data can be used to correlate the number of organisms extracted with specific habitat differences.

Because of the diversity found in many habitats, it is wise to separate extracted arthropods into easily recognizable forms such as springtails (Collembola), beetle mites

(Oribatida), nonbeetle mites (Acarina), spiders (Araneae), and beetles (Coleoptera). Other taxa can be added to the study depending on the student's knowledge.

Constructing a graph that depicts the groups of arthropods extracted, the physical data (pH, soil moisture, and temperature), the habitat type, and dates of collection will provide much valuable information on organismal biodiversity. The percent abundance for each group can be calculated by dividing the number of specific organisms collected by the number of all organisms in the sample and multiply by 100 to get a percentage. A cell phone photograph of the site will be useful for documenting the project and particularly valuable if creating a poster of the study.

1.6 Traps

A trap is any device that impedes or stops the progress of an organism. Traps are used extensively in entomology and may include devices used with or without baits, lures, or other attractants. The performance of a trap depends on its construction, location, time of day or year, weather, temperature, and kind of attractant. A little ingenuity coupled with knowledge of the habits of the insects or mites sought will allow for modifications or improvements in nearly any trap or may even suggest new traps. Only a few of the most useful traps are discussed here; however, Peterson (1964) and Southwood (1979) include extensive bibliographies on insect trapping.

Suggested reading

Separators and extractors
- Salmon (1946)
- Kevan (1955, 1962)
- Newell (1955)
- Kempson et al. (1962)
- Murphy (1962)
- Woodring (1968)
- Brown (1973)
- Merritt and Poorbaugh (1975)
- Gruber and Prieto (1976)
- Lane and Anderson (1976)
- Masner and Gibson (1979)
- Arnett (1985)
- Zolnerowich et al. (1990)

Traps
- Barber (1931)
- Williams and Milne (1935)
- Chamberlin (1940)
- Fleming et al. (1940)
- Huffacker and Back (1943)
- Broadbent et al. (1948)

- Johnson (1950)
- Williams (1951)
- Whittaker (1952)
- Dales (1953)
- Dodge and Seago (1954)
- Williams (1954)
- Eastop (1955)
- Woke (1955)
- Heathcote (1957a)
- Banks (1959)
- Dodge (1960)
- Race (1960)
- Fredeen (1961)
- Glasgow and Duffy (1961)
- Gressitt et al. (1961)
- Harwood (1961)
- Morris (1961)
- Taylor (1962b)
- Hollingsworth et al. (1963)
- Frost (1964)
- Hanec and Bracken (1964)
- Peterson (1964)

- Thorsteinson et al. (1965)
- Turnbull and Nicholls (1966)
- Bidlingmayer (1967)
- Kimerle and Anderson (1967)
- Everett and Lancaster (1968)
- Hartstack et al. (1968)
- Nijholt and Chapman (1968)
- Heathcote et al. (1969)
- Herting (1969)
- Thompson (1969)
- Catts (1970)
- Dresner (1970)
- McDonald (1970)
- Clinch (1971)
- Gojmerac and Davenport (1971)
- Hansens et al. (1971)
- Nakagawa et al. (1971)
- Emden (1972)
- Pickens et al. (1972)
- A'Brook (1973)

(Continued)

Suggested reading (Continued)

- Ford (1973)
- Klein et al. (1973)
- Hienton (1974)
- Wescloh (1974)
- Yates (1974)
- Acuff (1976)
- Hargrove (1977)
- Lammers (1977)
- Martin (1977)
- Rogoff (1978) *(includes extensive bibliography)*
- Stubbs and Chandler (1978, collecting and recording on pp. 1–37)
- Barnard (1979)
- Mcyerdirk et al. (1979)
- Southwood (1979)
- Hafraoui et al. (1980)
- Howell (1980)

- Sparks et al. (1980) *(includes pictures of several types of traps)*
- Banks et al. (1981)

Light traps
- Glick (1939, 1957)
- Frost (1957, 1958)
- Blakeslee et al. (1959)
- Meyers (1959)
- Goma (1965)
- Coon and Pepper (1968)
- Fincher and Stewart (1968)
- Stewart and Lam (1968)
- Callahan et al. (1972)
- Hocking and Hudson (1974)
- Roling and Kearby (1975)

Windowpane traps
- Chapman and Klinghorn (1955)

- Corbet (1965)
- Lehker and Dear (1969)
- Wilson (1969)

Interception nets and barriers
- Parman (1931, 1932)
- Gordon and Gerberg (1945)
- Leech (1955)
- Nielsen (1960)
- Merrill and Skelly (1968)
- Gillies (1969)
- Nielsen (1974) *(describes catches of insects in funnel traps on trunks of beech trees)*
- Steyskal et al. (1981)

The elevation or height above ground at which the trap is placed can be important in affecting the performance of traps, especially light traps. Optimum trap placement is a complex issue with many variables, including the behavior of the targeted insect, specific locality, as well as the size and color of the trap, influencing its performance.

1.6.1 Windowpane trap

This simple and inexpensive trap (Fig. 1.31) involves a barrier consisting of a windowpane held upright by stakes in the ground or suspended by a line from a tree or from a horizontal line. A trough filled with a liquid killing agent is placed so that insects flying into the pane drop into the trough and drown. They are removed from the liquid, washed with alcohol or other solvent, and then preserved or dried and pinned. This trap is not recommended for adult Lepidopterans or other insects that may be ruined if collected in a fluid.

Windowpane traps have been popular for many years because the traps passively intercept flying insects and require little effort on the part of the collector. Many different designs (Basset, 1988; Chapman, 1962; Hill and Cermak, 1997; Peck and Davies, 1980; Wilkening et al., 1981) exist for

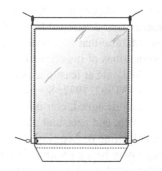

FIGURE 1.31

Windowpane trap.

windowpane traps, and the simplest trap is shown in Fig. 1.31.

Procedure. Construct a simple windowpane trap of glass or clear plexiglass with a base or trough that collects the stunned insects after they fly into it. The trough should be filled with a sufficient volume of soapy water to prevent evaporation (see also 1.6.4 Dish traps). Record the time and place where the trap was placed. Position the windowpane traps between two different habitats such as shrubs and the lawn or between a meadow and road to discover what insects move between the two habitats. To determine the direction of insect flight, be certain to place separate soapy trays on both sides of the trap. The study can be modified by measuring the times in which the traps are in place, altering the cardinal positions (north−south or east−west) of the traps, and even by separating by the different months of the year. To extend this project, colored plexiglass can be used to determine insect attraction to colors. The collected specimens can be identified to the appropriate taxonomic level of the researcher. (Note that this method is not recommended for those wishing to add Lepidopterans to their collections.)

A table or graph of the data will provide a quick overview of the project's results.

1.6.2 Interception nets and barriers

A piece of netting about 1.8 m high can be stretched between three trees or poles to form a V-shaped trap, with the wide end of the V kept open. A triangular roof should be adjusted to slope gently downward to the broad open side of the V. A device of this type will intercept many kinds of flying insects, particularly if the trap is situated with the point of the V toward the side of maximum light and in the direction of air movement. A pair of such nets set in opposite directions or a single net in a zigzag shape will intercept specimens from two directions. Insects flying into such a net tend to gather at the pyramidal apex, so they are easy to collect. The so-called funnel or ramp traps are interception devices that direct insects to a central point, where a retaining device or killing jar may be placed.

1.6.3 Malaise traps

A modification of the interception net led to the more complex trap developed by the Swedish entomologist René Malaise that now bears his name. Several modifications of his original design have been published, and at least one is available commercially (Townes, 1962, 1972; Steyskal, 1981). The trap, as originally designed, consists of a vertical net serving as a baffle, end nets, and a sloping canopy leading up to a collecting device (Fig. 1.32). The collecting device may be a jar with either a solid or evaporating killing agent or a liquid in which the insects drown. The original design is unidirectional or bidirectional with the baffle in the middle, but more recent types

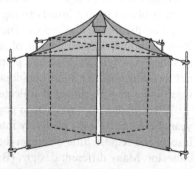

FIGURE 1.32

Malaise trap.

include a nondirectional type with cross-baffles and with the collecting device in the center. Malaise traps have been phenomenally successful, sometimes collecting large numbers of species that could not be obtained otherwise. Attractants increase the efficiency of the traps for special purposes.

1.6.4 Pitfall and dish traps

Another simple but very effective and useful type of interception trap consists of a dish, can, or jar sunk in the ground and is called a pitfall trap (Fig. 1.33). Wandering arthropods fall into the trap and are unable to escape from the partly filled container of 70% ethyl alcohol or ethylene glycol (automobile antifreeze). The latter is preferred because it does not evaporate. A cover must be placed over the open top of the container to exclude rain and small vertebrates while allowing insects and mites to enter. Pitfall traps may be baited with various substances, depending on the kind of insects or mites the collector hopes to capture. Although most specimens that fall into the trap remain there, the trap should be inspected daily because specimens can decompose or become damaged. Specimens should be removed and placed in alcohol or in a killing bottle while they are in their best condition.

Other pitfall trap modifications include placing a bait within the collection container or hanging it directly above. The cereal dish trap (another modified pitfall trap) represents a simple but effective device for obtaining insects attracted to dung. This trap consists of a small dish, preferably with a rim, set in the ground (Fig. 1.33C) and partly filled with 70%

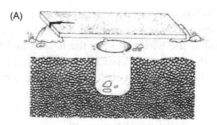

(A)

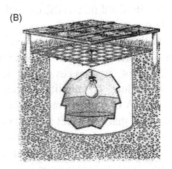

(B)

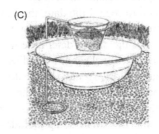

(C)

(D)

FIGURE 1.33

Pitfall and dish trap: (A) traditional pitfall trap; (B) baited pitfall trap; (C) cereal dish trap; and (D) PVC pitfall trap.

ethyl alcohol or ethylene glycol. A piece of stout wire, such as a coat hanger, is bent to form a loop at one end that holds the bait receptacle. A few zigzag bends in the other end of the wire will keep the looped end from swinging after the wire is pushed into the earth. The bait receptacle may be a small plastic or metal cup (such as that used for medicine doses), a coffee creamer, or a cup formed from aluminum foil. When baited with animal or human feces, this trap attracts beetles (typically Scarabaeidae and Staphylinidae), springtails, ants, earwigs, some parasitic Hymenoptera, and several families of flies, including Phoridae, Sepsidae, and Muscidae. The larger, strong-flying calliphorid and sarcophagid flies seldom fall into the liquid, although they are attracted to the bait. The alcohol fumes may cause smaller flies to drop into it. The trap is made of easily obtained materials, is easily transported, and provides excellent results. It deserves wide use.

Pitfall traps. There are several activities using pitfall traps that can be used to generate student science projects. Insects are attracted to different foods and if used as baits in pitfall traps, preferences of ground dwelling insects can be studied.

Procedure. To simplify a pitfall study, use a commercial bulb cutter to create a hole for the trap. A collecting container of the same diameter as the bulb cutter is placed in the hole such that it is level and even with the ground. To determine habitat and food preferences, place baited and nonbaited control traps in different habitats such as a meadow, woods, or pasture, and within each habitat select an open and a shady area. Insects collected from traps baited with various meats (beef, chicken, fish, etc.) can be compared against insects collected from plant-based baits. For each collection made be certain to record the date, the time the traps were placed and removed, the bait type, and habitat. In more advanced studies, traps with different baits can be monitored at different times of the day to determine the daily activity of the species collected.

Dish traps. Dish trap studies are based on the fact that some insects are attracted to certain colors. Aphids, pollinators, and parasitic wasps are attracted to yellow traps, whereas blue traps attract other insects. These simple and inexpensive dish traps can be used to determine the color preference of certain insects.

Procedure. The setup includes small red, blue, yellow, black, and white plastic plates, such as those used for birthday parties. Arrange the color plates in a circle and select color position at random. Fill each plate with enough soapy water to hold the insects.

Because some insect activity is dependent upon time of day, an experiment can be performed with dish traps to determine specific insect activity at selected times. Choose a habitat away from human interference and set out plates filled with soapy water. At predetermined times, empty the contents into individual containers marked with the colors of the plates and identify the type and number of insects captured. By repeating the procedure with the same set in the morning, afternoon, and at dusk, a student can determine differences in the activity by type of insects or even by trap color. To expand the project, the plates can be placed in different habitats and the same analyses can be performed.

Suggested reading

Pitfall and dish traps
- Broadbent et al. (1948)
- Broadbent (1949)
- Tretzel (1955)
- Greenslade (1964)
- Luff (1968, 1975)
- Alderz (1971)
- Briggs (1971)
- Greenslade and Greenslade (1971)
- Gist and Crossley (1973)
- Greenslade (1973)
- Schmid et al. (1973)
- Loschiavo (1974)
- Joossee (1975)
- Morrill (1973) *(bibliography)*
- Muma (1975) *(includes bibliography of can and pitfall trapping)*
- Newton and Peek (1975)
- Shubeck (1976)
- Smith (1976)
- Smith et al. (1977) *(describes low-cost carrying cages, pitfall traps, and rearing cages)*
- Thomas and Sleeper (1977)
- Houseweart et al. (1979)
- Reeves (1980)

Modifications to the basic pitfall trap include a section (up to a meter) of PVC pipe, capped at both ends, and where a small channel is cut lengthwise along the pipe (Fig. 1.33D). Burying the pipe such that the open channel is at the soil surface allows wandering insects to fall into the pipe.

1.6.5 Emergence traps and rearing cages

An emergence trap is any device that prevents adult insects from escaping when they emerge from their immature stages in any substrate, such as soil, plant tissue, or water (Yates, 1974). A simple canopy over an area of soil, over a plant infested with larvae, or over a section of stream or other water area containing immature stages of flying insects will secure the emerging adults (Lammers, 1977). If the trap is equipped with a retaining device (as in the Malaise trap), the adults can be killed and preserved shortly after emergence. Remember, however, that many insects should not be killed too soon after emergence because the adults are often teneral (soft bodied) and incompletely pigmented. For preservation, specimens must be kept alive until the body and wings harden completely and colors develop fully. Emergence traps and rearing cages (Fig. 1.34) enable the insects to develop naturally but ensure their capture when they mature or when larvae emerge to pupate.

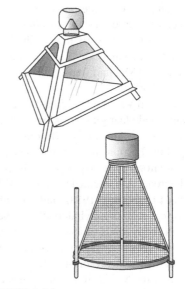

FIGURE 1.34

Emergence traps.

Suggested reading

Lobster eel traps
- Nicholls (1960)
- Doane (1961)
- Brockway et al. (1962)
- Morrill and Whitcomb (1972)

- Nielsen (1974) (*describes catches of insects in tunnel traps on trunks of beech trees*)
- Nakagawa et al. (1975)
- Reierson and Wagner (1975)

- Broce et al. (1977)
- Steyskal (1977)
- Becker et al. (1991)

1.6.6 Traps using the lobster or eel trap principle

This trap design can be made from any container that has its open end fitted with a truncated cone directed inward (Fig. 1.35). This trap design is commonly used to catch minnows, lobsters, and eels in water. An ordinary killing jar with a funnel fastened into its open end is an example. When the funnel is placed over an insect, the specimen

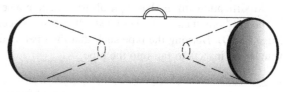

FIGURE 1.35

Eel trap.

will usually walk or fly toward the light and enter the jar through the funnel. Modified traps of this type include the Steiner and McPhail traps, which are used extensively in fruit fly surveys.

McPhail was the first entomologist to use an invaginated clear glass trap for evaluating the attraction of fruit flies to various baits (Burditt, 1982). The McPhail trap (Fig. 1.36), as it is known today, is used extensively for monitoring fruit fly populations in the United States and abroad. A homemade version of the McPhail trap was devised by Burditt (1988), and a number of baits, such as sucrose, yeast, or ethyl alcohol, can be used to attract and collect many kinds of insects. As an example, the McPhail trap was used in a survey to collect euglossine bees in tropical Amazonian forests (Becker et al., 1991). The inside of the Steiner trap usually has a sticky material containing a pheromone or other lure.

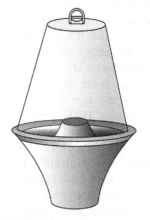

FIGURE 1.36

McPhail trap.

1.6.7 Light traps

Light traps take advantage of the attraction of many insects to a light source and their attendant disorientation. Using various wavelengths of light as the attractant, a great

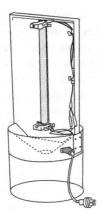

FIGURE 1.37

Blacklight trap.

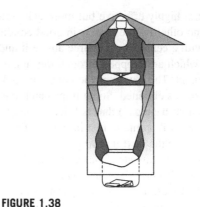

FIGURE 1.38

New Jersey trap.

variety of traps can be devised *using* a basic funnel and collection container design (Fig. 1.37).

Many traps can be constructed easily from materials generally available around the home. All wiring and electrical connections should be approved for outdoor use. Funnels can be made of metal, plastic, or heavy paper. Traps can be with or without a cover, but if they are operated for several nights, covers should be installed to exclude rain.

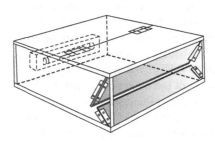

FIGURE 1.39

Wilkinson trap.

The New Jersey trap (Fig. 1.38) includes a motorized fan to force insects attracted to the light into a killing jar. This trap has been useful especially for collecting small, nonscaly insects, such as midges and gnats. This type of light trap, in which the insects fall directly into a killing jar, is not recommended for use with moths because such delicate specimens may be badly rubbed or torn. If small insects alone are desired, they may be protected from damage by larger insects by placing a screen with the proper-sized mesh over the entrance. The Minnesota trap is very similar to the New Jersey trap, but it does not include a fan or any motorized method of draft induction.

The Wilkinson trap (Fig. 1.39) requires somewhat more effort to construct than the preceding traps, but it has the advantage of confining, but not killing, the trapped insects. Moths, therefore, can be collected in good condition if the trap is inspected frequently and desirable specimens are removed quickly through the hinged top and placed in a killing jar.

Several highly effective but more elaborate devices have been made for collecting moths and other fragile insects in good condition. Basically, they all use the principle of a funnel with a central light source above it and vanes or baffles to intercept the approaching insects, which are dropped through the funnel into the container beneath that may hold a killing agent. The nature of the container and the type of killing agent affect the quality of the specimens obtained. Some traps catch the insects alive in a large collection chamber (such as a garbage can) that is filled or nearly filled with loosely arranged egg cartons. Most moths will come to rest in the cavities between the egg cartons and will remain there until removed the next morning.

To prevent rainwater from accumulating in the trap, place a screen-covered funnel inside the collection chamber to drain the water out through a hole in the bottom of the trap. Sometimes, a system of separators is added to guide beetles and other heavy, hard-bodied insects into a different part of the container than the moths and other delicate specimens.

The most effective light traps use lamps high in their output of ultraviolet light (Fig. 1.40). An example is the Robinson trap, which employs a 175 W mercury vapor lamp. Other trap designs use ultraviolet fluorescent tubes (most commonly 15 W), which are also effective and may be powered by a car battery or other portable source of electricity. High-wattage mercury vapor lamps emit more ultraviolet light than the more common fluorescent "blacklight" tubes, and many more insect specimens come to mercury vapor light setups. However, mercury vapor setups are not as portable and, in the field, require the use of a gasoline-powered generator. Self-ballasted mercury vapor lamps are not as effective on a watt-by-watt basis as those with a separate ballast.

Some collectors report that a few groups of insects respond best to incandescent lamps or even gas camping lanterns that do not emit ultraviolet light. However, for the majority of insects that may be taken with lights, high ultraviolet output is desirable.

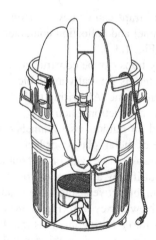

FIGURE 1.40

Mercury vapor light trap.

Suggested reading

Light traps
- Gui et al. (1942)
- Mulhern (1942)
- Pratt (1944)
- Williams (1948)
- Davis and Landis (1949)
- Frost (1952, 1957, 1958, 1964, 1970)
- Meyers (1959)
- Fredeen (1961)
- Graham et al. (1961)
- Hollingsworth et al. (1961)
- U.S. Department of Agriculture, Agriculture Research Service (1961)

(Continued)

Suggested reading (Continued)

- Barr et al. (1963)
- Beiton and Kempster (1963)
- Breyev (1963)
- Kovrov and Monchadskii (1963)
- Lowe and Putnam (1964)
- White (1964)
- Lewis and Taylor (1965)
- Belton and Pucat (1967)
- DeJong (1967)
- Bartnett and Stephenson (1968)
- Hardwick (1968)

- Stewart and Lam (1968)
- Powers (1969)
- Wilkinson (1969)
- Andreyev et al. (1970)
- McDonald (1970)
- Miller et al. (1970)
- Stanley and Dominick (1970)
- Barrett et al. (1971)
- Carlson (1971)
- Stewart and Payne (1971)
- Hollingsworth and Hartstack (1972)

- Tedders and Edwards (1972)
- Clark and Curtis (1973)
- Pérez and Hensley (1973)
- Morgan and Uebel (1974)
- Smith (1974)
- Nantung Institute of Agriculture (1975)
- Apperson and Yows (1976)
- Onsager (1976)
- Stubbs and Chandler (1978)
- Burbutis and Stewart (1979)
- Howell (1980)
- Hathaway (1981)

Except in special cases, light sources should be placed near the ground. This is counterintuitive to some collectors who try to place lights as high as possible. However, unpublished studies suggest that more specimens come to ground-level light sources, and species that tend to be "flighty" more reliably remain near these lights. Of course, specialized light collecting may require placing of a light trap high in a tree or in some other location.

A lightweight, spill-proof 12 V battery, in which the acid electrolyte is a gel rather than a liquid, is far superior to the standard automotive battery for powering light traps. These new batteries are fairly expensive and require a special charger. Special lightweight, nickel—cadmium battery packs, used to power blacklights for collecting, are marketed by some dealers of entomological equipment.

1.6.8 Light sheets

Another highly effective method of employing light to attract moths and other nocturnal insects involves a "light sheet" (Fig. 1.41). A particularly effective and efficient light sheet or "light net" design is one made of sturdy netting material with edges of ripstop nylon and having a rectangular cutout near the bottom in which a mercury vapor or other light source is placed. The netting and cutout greatly reduce flapping caused by wind, and the cutout allows one light source to shine in all directions. Ground sheets made of ripstop nylon also are useful, lightweight, easy to clean, and quick-drying.

An alternative light sheet design employs light sources (ultraviolet

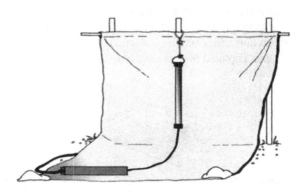

FIGURE 1.41

Light sheet.

fluorescent tubes, gasoline lanterns, or automobile headlights) placed in front of the light sheet. As insects are attracted and alight on the sheet, they are easily captured in cyanide bottles or jars by the collector, who stands in attendance or at least checks the sheet frequently. The sheet may be pinned to a rope tied between two trees or fastened to the side of a building, with the bottom edge spread out on the ground beneath the light. Some collectors use supports to hold the bottom edge of the sheet several centimeters above the ground so that specimens cannot walk into the vegetation under the sheet and be overlooked. Other collectors turn up the edge to form a trough into which insects may fall as they strike the sheet.

The light sheet remains unsurpassed as a method of collecting moths in flawless condition or of obtaining live females for rearing purposes. Its main disadvantage is that species that fly very late or those that are active only in the early morning hours may be missed unless the collector is prepared to spend most of the night at the sheet. Many other insects besides moths are attracted to the sheet, and collectors of beetles, flies, and other kinds of insects would do well to collect with this method.

It should be emphasized that the phases of the moon profoundly influence the attraction of insects to artificial light. Attraction is inhibited by a bright moon. The best collecting period in each month extends from the fifth night after the full moon until about a week before the next full moon.

1.6.9 Color traps

Colored objects also serve as attractants for insects. A bright yellow pan (Fig. 1.42) containing water is used to collect winged aphids and parasitic hymenopterans. These insects are attracted by the color of the pan and get drowned in the water. The trap becomes more effective when a little detergent has been put in the water to reduce surface tension and thereby cause insects to drown more quickly. Yellow seems to be the best color for traps, but various kinds of insects react differently to different colors. The Manitoba trap (Fig. 1.43) has a black sphere to attract horse flies (family Tabanidae), which are then captured in a canopy-type trap.

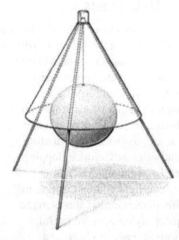

FIGURE 1.42

Pan trap.

FIGURE 1.43

Manitoba trap.

Suggested reading

Color traps
- Gui et al. (1942)
- Hottes (1951)
- Brown (1954)
- Granger (1970)
- Kring (1970)
- Beroza (1972)
- Emden (1972)
- Kieckhefer et al. (1976)
- Hendrix and Showers (1990)
- McClain et al. (1990)

Sticky traps
- Golding (1941, 1946)
- Johnson (1950)
- Moreland (1955)
- Heathcote (1957b)
- Still (1960)
- Murphy (1962, pp. 226–227)
- Taylor (1962b)
- Maxwell (1965)

- Gillies and Snow (1967)
- Lambert and Franklin (1967)
- Prokopy (1968, 1973)
- Dresner (1970)
- Harris et al. (1971)
- Mason and Sublette (1971)
- Emden (1972)
- Buriff (1973)
- Chiang (1973)
- Goodenough and Snow (1973)
- Williams (1973)
- Evans (1973)
- Edmunds et al. (1976, methods of collecting and preservation on pp. 8–26)
- Harris and McCafferty (1977)
- Murphy (1985)
- Bowles et al. (1990)
- Knodel and Agnello (1990)

- Vick et al. (1990)
- Jenkins (1991)
- Miller et al. (1993)

Artificial refuge traps
- Glen (1976)
- Shubeck (1976)

Electrical grid traps
- Graham et al. (1961)
- Hollingsworth et al. (1963)
- Mitchell et al. (1972, 1973, 1974)
- Goodenough and Snow (1973)
- Hienton (1974)
- Glen (1976)
- Rogers and Smith (1977)
- Stanley et al. (1977)
- Kogan and Herzog (1980)

1.6.10 Sticky traps

In its simplest design, this type of trap involves a board, a piece of tape, a pane of glass, a piece of wire net, and a sphere, a cylinder, or some other object (often painted yellow). The trap is coated with a sticky substance and suspended from a tree branch or other convenient object. Insects landing on the sticky surface cannot extricate themselves. The sticky material can be dissolved with a suitable solvent, usually toluene, xylene, ethyl acetate, or various combinations of these solvents. Later, the insects are washed in Cellosolve and then in xylene. Sticky traps should not be used to collect specimens such as Lepidopterans, which are ravaged by the sticky substance and cannot be removed without being damaged.

Various sticky-trap materials are available commercially, some with added attractants, including lights. However, caution should be exercised in selecting a sticky substance, because some are difficult to dissolve. General techniques for cleaning specimens taken with sticky traps have been outlined (Murphy, 1985), and a cleaning procedure involving an ultrasonic cleaner has been developed (Williams and O'Keeffe, 1990).

1.6.11 Snap traps

Two kinds of traps designed for quantitative sampling may be termed "snap traps." One (Menzies and Hagley, 1977) consists of a pair of wooden or plastic disks, slotted in the center to fit on a tree branch and connected to each other by a pair of rods. A cloth cylinder is affixed at one end to one of the disks and at the other end to a ring sliding on the rods. After the cloth cylinder has been pulled to one end and secured in place, the ring is held by a pair of latches. When insects have settled on the branch, its leaves, or flowers, the latches are released by pulling on a string from a distance, and the trap is snapped shut by a pair of springs on the rods, capturing any insects present. One of the canopy traps (Turnbull and Nicholls, 1966) operates in a similar manner. When a remotely controlled latch is pulled, a spring-loaded canopy is snapped over an area of soil, and insects within the canopy are collected by a suction or vacuum device. This trap was designed for use in grasslands.

1.6.12 Artificial refuge traps

Many insects, especially beetles, are found under stones, planks, or rotten logs. Providing refuges (such as pieces of wood, cardboard, or complex traps) is also an effective form of trapping.

1.6.13 Electrical grid traps

In recent years, electrocuting insects has been used extensively in pest-control works. The insects are lured to a device by a chemical (see Section 1.7), light, or other substance placed in a chamber protected by a strongly charged electrical grid (Fig. 1.44). This method is not effective in preserving insect specimens but may be useful for purposes such as surveying the arthropod fauna of an area.

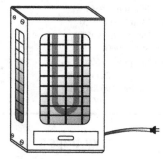

FIGURE 1.44

Electrical grid trap.

1.6.14 Combination traps for student studies

Traps previously discussed collect only a subset of all insects in a specific habitat. Scientists often use several types of traps when studying the biodiversity of a specific habitat even if the study involves a single taxon. To reduce the time and effort required to set up Malaise, intercept, pan, and cone traps, Russo et al. (2011) developed a composite insect trap that incorporates the catch characteristics of each insect trap. A composite trap incorporates the concepts of intercept and Malaise traps and a yellow pan trap to collect insects that are attracted to yellow and to water. While the composite trap will not collect all flying insects in an area, the trap is more general and can capture a wide diversity of insects. For those interested in constructing a composite trap, directions and illustrations can be found at Russo et al. (2011).

1.7 Baits, lures, and other attractants

Any substance that attracts insects may function as a bait. Natural products, chemicals derived therefrom or synthesized, and secretions of insects may all be attractants

(Jacobson and Beroza, 1964; Jacobson, 1972). Exposure of attractive substances is found in a variety of constructed traps.

Suggested reading

Baits and lures
- Walsh (1933)
- Golding (1941)
- Dethier (1955)
- Macleod and Donnelly (1956)
- Atkins (1957)
- Colless (1959)
- Rennison and Robertson (1959)
- Beroza and Green (1963)
- Hocking (1963)
- Mason (1963)
- Wilton (1963)

- Jacobson and Beroza (1964)
- Coffey (1966)
- Newhouse et al. (1966)
- Sanders and Dobson (1966)
- Strenzke (1966)
- DeJong (1967)
- Aeree et al. (1968)
- Everett and Lancaster (1968)
- Fincher and Stewart (1968)
- Morris and DeFoliart (1969)
- Beroza (1970, 1972)
- Nakagawa et al. (1971)
- Wellso and Fischer (1971)
- Beavers et al. (1972)

- DeFoliart (1972)
- Knox and Hays (1972)
- Roberts (1972)
- Beroza et al. (1974)
- Debolt et al. (1975)
- Howell et al. (1975)
- Pinniger (1975)
- Shorey and McKelvey (1977)
- Bram (1978)
- Zimmerman (1978)
- Howell (1980)
- Laird (1981)

1.7.1 Sugaring for insects

One of the oldest collecting methods involves the use of a specially prepared bait in which some form of sugar is an essential component. The bait may contain refined or brown sugar, molasses, or syrup. Such substances are often mixed with stale beer or fermented peaches, bananas, or other fruit. Sugar-baited traps are most often used for moths and butterflies, but they are also effective for some flies (Dethier, 1955) and caddisflies (Bowles et al., 1990).

One particularly effective recipe for Lepidopterans uses fresh, ripe peaches, culls, or suitable windfalls. Remove the seeds but not the skins, mash the fruit, and then place it in a 4 L (1 gal) container with a snugly fitting but not airtight cover. Avoid using metal containers that may rust or corrode. Fill each container only one-half to two-thirds full to allow space for expansion. Add about a cup of sugar and place in a moderately warm place for the mixture to ferment. The bubbling fermentation reaction should start in a day or so and may continue for 2 weeks or more, depending on the temperature. During that time, check the fermentation every day or every other day and add sugar until fermentation appears to have subsided completely. As the added sugar is converted into alcohol, the growth of yeast slows and eventually ceases.

If the mixture is allowed to run low in sugar during the fermentation process, vinegar will be produced instead of alcohol. Remember to smell the bait periodically and to add plenty of sugar to avoid vinegar formation. The amount of sugar consumed will be surprising, usually over 0.4 kg/L (3.3 lb/gal). After fermentation ceases, the bait should remain stable and should be kept in tightly sealed containers to prevent contamination and evaporation. Canned fruit, such as applesauce, may also be used to make the bait, but such products are sterile and a small amount of yeast must be added to start fermentation. The bait should have a sweet, fruity, wine-like fragrance. Although the bait may seem troublesome to prepare, it keeps for years and is available at any time, even when fruit is not in season.

Immediately before use, the bait may be mixed with 30%−50% molasses, brown sugar, honey, or a mixture of these ingredients. This thickens the bait, retards drying, and makes the supply last longer. Set out the sugar bait during the early evening before dark. Apply the bait with a paintbrush in streaks on tree trunks, fence posts, or other surfaces. Choose a definite route, such as along a trail or along the edge of a field, so that later you can follow it in the dark with a lantern or flashlight. Experienced collectors approach the patches of bait cautiously, with a light in one hand and a killing jar in the other to catch moths before they are frightened off. Some collectors prefer to wear a headlamp, leaving both hands free to collect specimens. Although some moths will escape, a net usually is regarded as an unnecessary encumbrance because moths can be directed rather easily into the jar. Sugaring is an exceptionally useful way to collect noctuid moths, and the bait applied in the evening often will attract various diurnal insects on the following days. This bait has been used in butterfly traps with spectacular results. However, collecting with baits is notoriously unpredictable, being extremely productive on one occasion and disappointing on another, under apparently identical conditions.

Suggested reading

Feces
- Steyskal (1957)
- Coffey (1966)
- Fincher and Stewart (1968)
- Merritt and Poorbaugh (1975)

1.7.2 Feces

Animal feces (including human fees) attract many insects. A simple but effective method of collecting such insects is to place fresh feces on a piece of paper on the ground and wait a few minutes. When a sufficient number of insects have arrived, a net with its bag held upward can be brought carefully over the bait about 1 m above it. This will not disturb the insects nor will they be greatly disturbed when the net is lowered gently about two-thirds of the distance to the bait. At this point, the net should be lowered quickly until its rim strikes the paper. The insects, mostly flies, will rise into the net, which may then be lifted a short distance above the bait and quickly swung sideways, capturing the insects at the bottom of the bag. Feces are most attractive to insects during the first hour after deposition, and many flies can be caught during that relatively short period of time. Because of this, the "baiting with feces" method may be used for quantitative studies.

Other insects may continue coming to the feces for a more extended period and may be captured by placing a canopy trap over the feces or by using the feces with the cereal dish trap (see Section 1.6.4). Emergence traps placed over old feces will capture adult insects emerging from immature forms feeding there. The same methods may also be used with other baits, such as decaying fruit, small dead animal carcasses, and a wide variety of other substances.

1.7.3 **Oatmeal**

Hubbell (1956) showed that dry oatmeal scattered along a path will attract insects such as crickets, camel crickets, cockroaches, and ants. Some of these insects feed only at night and may be best located by using a flashlight or headlamp. The specimens may be hand-collected, aspirated, or netted and placed into a killing jar.

1.8 **Pheromones and other attractants**

Insects rely heavily upon chemical communication. Over the last several years, the research into insect chemical communication has been phenomenal. New terms, new concepts, and the refinement of old ideas have been necessary. For instance, the term *allelochemical* has been proposed for chemicals secreted outside the body of an organism. *Semiochemicals* are compounds that influence insect behavior or that mediate the interactions between organisms. Currently, at least three categories of semiochemicals have been identified. A *synomone* is an interspecific chemical messenger that is beneficial to emitter and receiver. Examples include secondary plant chemicals that attract entomophagous insects to the plant and subsequently to prey or hosts. A flower fragrance that is attractive to a bee through nectar and pollen gained is also beneficial to the flower through pollination. In contrast, an *allomone* is an interspecific chemical that, in a two-species interaction, is beneficial only to the emitter. A *kairomone* is an interspecific chemical that is beneficial only to the receiver. Insect collectors are beginning to appreciate the role that interspecific attractants play in capturing insects, and in the future, we will probably rely on them even more often. For instance, "Spanish fly" (cantharidin) is an irritant to vertebrates that is produced by meloid beetles but has recently come into use as an extremely effective attractant for other beetles (such as pedilids) and bugs (such as bryocerines).

Suggested reading

Pheromones
- Beaudry (1954)
- Holbrook and Beroza (1960)
- Jacobson and Beroza (1964)
- Howland et al. (1969)
- Beroza (1970)
- Campion (1972)
- Jacobson (1972)
- Goonewardene et al. (1973)
- Pérez and Hensley (1973)
- Shorey (1973)
- Beroza et al. (1974)
- Birch (1974)
- Campion et al. (1974)

- Peacock and Cuthbert (1975)
- Weatherston (1976)
- Shorey and McKelvey (1977)
- Steck and Bailey (1978)
- Neal (1979)
- Howell (1980)
- Sparks et al. (1980)
- Hathaway (1981)
- Vite and Baader (1990)
- Gray et al. (1991)
- Mullen (1992)
- Mullen et al. (1992)
- Bartelt et al. (1994)

Carbon dioxide
- Reeves (1951, 1953)
- Rennison and Robertson (1959)
- Takeda et al. (1962)
- Whitsel and Schoeppner (1965)
- Newhouse et al. (1966)
- Snoddy and Hays (1966)
- Wilson et al. (1966)
- Gillies and Snow (1967)
- Morris and DeFoliart (1969)
- Hoy (1970)
- Stryker and Young (1970)

(Continued)

Suggested reading (Continued)

- Davies (1971)
- Batiste and Joos (1972)
- Blume et al. (1972)
- Roberts (1972)
- Wilson et al. (1972)
- Debolt et al. (1975)
- Gray (1985)
- Carroll (1988)

Substances naturally produced and emitted by insects that cause a behavioral response in individuals of the same species are known as *pheromones*. Sex pheromones, in which the pheromone is emitted by one sex and attracts individuals of the opposite sex, are often used in traps to aid in controlling pest species. They also are emitted into the air from controlled-release dispersers throughout a particular crop to disrupt mating by misdirecting male insects to these synthetic sources as well as habituating males to the scent of their own females. Most pheromones are blends of several components; the blend is highly specific, attracting only one species. Pheromones are known to attract members of one sex, usually males, from great distances.

1.8.1 Carbon dioxide

Host animals likewise may be used as bait for various bloodsucking insects, with or without constructed traps. Carbon dioxide (Fig. 1.45) in the form of dry ice, cylinder gas, or marble chips treated with an acid such as vinegar serves as an attractant for certain insects and has been very successful in attracting horse flies to Malaise and Manitoba traps. Dry ice is also an effective attractant for adult mosquitoes, horse flies, and parasites such as ticks.

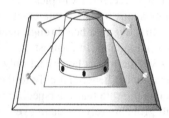

FIGURE 1.45

CO_2 trap.

1.8.2 Sounds

Sounds are produced by many insects to attract other members of the same species. These sounds are very specific in pitch, tempo, and duration. Sound sources once tape-recorded are now often synthesized and broadcast using high-tech electronic amplifiers. Photocells often serve as timing devices to turn sounds on or off, rather than manual switches. Recordings of such sounds played at the proper volume have been effective in luring grasshoppers, mole crickets, and other kinds of insects.

1.9 Collecting aquatic insects

Insects and mites emerging from water may be collected by some of the same methods as terrestrial insects. Aquatic-insect-collecting dip nets and heavy-duty aquatic nets can be

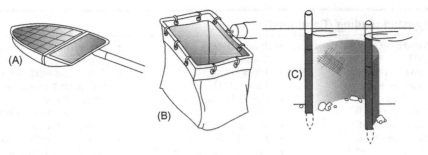

FIGURE 1.46

Aquatic nets: (A) aquatic dip net; (B) aquatic canvas net; and (C) kick screen.

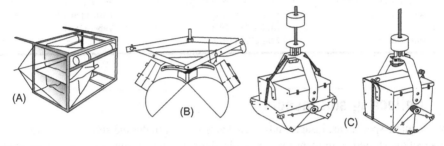

FIGURE 1.47

Specialized aquatic-insect-collecting techniques: (A) tow trap; (B) dredge; and (C) modified Birge–Ekman dredge.

purchased commercially (Fig. 1.46A and B). Kick screens (Fig. 1.46C) can be constructed easily from a fine-mesh window screen attached to two stout poles. Holding the screen on the bottom of the stream immediately downcurrent from where rocks and other debris are dislodged can yield a large number of aquatic organisms. Flume collecting is an old technique that has recently been revived. The extraction of insects and mites from flumes involves glass-topped sleeve cages, modified Malaise and tent traps, and Berlese photoattractive traps positioned directly above water. Other types of aquatic collecting techniques, such as tow traps and dredges (Fig. 1.47), require specialized equipment that we do not describe here. The collection of aquatic insects is of great importance in public health and general ecological studies.

Suggested reading

Sound traps
- Belton (1962)
- Cade (1975)
- Walker (1982)
- Bailey (1991)

- Dethier (1992)
- Parkman and Frank (1993)
- Thompson and Brandenburg (2004)

Aquatic insects
- Hodgson (1940)
- Welch (1948) (general)
- Jonasson (1954)

(Continued)

Suggested reading (Continued)

- Mundie (1956, 1964, 1966, 1971)
- Gerking (1957)
- Essig (1958)
- Lindeberg (1958)
- Grigarick (1959)
- Pieczynski (1961)
- Sladeckova (1962)
- Morgan et al. (1963)
- Cushing (1964)
- Macan (1964)
- Corbet (1965)
- Wood and Davies (1966)
- Kimerle and Anderson (1967)
- Tarshis (1968a,b)

- Waters (1969)
- Coulson et al. (1970)
- Elliott (1970)
- Carlson (1971)
- Edmondson and Winberg (1971)
- Mason and Sublette (1971)
- Fahy (1972)
- Finch and Skinner (1974)
- Langford and Daffern (1975)
- Murray and Charles (1975)
- Apperson and Yows (1976)
- Edmunds et al. (1976)
- Landlin (1976)
- McCauley (1976)
- Masteller (1977)

- Edmunds and McCafferty (1978)
- LaGasa and Smith (1978)
- Pennak (1978) (general)
- Ettinger (1979)
- Lawson and Merritt (1979)
- LeSage and Harrison (1979)
- Wood et al. (1979, methods of collecting, preserving, and rearing on pp. 45−53)
- McCafferty (1981)
- Merritt et al. (1984) (general)
- Halstead and Haines (1987)
- Weber (1987)

1.10 Collecting soil insects

As with aquatic specimens, insects and mites that live on or under the soil surface require special techniques and equipment for their collection and study. Often separators, extractors, and pitfall and Berlese traps can be utilized to collect soil-inhabiting insects. Various flotation techniques have been used to collect small insects and mites from soils and other substrates as well. Many soil-inhabiting species are of great economic importance because they devour the roots of crops. Many of these insects spend their immature stages in soil but emerge and leave the soil as adults. A considerable amount of literature on collecting soil insects has been published, the most useful of which is cited here.

1.11 Collecting ectoparasites

Ectoparasites are organisms that live on the body of their host. Examples include lice and fleas. Some ectoparasites, particularly those that fly, may be collected in some of the traps discussed previously, using their hosts as bait; others may be collected either by hand or with special devices described in the listed readings.

1.12 Collecting regulated insects

Collecting insects or other arthropods at points of entry into the state or country or by way of specific and intensive surveys is an important part of regulatory officials' duties. Extreme care must be exercised when collecting, preserving, and identifying the invasive organisms. Care must be taken to preserve the sample together with its collection information, in a manner that will allow others to also verify species identity. Both shipping and rearing insects (sections found in this book) are critical for regulatory entomologists to understand and utilize.

1.13 Collecting insects for pest management audits

Collecting tools and techniques are an important part of a pest management audit, regardless of the type of building or the commodities stored or manufactured in the facility. Intensive inspections require a solid understanding of how and why pests may enter a building. Understanding the importance of food, water, and harborage to pests' establishment is crucial in conducting pest audits. Understanding building construction and maintenance is also invaluable. Insect pest managers can take pest information observed during an audit and subsequently describe a correct course of action to eliminate existing and exclude other potential pests from facilities.

Suggested reading

Collecting soil insects
- Davidson and Swan (1933)
- Barnes (1941)
- Salt and Hollick (1944)
- Kevan (1933, 1962)
- Newell (1933)
- Turnock (1937)
- MacFadyen (1962)
- Murphy (1962)
- Teskey (1962)
- Brindle (1963) *(includes collecting methods)*
- Evans et al. (1964, techniques on pp. 61–88)
- Kühnelt (1976, observation and collecting techniques on pp. 35–65, bibliography on pp. 385–466)
- Lane and Anderson (1976)
- Akar and Osgood (1987)

Collecting ectoparasites
- Banks (1909) (mostly of historical interest, but describes the old methods and contains much general information about insects)
- Klots (1932)
- Cantrall (1939–1940, 1941)
- Comstock (1940)
- Chu (1949)

- Grandjean (1949)
- Lipovsky (1951) (mites)
- Cook (1954)
- Williamson (1954)
- Wagstaffe and Fidler (1955)
- Balogh (1958)
- Lumsden (1958)
- Morgan and Anderson (1958)
- Oldroyd (1958)
- Fallis and Smith (1964)
- Peterson (1964)
- Urquhart (1965, elementary directions for making insect collections, pp. 1–19)
- Knudsen (1966, 1972, pp. 128–176)
- Norris (1966)
- U.S. Department of Agriculture, Plant Pest Control Division (1966–1970)
- Watson and Amerson (1967)
- Lehker and Deay (1969)
- Nicholls (1970)
- Brown (1973)
- Ford (1973)
- Cogan and Smith (1974, p. 152)
- Thompson and Gregg (1974)
- Cheng (1975)

- McNutt (1976)
- Service (1976) (includes extensive bibliographies)
- Stein (1976)
- Martin (1977)
- Bland and Jacques (1978)
- Edmunds and McCafferty (1978)
- Lincoln and Sheals (1979)
- Southwood (1979) (includes extensive bibliographies)
- Banks et al. (1981)
- Eads et al. (1982)
- Pritchard and Kruse (1982)
- Arnett (1985)
- Stehr (1987, 1991)
- Borror et al. (1989)
- Dindal (1990)
- Stehr (1991)
- Dryden and Broce (1993)
- Kasprzak (1993)

Forensic entomology
- Erzinelioglu (1983)
- Anderson (1999)
- Castner and Byrd (2001)
- Greenberg and Kunish (2002)
- Haglund and Sorg (2002)
- Saferstein (2004)
- James and Nordby (2005)

Using insect glue traps as monitors is very important to pest auditors. Because inspections occur only over the course of 1 or 2 days, they represent a snapshot in time of what is occurring in the building. Monitors, on the contrary, work 24 hours a day, 7 days a week, and provide a record of events over a much longer period of time. Flashlights, screwdrivers, probes, flushing agents, scrapers, mirrors, collecting vials, clipboards, blueprints of the building, knee pads, and bump caps are all part of an auditor's routine inspection tool bag. On occasion, more sophisticated equipment is required to listen for or observe pests in areas where they hide. Auditors must be ever alert to signs of pest infestations, even when the pest cannot be found. Often alerting the owners to "conducive conditions" (those conditions that might enable a pest to enter or to flourish if introduced) can be of as much value as finding the pest.

1.14 Collecting insects for forensic or medico-criminal investigations

Use of arthropods in forensic or legal investigations has become standard practice in recent years. Insects and mites are often the subject of lawsuits especially where contamination of products, misapplication of pesticides, transmission of disease, or human myiasis is concerned. Collection and documentation of specimens as evidence requires great attention to procedure and protocol.

Many different collecting techniques can be employed in death-scene investigations. Direct observation or sweep net sampling can often provide important information of the adult stages of insects above the body if taken before the body is disturbed. These data are often lost evidence if not taken immediately. Direct collecting of insects from on, in, and around the body is one of the most effective means of obtaining reliable data. Often, collecting various stages of insects and recording the exact location of the collections are vital. Rearing a portion of the immature stages to adulthood can provide important times and identifications of insects. For detailed protocols about collecting and preserving insects and related organisms in forensic cases, please consult the cited references.

Vehicles travel through many different insect habitats. Most insects are found in specific habitats such as near water sources, high mountains, desert areas, woods, and forests. A project that incorporates these two concepts may illustrate the procedures followed by forensic entomologists. The project involves the examination of insects found on the front of a vehicle to determine where and when the vehicle traveled. The project can be configured to be as detailed or as simple as the researchers' needs and circumstances warrant.

Procedure: Forensic scientists carefully gather, record, and maintain evidence for future use by using a standard protocol. In this project, the student investigator will use a standard evidence sheet or form to record insect evidence of the vehicle passing through various habitats. To begin the study, the vehicle grill and radiator should be washed to remove all previous insects. A case number for each trip is created and the following data are recorded: vehicle's model, make, and year; a photograph of the front of the vehicle before and after the project to serve as a comparison; the vehicle's mileage at the start and finish of the project; the start and end dates of the project; weather conditions during the project; and habitat visited. To collect and preserve the insect specimens from the vehicle, take a piece of clear cellophane tape and touch it to the specimens then tape the cellophane tape to the evidence sheet. Repeat until representative samples have been collected and recorded.

Identify the collected specimens to the level (order, family, genus, or species) most comfortable to the researcher and compare to specimens known to exist in each of the habitats. To extend the study for more advanced students, travel through the same habitat once a month, or during early morning, midday, and at night to compare the differences in taxon diversity and occurrence as well as numbers of insects collected. This procedure can be repeated using other variables such as traveling speeds (keeping safety issues in mind) and comparing the number of insects by speed traveled.

1.15 Rearing

Collectors should take every opportunity to rear insects and mites. Reared specimens generally are in the best condition. Further, rearing provides life stages that otherwise might be collected only rarely or with great difficulty. With respect to parasitic insects, such as some hymenopterans and dipterans, rearing provides unambiguous proof of host associations and other scientifically important information.

By preserving one or more specimens from each of the stages as they are reared, the collector can obtain a series of immature stages along with associated adults (Clifford et al., 1977). Such series are desirable, especially for species in which the adult is known but the immature stages are unknown or difficult to identify. Often, the converse is also true. Some species of insects, such as stem-mining flies, are fairly abundant in the larval stage. Sometimes we do not know whether the specimens represent a species that has been described and named from an adult whose life history is unknown. Because adults of these flies are seldom found, the easiest way to obtain the stage necessary for specific determination is to rear the larvae or pupae.

If only a few specimens are reared, then the shed "skins" (exuviae) and pupal cases of puparia should be preserved because they are valuable when properly associated with the reared adult. Do not preserve a pupa or puparium with an adult unless you are positive that the association is correct. Put pupae in separate containers so that adults or parasites that emerge are associated with certainty. When feasible, the parasite's host should be preserved for identification. Keep careful notes throughout the rearing so that all data relative to the biology of the species are properly correlated.

1.15.1 Containers for rearing

To rear specimens successfully, rearing cages must simulate closely the natural conditions in which the immatures were found. A screen cylinder capped on both the top and bottom can be an effective rearing container (Fig. 1.48A). Almost any container will serve as a temporary cage for living insects or mites. A paper bag is a simple temporary cage that is very handy on field trips. Plant material or a soil sample containing insects or mites is placed in the paper bag, which is then sealed. A paper bag also can be placed over the top of a plant on which insects or mites are found. The bottom edge of the bag is tied tightly around the exposed stems. Stems are then cut and placed in a jar of water. Paper bags are not transparent and must be removed to observe the specimens or to determine when the foliage needs to be changed. Clear plastic bags are better suited to such viewing. However, they are not recommended for more than short-term use because they are airtight and specimens may be damaged by drowning in condensed water inside the bag.

A glass jar whose lid is replaced by a piece of organdy cloth or gauze held in place by a rubber band forms a simple temporary cage (Fig. 1.48B). A few such jars in a collecting kit are useful for holding live insects. For aquatic species, a watertight lid on the jars is advisable. Aquatic insects often die when transported over a considerable distance in water that sloshes in the container. Fewer specimens will die if the jar is packed with wet moss or leaves. After arrival at your destination, release the insects into a more permanent rearing container.

Certain aquatic insects may be reared readily indoors in an aquarium or a glass jar. The main goal is to duplicate their natural habitat. If the specimen was collected from a rapidly flowing stream, it probably will die indoors unless the water is aerated. Other insects do well in stagnant water. Aquatic vegetation usually should be provided in the aquarium, even for predaceous specimens, such as dragonfly nymphs, which often cling to underwater stems. Keep sufficient space between the surface of the water and the aquarium cover to allow the adult insect to emerge. The amount of space will vary according to the insect being reared. For example, a dragonfly needs considerable space and a stick, rock, or other object above the water on which to perch after emerging so that the wings will expand fully.

Aquatic insects can be reared in their natural habitat by confining them in a wire screen or gauze cage. Part of the cage must be submerged in water and anchored securely. The

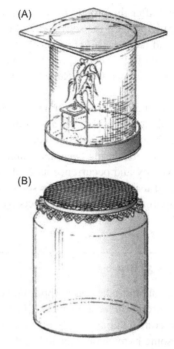

FIGURE 1.48

Rearing chambers: (A) screen cylinder capped on both ends and (B) glass jar with ventilated lid.

screen used in aquatic cages should be coarse enough to allow food through yet fine enough to retain the insects being reared.

Most adult insects, both terrestrial and aquatic, are teneral when they first emerge and should not be killed until the body and wings harden and the colors develop fully. Hardening may require a few minutes, hours, or days. Keep even small flies alive for 1 full day after they emerge. Specimens that are killed while being teneral will shrivel when mounted. Some insects, especially butterflies and moths, will beat their wings against the cage and lose many scales or tear their wings if kept in cages too long after emerging. Providing adequate space in which emerging insects may expand their wings fully and move about slightly is therefore critical in the design of rearing cages.

Some beetles and other boring insects often are abundant in bark and wood. Excellent adult specimens may be obtained if pieces of infested wood are placed in glass or metal containers. However, remember that rearing these insects sometimes requires considerable time. Cages made of wood or cardboard are not suitable for such insects because those found in wood or bark usually are well equipped, both in immature and adult stages, to chew their way through a cage made of such material and thus escape.

A flowerpot cage is one of the best containers for rearing plant-feeding species over an extended period. The host plant, if its size and habitat permit, is placed in a flowerpot and a cylinder of glass, plastic, or wire screen placed around the plant (Fig. 1.49). Another type of flowerpot cage is made by inserting a cane or stick, taller than the plant, into the soil in the pot. One end of a net or muslin tube is fitted over the edge of the pot and is held in place by a string. The other end of the tube is tied around the top of the stick. An advantage of the flowerpot cage is that the plant is living, and so it is not necessary to add fresh plant material daily.

FIGURE 1.49

Flowerpot cage for long-term rearing.

Plant-feeding mites will not wander far as long as suitable host material is available for them. Because mites are wingless even as adults, they can be confined in an open rearing container by making a barrier around the top edge or upper inner sides of the container with petroleum jelly or talcum powder.

Emergence cages are essentially rearing cages that are used when it is impractical or impossible to bring specimens indoors. Emergence cages may also be considered as traps and are discussed under that heading (see Section 1.6.5). With plant-feeding insects, a sleeve consisting of a muslin tube with open ends is slipped over a branch or plant and tied at one end (Fig. 1.50). The insects are then placed in the tube, and the loose end of the tube is tied. This cloth tube can be modified to allow observation of the insects by replacing the midsection with a "window" of clear plastic or

FIGURE 1.50

Muslin sleeve cage.

wire screen. If the insects in the tube require duff or debris in which to pupate, the tube should be placed perpendicular to the ground and duff or debris placed in the lower end.

1.15.2 Rearing conditions and problems

1.15.2.1 Moisture

The moisture requirements of insects and mites are varied. Examination of the habitat from which specimens were collected should provide clues about their moisture requirements in captivity. Many insects in the pupal stage resist desiccation. Species that normally infest stored foods also require very little moisture. In fact, many produce their own metabolic water. Most species found outdoors require higher levels of humidity than generally exist indoors. Additional moisture can be added to indoor rearing cages in several ways. To increase the humidity in a cage, keep a moist pad of cotton on top of the screen cover of the cage or place a moist sponge or a small glass vial filled with water in the cage. The mouth of the vial is plugged with cotton and the vial laid on its side so that the cotton remains moist. Pupae may be held for long periods in moist sawdust, vermiculite, sphagnum, or peat moss. In a flowerpot cage, the water used to keep the plant alive and moisture released through transpiration should provide sufficient moisture for the plant-feeding insects and mites. Spraying the leaves daily also may supplement moisture requirements in rearing cages. Too much moisture may result in water condensation on the sides of the cage, which may trap the specimens and damage or kill them. Excess moisture also enhances the growth of mold and fungus, which is detrimental to the development of most insects and mites. A 2%−3% solution of table salt sprayed regularly in the cage will help prevent mold and fungus growth.

1.15.2.2 Temperature

Of all the environmental factors affecting the development and behavior of insects and mites, temperature may be the most critical. Because arthropods are cold-blooded, their body temperature is usually close to the temperature of the surrounding environment, and their metabolism and development are directly affected by increases and decreases in temperature. Each stage of an insect or mite species has a low and a high point at which development ceases. These are called *threshold* temperature levels.

Most species that are collected and brought indoors for rearing can be held at normal room temperature; the optimum temperature for rearing varies from species to species and with different stages of the same species. As with all rearing techniques, every attempt should be made to duplicate optimum natural conditions. Specimens that normally would overwinter outdoors should be kept during the winter in rearing cages placed in an unheated room, porch, or garage. Never place an enclosed rearing cage in direct sunlight; the heat becomes too intense and may kill the specimens.

1.15.2.3 Dormancy and diapause

Insects and mites cannot control the temperature of their environment. Instead, they make physiological adjustments that allow them to survive temperature extremes. In regions with freezing winters, insects and mites have at least one stage that is resistant to low temperatures. The resistant stage may be egg, larva, nymph, pupa, or adult. When winter arrives, only the resistant stage survives. *Dormancy* is the physiological state of an insect or mite during a period of arrested development. *Diapause* is the prolonged period of arrested development brought about by such adverse conditions as heat, drought, and cold. This condition can be used to advantage in rearing. For example, if rearing cages must be unattended for several days, then many specimens can be refrigerated temporarily to slow down their activity and perhaps force diapause. This measure should be used with caution because the degree and duration of cold tolerated by different species vary.

> Dormancy:
>   The physiological state of an insect or mite during a period of arrested development.
> Diapause:
>   The prolonged period of arrested development brought about by such adverse environmental conditions as heat, drought, and cold.

The reverse situation that of causing diapause to end is equally useful. Overwintering pupae that normally would not develop into adults until spring can be forced to terminate diapause early by chilling them for several weeks or months, and then bringing them to room temperature so that normal activity will resume. Often, mantid egg cases are brought indoors accidentally with Christmas greenery. The eggs, already chilled for several months, hatch when kept at room temperature, often to the complete surprise and consternation of the unsuspecting homeowner.

1.15.2.4 Light

Most species of insects and mites can be reared under ordinary lighting conditions. However, artificial manipulation of the light period will control diapause in many species. If the light requirements of the species being reared are known, then the period of light can be adjusted so that the specimens will continue to develop and remain active instead of entering diapause. Light and dark periods can be regulated with a 24 hours timing switch or clock timer. The timer is set to regulate light and dark periods to correspond with the desired lengths of light and darkness. For example, providing 8−12 hours of continuous light during a 24-hour cycle creates short-day conditions that resemble winter; providing 16 hours of continuous light during a 24-hour cycle creates long-day conditions that resemble summer. Remember that many insects and mites are very sensitive to light; sometimes even a slight disturbance of the photoperiod can disrupt their development.

Artificial manipulation of the light period will control diapause in many species.

1.15.2.5 Food

The choice of food depends upon the species being reared. Some species are detritivores and will accept a wide assortment of dead or decaying organic matter. Examples include most ants, crickets, and cockroaches. Other insects display food preferences so restricted that only a single species of plant or animal is acceptable. At the time of collection, carefully note the food being consumed by the specimen and provide the same food in the rearing cages.

Carnivorous insects should be given prey similar to that which they normally would consume. This diet can be supplemented when necessary with such insects as mosquito larvae, wax moth larvae, mealworms, maggots, and vinegar flies or other insects that are easily reared in large numbers in captivity. If no live food is available, a carnivorous insect sometimes may be tempted to accept a piece of raw meat dangled from a thread. Once the insect has grasped the meat, the thread can be gently withdrawn. The size of the food offered depends on the size of the insect being fed. If the offering is too large, the feeder may be frightened away. Bloodsucking species can be kept in captivity by allowing them to take blood from a rat, mouse, rabbit, or guinea pig. *A human should be used as a blood source only if it is definitely known that the insect or mite being fed is free of diseases that may be transmitted to the person.*

Stored-product insects and mites are easily maintained and bred in containers with flour, grains, tobacco, oatmeal or other cereal foods, and similar products. Unless leaf-feeding insects are kept in flowerpot cages where the host plant is growing, fresh leaves from the host plant usually should be placed in the rearing cage daily and old leaves removed.

1.15.2.6 Artificial diets

Some species can be maintained on an artificial diet. The development of suitable artificial diets is complex, involving several factors besides the mere nutritional value of the dietary ingredients. Because most species of insects and mites have very specific dietary requirements, information regarding artificial diets is found mainly in reports of studies on specific insects or mites.

Suggested readings

Rearing
- Needham (1937)
- Hodgson (1940)
- Gerberich (1945)
- Chu (1949)
- Hopkins (1949)
- Harwood and Areekul (1957)
- Levin (1957)

- Fischer and Jursic (1958)
- Lumsden (1958)
- Peterson (1964)
- Krombein (1967)
- U.S. Department of Agriculture, Extension Service (1970)
- Ford (1973)

- Smith (1974)
- Sauer (1976)
- Clifford et al. (1977)
- Smith et al. (1977)
- Barber and Matthews (1979)
- Banks et al. (1981)
- Edwards and Leppla (1987)
- Gray and Ibaraki (1994)

1.15.2.7 Special problems and precautions in rearing

Problems may arise in any rearing program. Cannibalism, for instance, is a serious problem in rearing predaceous insects and necessitates rearing specimens in individual containers. Some species resort to cannibalism only if their cages become badly overcrowded. Disease is also a problem and can be caused by introducing an unhealthy specimen into a colony, poor sanitary conditions, lack of food, or overcrowding.

Cages should be cleaned frequently and all dead or clearly unhealthy specimens removed. Exercise care not to injure specimens when transferring them to fresh food or when cleaning the cages. Mites and small insects can be transferred with a camel's-hair brush.

Attacks by parasites and predators can also be devastating to a rearing program. Carefully examine the host material when it is brought indoors and before it is placed in the rearing containers. This reduces the possibility of predators and parasites being introduced accidentally into the cages. Also, place rearing cages where they will be safe from ants, mice, the family cat, and other predators.

1.16 **Collecting insects for molecular research**

Molecular studies of insects continue to expand our knowledge of phylogenetic relationships among insect taxa confirm the identification of a known species, support the description of new species, and elucidate the boundaries of morphologically similar species. Molecular analyses serve to correlate instars with their respective adults, a critical need in forensic entomology. In addition, molecular studies can provide data on host−insect interactions (Lehmann et al., 2012).

Collection of insect or mite specimens for molecular research requires few special techniques. A discussion of the detailed molecular procedures in the papers reviewed in this chapter is beyond the scope of this manual and interested readers are directed to the article(s). Asghar et al. (2015) provide an overview of different extraction methods for DNA from insets.

In most molecular studies, arthropods may be collected in much the same manner as discussed for general collections. Clean, contaminant-free specimens must be ensured; however, keep in mind that some insects may carry foreign debris on their bodies, such as pollen or other potential contaminants that may confound some studies. Some arthropods also carry parasites (either externally or internally) that may pose confusion when DNA or RNA extractions are done. Some mites commonly conceal themselves under wing covers of insects or other locations and may potentially go unnoticed.

Specific preservation techniques of the sample must follow the protocol determined by the procedure set out in Chapter 2, Agents for killing and preserving. It is imperative to know the ultimate and specific use of the specimen beforehand; however, because the preservation techniques may affect the molecular technique being performed.

Common methods include preservation in ethanol, freezing on dry ice and then holding in an ultralow freezer, and placement in liquid nitrogen or other specific products, such as RNALater, which helps preserve the more fragile RNA materials.

Packaging and labeling must also be done with care when specimens are meant for molecular studies. Using a soft pencil to label specimens in liquids is still the best method. Labeling specimens both inside and outside of the package or vial should be standard operating procedure because the extreme environment in which the specimens are held often causes labels, even with the best adhesive, to detach. Voucher specimens are often valuable. These may consist of properly preserved whole insects or parts thereof (e.g., wings of larger insects) to positively link the insect identification to the molecular data.

1.17 Preparation of insects for molecular research

1.17.1 Killing agents

Insect killing agents can impact the success of DNA extraction. In addition, the species under study can impact DNA analysis because of the specific cuticle's differential ability to absorb the killing solution. Table 1.1 lists the killing agents and the preservatives currently used by researchers. Note: some killing agents can also serve as preservatives.

1.17.1.1 Cyanide

Dean and Ballard (2001) measured DNA success by (1) the size of extracted DNA, (2) extraction yield, and (3) ability to amplify from four target regions of *Drosophila simulans* (Diptera: Drosophilidae). Based on these criteria, they extracted the highest amount of mitochondrial DNA from *D. simulans* killed with cyanide and the lowest DNA from flies killed in 70% ethanol. Storage time showed significant ($P < .001$) effect on amplification with fresh specimens showing greater success than specimens stored for 2 years.

1.17.1.2 Ethyl acetate

Ethyl acetate has long been used as a killing agent supplanting the previous use of hydrogen or potassium cyanide. Dillon et al. (1996) reported inconsistent results from Ichneumonidae (Hymenoptera) killed with ethyl acetate and questioned if the wide use of ethyl acetate as a killing agent may have led to inconsistent results, especially with museum specimens. Willows-Munro and Schoeman (2014) found that adult Lepidopterans killed with ethyl acetate, cyanide, and freezing at $-20°C$ produced cytochrome c oxidase (COI) with a 658 bp sequence suitable for barcoding. Their extraction success with ethyl acetate runs counter to that of Dillon et al. (1996).

For large Hymenopterans (Ichneumonidae) with extensive fat reserves, Quicke (1999) reported that specimens killed with ethyl acetate and air-dried revealed fungal DNA because of the invasion of fungi on the decayed tissues.

Table 1.1 Comparison of DNA extraction success influenced by collecting methods, preservation techniques, and taxon.

Taxon	Procedure	Killing agent	Preservative	Contentration preservative	Treatment	Procedure before analysis	DNA results	Citation
General review	Technique review	Killed in preservative	1. Ethanol 2. Ethanol 3. Alive then frozen	Prefix no. refers to preservative 1. 70% 2. 95%–99%	Not stated	1. Kill then dry quickly	2. Good option 3. Good option	Krogman and Holstein (2010)
Coleoptera Carabidae	Preservative	1. Ethyl acetate 2. Ethanol 3. Carnoy's Solution 4. DNA isolation buffer 5. Liquid nitrogen	5. –80°C	2. 95% 3. 3:1 methanol: acetic acid (see citation)	Homogenized or whole	11–144 days Stored in 95% ethanol except those in liquid N	Ethanol, DNA isolation buffer and cryopreservation all yielded intact DNA	Reiss et al. (1995)
Coleoptera Carabidae Passinae 14 species	Natural history	Not stated	Between 2 and 94 years ago before extraction	Dried	Whole	Dried	Recovered DNA from 11 specimens, but not from 2 collected in 1910 and 1980	Gilbert et al. (2007)
Coleoptera Gyrinidae Dysticidae Carabidae	Preservative	100% ethanol	1. Ethanol 2. Freeze killed at –80°C	1. 100%	Whole	1. Stored at –20°C in 100% ethanol	Increased number of loci available for beetle	Wild and Maddison (2008)
Coleoptera Tenebrionidae *Tenebria molitor*	Preservative	Killed in preservative	1. Acetone 2. Carnoy's solution 3. Ethanol 4. Ethanol 5. Ultracold freezer 6. Liquid nitrogen 7. Dried specimens	1. Standard grade 2. Standard grade 3. 75% 4. 100%	Whole	All stored for 1 day to 18 months at room temperature except frozen specimens	Acetone, 100% ethanol quality DNA	Bisanti et al. (2009)

(Continued)

Table 1.1 Comparison of DNA extraction success influenced by collecting methods, preservation techniques, and taxon. *Continued*

Taxon	Procedure	Killing agent	Preservative	Concentration preservative	Treatment	Procedure before analysis	DNA results	Citation
Coleoptera Curculionidae *Ips typographus*	Preservative	Killed in preservative	1. Acetone 2. Formalin 3. Sodium benzoate 4. Copper sulfate 5. Sodium chloride 6. Ethylene glycol 7. Renner solution 8. Water with fungicide 9. Glycerin	1. 30% 2. 30% 3. 3% 4. 3% 5. Saturated 6. Pure 7. Glycerin, acetic acid, water in 40:20:10:30 ratio 8. Euxyl K100	Whole wings	4 weeks at room temperature	Positive PCR for: Renner's solution, ethylene glycol, glycerin, NaCl DNA quantity from Renner solution, sodium chloride, glycerin, NaCl	Stoecke et al. (2010)
Coleoptera Carabidae *Cylindera lemniscata* Staphylinidae *Arhetini* sp.	Killed and preserved	100% ethanol	1. Propylene glycol	20%, 40%, 60%, 80%, and 100%	Legs, heads, thoraces	Stored in propylene glycol at 5 concentrations tested at 2 weeks, 3 months, and 6 months for DNA	All concentrations propylene glycol yielded ≈ 800 bp	Ferro and Park (2013)
Coleoptera Carabidae *Amara alpina*	Natural history (pinned)	Not stated	1. Less than 10 years old 2. Greater than 10 years old 3. Ancient (permafrost deposits) Collection age range: 9–136 years before DNA extraction	Pinned ancient specimens greater than 560,000–590,000 years old	Hind legs or sclerites from ancient specimens	Ancient specimens stored in −20°C	Museum specimens 46%–100% DNA recovery Ancient specimens yielded 26% mtDNA and nuDNA ≈ 10 yielded either mt- or nuDNA, 54% yielded neither	Heintzman et al. 2013

	Preservative	Solution	Concentration	Part	Storage	Results	Reference
Coleoptera Buprestidae *Lamprodila rutilans* Nitidulidae *Meligethes aeneus*	Not stated	Sodium dodecyl sulfate (SDS) and eDTA 1. SDS and eDTA 2. SDS and eDTA 3. SDS and eDTA	1. 2% and 100 mM 2. 1% and 50 mM 3. 0.66% and 33 mM	Whole	1, 4, and 8 weeks	All solutions effective	Pokluda et al. (2014)
Coleoptera Curculionidae Ambrosia beetle *Xylosandrus compactus*	Killed in preservative	1. Ethanol 2. Purell hand sanitizer 3. Prestone low-tox antifreeze 4. Super tech antifreeze	1. 95%	Whole	2 and 7 days All specimens rinsed and stored in 95% ethanol at −20°C	All successfully amplified DNA	Steininger (2015)
Diptera Simuliidae, *Simulium damnosum*	Killed in preservative	1. Liquid nitrogen 2. Ethanol 3. Silica gel 4. Carnoy's solution 5. Formal saline 6. Methanol 7. Ethanol 8. Propan-2-ol	4. Ethanol-acetic acid 5. 10% formaldehyde 6. 60%, 80%, and 100% 7. 60%, 80%, and 100% 8. 60%, 80%, and 100%	Whole	Stored at 4°C Stored at 4°C 6. −2°C, 4°C, or ambient 7. −2°C, 4°C, or ambient 8. −2°C, 4°C, or ambient	Liquid nitrogen, silica dried specimens yielded high DNA concentrations. Carnoy's, ethanol, methanol, propan-2-ol DNA fragmented formal saline. Pinned specimens undetectable results	Post el al. (1993)
Diptera Culicidae *Anopheles gambiae*	Natural history (pinned)	15—93 years prior to analysis Total: 20 pinned		Abdomen and legs of some		Sizes of DNA products extracted same as those of fresh specimens. 17 of 20 mosquito species yielded amplifiable DNA, 12 PCR identifications matched label ids.	Townson (1999)

(Continued)

Table 1.1 Comparison of DNA extraction success influenced by collecting methods, preservation techniques, and taxon. *Continued*

Taxon	Procedure	Killing agent	Preservative	Contentration preservative	Treatment	Procedure before analysis	DNA results	Citation
Diptera Drosophilidae *Drosophila simulans*	Natural history (pinned)	Killed in preservative	Pinned for 2 years	Killed in: 1. Cyanide 2. Ethyl acetate 3. Freezing at −20°C 4. 70% ethanol	Stored with and without naphthalene	Two years after killing	Naphthalene: no significant effect on extraction or amplification. Killing method significantly affect extraction but not amplification. Time of storage did not affect extraction, but amplification impacted by storage time but killing method had no significant effect	Dean and Ballard (2001)
Diptera Sarcophagidae *Sarcophaga Peckia Biaesoxipha Rowinia Wohlfahrtia Brachicoma Mucscidae Musca*	Natural history (pinned)	Frozen −70°C;	Fresh and natural history collection	Fresh at −70°C 95% ethanol dried	Pinned specimens entire thorax, frozen thoracic muscles	Fresh frozen: 1.2–15 months Ethanol: 1.2–8 months Pinned: 1–15 years	Extracted DNA from 15-year-old specimens	Wells et al. (2001)
	Biaesoxipha Rowinia	*Biaesoxipha Rowinia*	*Rowinia*	*Rowinia*				
Diptera: Calliphoridae, 8 species *Cochliomyia* spp. *Chrysomya* spp. *Lucilia* sp. *Hemilucilia* sp. *Dermatobia* sp.	Natural history (pinned)	Not stated	Pinned 1–57 years prior to analysis		Thorax, abdomen, legs, and wings removed and ground		All 35 samples amplified 137 bp COII, 43% of 155bp COII, 34% of 305bp COII, 63% of 315bp of COII, and 40% of 357bp of COII	Junqueira et al. (2002)

Taxon	Specimen source	Preservation	Method / sample	Age / frequency	Portion used	Storage	Results	Reference
Diptera Simuliidae *Simulium posticatum* *S. erythrocepalum*	Nondestructive	95% ethanol	Sonication	Frequency: 50/60 Hz	30 s to 1 min	Fixed in 95% ethanol Stored in 80% ethanol at −20°C	Sonicated larvae had higher DNA than nonsonicated larvae	Hunter et al. (2008)
Diptera Culicidae *Culicoides* spp.	Preservative light trap	Not stated	Ethanol	75%	Whole, some homogenized	Not stated	Differentiated three species in Obsoletus group	Lehmann et al. (2012)
Diptera Simuliidae *Gigantodax* spp. *Simulium* spp.	Natural history (pinned)	Not stated	271 pinned specimens, 2 genera, and 36 species	Most specimens over 10 years old with mean of 26 years, median of 28 years	2 or 3 legs per specimen		Recovered high-quality DNA from 80% (215) of the specimens: 18% had full length 500–658 bp, 36% had 201–499 bp, 46% less than 200 bp. Recovered barcodes from 51-year-old specimens	Hernandez-Trianna et al. (2014)
Diptera Calliphoridae 18 species Sarcophagidae 3 species	Nondestructive	Freezing	Sonication	Frequency: 42 kHz	1–2 min	Held −20°C	Sequence length reduced and low DNA quality	Stamper et al. (2017)
Hemiptera: Cydnidae Dinidoridae Tessaratomidae Thyrecoridae Parastrahiidae	Natural history (pinned)	Not stated	48 pinned of 46 species of Pentatomoidea	Collected 15 years ago to specimen collected in 1894	Thorax and abdomen		PCR amplified mitochondrial DNA from 10 of 48 specimens	Lis et al. (2011)
Hemiptera Heteroptera (Most Miridae) 1689 specimens	Natural history (pinned)	Not stated	Collected less than 40 years ago, median age 11	Most pinned, a few in 95% ethanol	Single leg		Barcodes for 80%, for specimens up to 35 years old. 380 species in 191 genera in 30 families	Park et al. (2011)

(Continued)

Table 1.1 Comparison of DNA extraction success influenced by collecting methods, preservation techniques, and taxon. *Continued*

Taxon	Procedure	Killing agent	Preservative	Contentration preservative	Treatment	Procedure before analysis	DNA results	Citation
Hymenoptera Halictidae *Augochlorella striata*	Preservative: pan trap	Collected in water and dish detergent	1. Methanol 2. Methanol 3. Ethanol 4. Ethanol 5. Ethanol 6. Ethanol-methanol 7. DMSO	1. 50% 2. 95% 3. 50% 4. 70% 5. 95% 6. 70:30 7. Pure	Whole	Stored room temperature 1–12 months	DMSO was best but has issues, 95% ethanol good alternative to DMSO	Frampton et al. (2008)
Hymenoptera Ichneumonidae *Ventura canescens* Encritidae *Leptomastix dactylopii*	Preservative	Killed in preservative	1. –80°C 2. Ethanol 3. Killed with CO_2 4. Killed with CO_2 5. Critical point dried 6. Museum specimens 7. Killed in ethyl acetate then air-dried 8. Killed in 100% ethanol then air-dried 9. Killed, stored in ethylene glycol 10. Killed, stored in formalin formaldehyde 3. ethanol + 10% glycerin	2. 100% at 4°C 6. 5 or 45 years 10. 4% formaldehyde 3. 90% ethanol	All specimens stored for 7 days to 18 months	4. Stored in silica gel 5. Stored in 100% ethanol and stored in silica gel desiccator, or stored under naphthalene	Freezing at –80°C or 100% ethanol 4°C Ethyl acetate and air-dried low yields Efficiently preserved DNA for extraction	Dillon et al. (1996)

Taxon	Source	Killing	Storage method	Concentration	Body part	Temperature	Results	Reference
Hymenoptera Formicidae *Atta* spp.	Preservative	Killed in 95% ethanol	1. Cryofreezing 2. Ethanol 3. Ethanol 4. Ethanol 5. Silica gel 6. Buffer	2. 95% 3. 95% 4. 95% 6. 0.25 M eDTA, 2.5%, SDS, 0.5 M Tris-HCl, pH 9.2	Head crushed	−70°C −20°C −4°C Room temperature Room temperature Room temperature	Cryofreezing best, next 95% ethanol at −20°C and 4°C were effective	Carvalho and Vieira (2000)
Hymenoptera Ichneumonoidea 7 subfamilies Chalcidoidea 7 subfamilies	Preservative	Killed in preservative	1. Liquid nitrogen 2. Ethyl acetate, slow dried 3. Rapid drying, silica gel 4. Ethanol 5. Ethanol then frozen 6. Ethanol 7. Ethanol 8. Ethanol 9. Ethanol then frozen	4. 70% 5. 70% 6. 70% with CPD or HMS 7. 96% at room temperature 8. 96% with CPD or HMS 9%	Whole	Not stated in silica gel	1. Good for chalcids 2. Variable 3. Fair 4. Fair to good 5. Fair to good 6. Fair to good 7. Good 8. Good 9. Good	Quicke et al. (1999)
Hymenoptera Apidae *Bombus apposites* *Bombus huntii* *Bombus occidentalis*	Natural history (pinned)	Unknown	Natural history collection		Mesothoracic leg		Microsatellite alleles amplified from 101-year-old specimens but amplification success in 60-year-old specimens	Strange et al. (2009)
Hymenoptera Formicidae 7 subfamilies	Preservative	Killed in preservative	1. Ethanol 2. DMSO 3. Propylene glycol 4. RNALater 72 pinned specimens 19–115 years old	1. 95%		Stored in solution at room temperature until extraction	95% ethanol best propylene glycol can be used in lieu of 95% ethanol	Moreau et al. (2013)
Hymenoptera Apidae *Xylocopa* spp.	Natural history	Not stated	Washed in 95% ethanol, dried, and frozen		Middle legs	Frozen for several hours	60 samples successfully amplified DNA, 12 extractions failed to produce quantifiable DNA	Blaimer et al. (2016)

(Continued)

Table 1.1 Comparison of DNA extraction success influenced by collecting methods, preservation techniques, and taxon. *Continued*

Taxon	Procedure	Killing agent	Preservative	Contentration preservative	Treatment	Procedure before analysis	DNA results	Citation
Lepidoptera	Preservative	Potassium cyanide	Ethyl acetate	Held in ethyl acetate for 20–30 min	Right leg	Stored 20–30 min	More than 500 bp detected	Schmidt et al. (2017)
Geometridae		Ammonium chloride	Silica gel					
			6. Ethanol	6. 70% with CPD or HMS			6. Fair to good	
			7. Ethanol	7. 96% at room temperature			7. Good	
			8. Ethanol	8. 96% with CPD or HMS 109%			8. Good	
			9. Ethanol 1924–89				9. Good Between 1954 and 1989	
Psocoptera Caeciliusidae *Xanthoscelius sommermanae* Amphipsocidae *Polypsocus corruptus* Microsporidia Microsporia	Killed and preservative	Propylene glycol	Propylene glycol	Commercial grade	Abdomen whole	Stored in 100% ethanol for 2 months	Two new species microsporidia described from sequence and morphological data	Sokolova et al. (2010)

Mixed taxa

Taxon	Procedure	Killing agent	Preservative	Contentration preservative	Treatment	Procedure before analysis	DNA results	Citation
Hemiptera	Preservative	Killed in preservative	1. Acetone	All pure grade	Homogenized some whole	13–36 months	Acetone most effective, methanol and chloroform no DNA visible	Fukatsu (1999)
Aphididae, Psyllidae			2. Ethanol					
Aleyrodidae						18 months		
Delphacidae			3. 2-Propanol			9 months		
Deltocephalidae			4. Diethyl ether			9 months		
Lepidoptera			5. Methanol					
Pyralidae			6. Chloroform					
Isoptera: Termitidae						27–63 months		
Rhinotermitidae						3–67 months		
Blattodea								
Epilampridae						18 months		
Panaesthilidae						48 months		

Taxa	Method	Killing	Preservative	Concentration	Extraction	Treatment	Results	Reference
Arachnida: Acarina, Aranea; Coleoptera: Coccinellidae; Dietera; Hymenoptera: Eurytomidae	Nondestructive	Killed in 80% ethanol	Ethanol	80%	GuSCN-based extraction buffer	60°C waterbath for 1, 2, or 4 h	Slight discoloration to significant surface distortion; DNA amplified from all species; Between 1954 and 1990	Rowley et al. (2007)
Coleoptera: Dermestidae, Buprestidae, Cerambycidae; Hemiptera: Aphididae; Diptera: Tephritidae	Nondestructive	Killed in preservative	1. Propylene glycol 2. Air-dried 3. Ethanol 4. Fresh	1. 20%, stored in ethanol 3. Not given	Whole specimen	Storage time not given	DNA extracted with ANDE. All preservative showed at least 500 bp	Castalanelli et al. (2010)
Coleoptera: Scarabaeidae; Lepidoptera: Noctuidae	Natural history (pinned)	Not stated	Beetles 2–53 years old at start of analysis, mean age 32.8 and median age 34.0; Lepidoptera Mean age = 16.0 and 21.3 years, median age = 15 years		Legs		PCR amplicons yielded 559 bp of COI gene from Coleoptera, Lepidoptera, Diptera, Hemiptera, Odonata. 67% for specimens up to 25 years old and 51% up to 55 years old	Mitchell (2015)

1.17.1.3 Freezing

Freezing live specimens at $-20°C$ and storing for 2 years with and without the presence of naphthalene produced DNA from *D. simulans* (Diptera: Drosophilidae) that were intermediate between the cyanide treatment and the 70% ethanol treatment. The 2-year storage time had a significant ($P < .001$) effect regardless of the killing method (Dean and Ballard, 2001). For nocturnal Lepidoptera studied by Willows-Munro and Schoeman (2015), freezing at $-20°C$, along with ethyl acetate and cyanide produced satisfactory DNA.

1.17.1.4 Ethanol

The hymenopterans Ichneumondidae and Chalcidae killed with 70% ethanol and subsequently frozen showed fair to good extraction of DNA as did critical point drying (CPD) or storage in hexamethyldisilazane (HMDS). The success rate increased to good when specimens were killed in 96% ethanol and stored in a freezer or CPD, or stored in HMDS (Quicke et al., 1999).

1.17.2 Preservation of specimens

DNA sequencing techniques have become so advanced that extraction and amplification success depend greatly on how well the specimens are preserved (Schmidt et al., 2017). The ideal preservative should remove water and oxygen to minimize DNase activity, reduce hydrolytic processes, and prevent contamination by other organisms, including microorganisms. Long-term storage of preserved specimens for DNA analysis must prevent further denaturation and degradation of proteins and all oxidative processes must be slowed or stopped. In addition to the preservation technique, degradation of DNA in specimens can be attributed to time, heat, and the age of the specimen. If preservatives used are 70% ethanol and other liquids that slowly penetrate the cuticle such as methanol or Carnoy's solution, the resulting DNA extractions are less than if no preservative was used. Choice of suitable preservative can minimize the uncertainty of extracting suitable DNA (Nagy, 2010). According to Quicke et al. (1999), who reviewed different preservation techniques for hymenopteran specimens, advantages and disadvantages differ by the technique used. Contradictory results are due to a lack of detailed data on methods and materials used by previous researchers.

Selection of appropriate preservation methods depends on the collecting techniques used, the availability of drying and buffering materials, ability to immediately extract DNA after collecting, and access to freezing equipment.

1.17.2.1 Freezing

Freezing at $-20°C$, $-80°C$, or in liquid nitrogen produces DNA results similar to DNA from fresh specimens (Dillon et al., 1996; Reiss et al., 1995; Willows-Munro, 2014). Deep freezing and rapid drying options are often unavailable in field work so other preservation techniques are needed.

In a study of random amplified polymorphic DNA extraction of DNA from *Atta* spp. (Hymenoptera: Formicidae), Carvalho and Viera (2000) tested cryofreezing, storage in 95% ethanol held at different temperatures, a buffer solution, and a silica gel treatment. All methods tested yielded DNA for up to 210 days but at 360 days, DNA was degraded in specimens stored in 95% ethanol at room temperature. Thus low-temperature storage is recommended for long-term storage, whereas silica gel and buffers can serve as viable alternatives.

1.17.2.2 Ethanol

For long-term storage in the absence of freezing, specimens can be stored in 95%−99% ethanol but long-term storage in 70% ethanol is not recommended. If specimens are stored in 70% ethanol, the specimens should be placed in a dark, cool place (Krogmann and Holstein, 2010). The use of 100% ethanol is not recommended because of the presence of trace amounts of benzene in absolute ethanol. Storage of specimens in small vials capped with a stopper and placed into a canning jar filled with 95% ethanol mitigate the need to constantly replace ethanol. However, Moreau et al. (2013) found that in Formicidae, extracted concentrations of DNA were best using ethanol versus dimethyl sulfoxide (DMSO) or RNALater. As cautioned by Moreau et al. (2013), ethanol may show inconsistent results. They cite work by Fukatsu (2002) showing that acetone-preserved specimens more effectively extracted DNA than absolute ethanol. The difference may be that Moreau et al. (2010) used 95% ethanol. Wild and Madison (2008) increased the number of loci available for beetle systematics in their work on three families of Coleoptera by storing beetles in absolute ethanol at -20°C.

For Bisanti et al. (2009), absolute ethanol- and acetone-preserved Tenebrionidae (Coleoptera) had higher quality DNA than specimens preserved in 75% ethanol. However, acetone-preserved specimens showed lower DNA degradation than those preserved by absolute ethanol perhaps because of the rapidity of acetone penetration into the cuticle.

In the field, specimens collected in 70% ethanol should be dried quickly and stored in a freezer or high percentage of ethanol. If collected specimens are to be stored in 95%−99% ethanol the ethanol−specimen ratio should be greater than 5:1. For small mites, spiders, and insects, Rowley et al. (2007) stored all specimens in 80% ethanol to prevent distortion of the exoskeleton.

1.17.2.3 Hand sanitizer

Where high concentrations of ethanol may not be readily available, hand sanitizers can serve as an effective alternative to other preservatives especially for citizen science projects. Ambrosia beetles, *Xylosandrus compactus* (Coleoptera: Curculionidae), stored in a high-concentration of alcohol-based hand sanitizer, had highly amplified DNA as did specimens preserved in 95% ethanol, commercial propylene glycol, and ethylene glycol (Steininger, 2015).

1.17.2.4 Methanol

The Halictidae (Hymenoptera), *Augochlorella striata*, was preserved in 50% and 95% methanol, and both preservatives failed to amplify polymerase chain reaction (PCR) using mitochondrial primers (Frampton, 2008).

1.17.2.5 Drying fresh specimens

If liquid preservatives or freezing equipment are unavailable in the field, silica gel can be a viable alternative. Post et al. (1993) examined the viability of DNA extracted from adult Simuliidea (Diptera) preserved with liquid nitrogen, ethanol, silica gel, Carnoy's solution, formal saline, methanol, and propan-2-ol. Carnoy's solution, methanol, and propano-2-ol failed to yield results suitable for DNA fingerprinting. Silica gel-dried specimens yielded the same quality of DNA as specimens stored in liquid nitrogen and the greatest amount of extracted DNA came from materials stored in absolute ethanol held at 4_oC (Post et al., 1993). Expended silica gel can be reconstituted in a drying oven, making this an economical preservative.

1.17.2.6 Critical point drying

To avoid inconsistent results from museum specimens that fail to state the killing agent, CPD drying of specimens may offer a degree of success. Dillon et al. (1996), in a study of Hymenoptera, killed specimens with CO_2 and critical point dried the specimens. Subsequent DNA analysis indicated that CPD specimens yielded DNA as did specimens stored at -80_oC, 100% ethanol, fresh specimens, and rapid air-dried, whereas museum specimens, ethyl acetate, and formalin showed degraded DNA.

1.17.2.7 Acetone

PCR confirmed that acetone, ethanol, 2- propanol, diethyl ether, and ethyl acetate preserved RNA, DNA, and proteins in the pea aphid, *Acyrthosiphon pisum* (Hemiptera: Aphididae) and the aphid's bacterial endosymbionts, *Buchnera* sp., mycetocyte symbiont, *Wolbachia*, and gut bacteria (Fukatsu, 1999). According to Fukatsu (1999), acetone and ethanol were more vulnerable to water contamination than DNA so if RNA is to be preserved, water contamination of the specimens must be kept at a minimum. Thus after initial preservation, the hydrated acetone should be replaced with fresh acetone several times, which will preserve specimens for several years.

1.17.2.8 Sodium dodecyl sulfate

Pokuda et al. (2014) stored small (≈ 72 mm long) nitidulid and large buprestid (≈ 714 mm long) beetles in three different solutions of an aqueous 2% sodium dodecyl sulfate (SDS), and all concentrations were mixed with a 100 mM detergent and ethylenediaminetetraacetic acid (eDTA) solution. Beetles were stored for 1, 4, and 8 weeks prior to analysis, and all solutions and all storage times effectively isolated DNA.

1.17.2.9 Dimethyl sulfoxide

DMSO was the best preservative among methanol and ethanol at concentrations ranging from 50% to 100%, an ethanol−methanol mixture (70:30) (Frampton et al., 2008). If DMSO is not available, 95% ethanol could be used as a substitute. DMSO alters morphological characters, making specimens unsuitable for morphological identification.

1.17.3 Preservatives in traps

Prior to starting a molecular analysis of insects, questions regarding preservation methods and the duration of storage time before analysis should be answered. Will specimens be collected and maintained live until DNA extraction? Are there facilities and equipment available to dry, cool, or freeze specimens in the field, and what are the costs, if any, of transporting the specimens to the laboratory? And are there restrictions to transporting specimens especially if hazardous chemicals are involved in shipping?

Traps are often placed in remote sites requiring travel time and are often emptied and fresh preservatives added once a week or at longer intervals. If the traps are emptied once a week, the preservative must retain sufficient volume to cover the specimens because evaporation and dilution of the killing/preservatives are major issues when preserving DNA quality. The preservative must be nontoxic to the researcher, environmentally benign, kill specimens quickly to avoid DNA degradation, and the preservative must be readily available even in remote areas. If large samples are to be collected, the cost of the preservative must be considered.

1.17.3.1 Water and detergent

Frampton et al. (2008) preserved bees collected in pan traps filled with water and a commercial dish detergent. The traps were placed in the sun and shade between 6 and 24 hours, after which the bees were stored in various concentrations of methanol, ethanol, a mixture of methanol and ethanol, and pure DMSO. After drying, the specimens were pinned and saved until DNA extraction. Pure DMSO maintained overall genomic quality, whereas the other preservatives exhibited DNA degradation. DMSO, however, caused morphological changes to the specimens that made traditional identification difficult.

1.17.3.2 Ethanol fuel

Ethanol fuel proved to be the most efficient killing solution in pitfall traps for Orthoptera and permitted DNA to be successfully characterized in a study conducted in Brazil (Szinwelski et al., 2012). Other killing solutions tested that failed to keep DNA intact were ethanol−glycerin−10% formaldehyde mixture and a 100% ethanol−10% glycerin solution. The benefits of ethanol fuel as a killing and preservative in pitfall traps are its availability and minimal cost when used in large volumes. Szinwelski et al. (2012) recommends that the volume of ethanol fuel be sufficient to allow for 2 days of evaporation.

1.17.3.3 Sodium dodecyl sulfate and ethylenediaminetetraacetic acid

Pokulda et al. (2014) tested an aqueous mixture of the chelator SDS and the detergent eDTA in three different concentrations for efficacy on preserving DNA in two families of Coleoptera. All solutions effectively preserved DNA. The SDS and eDTA solutions may solve the issue of using ethanol in traps not regularly attended.

1.17.3.4 Propylene glycol

To overcome the inherent disadvantages of ethanol use in various traps, propylene glycol has been used successfully in pitfall traps to extract DNA from Carabidae (Coleoptera) and Staphylinidae (Coleoptera) (Ferro and Park, 2013). Microsporidia, associated with Psocoptera captured in a canopy Malaise trap containing propylene glycol, produced DNA to permit the discovery of four new species of microsporidia (Sokolova et al., 2010). Fixation and storage of arthropods to be stored at high temperatures ($>20°C$) for a considerable amount of time are preferable to 70%−90% ethanol under similar storage conditions (Sokolova et al., 2010).

1.17.3.5 Formaldehyde

Formalin has been used in traps because it does not evaporate as rapidly as ethanol. Dillon et al. (1996) tested ten preservation techniques for their DNA extraction efficiency from two parasitic wasps (Hymenoptera: Ichneumonidae and Encrytidae). Formalin failed to yield DNA that could be PCR amplified. In addition, Stoecke et al. (2010) recorded that 3% formalin along with acetic acid had the highest error rate and lowest PCR rate of the nine trap preservatives tested.

1.17.4 Pinned specimens (Natural History Museum Collections)

Museum specimens (dried) have been an underutilized source of DNA material because of the impression that specimens older than 2 years have degraded DNA and that existing PCR techniques fail to extract DNA from old specimens (Mitchell, 2015).

   If successful extraction of DNA from old museum specimens can be accomplished, data generated can provide a wide range of inter- and intraspecific species diversity based on environmental changes over time. With the increasing political and economic restrictions on collecting insects, researchers frequently have limited access to fresh specimens. Insect labels often lack detail on the type of killing agent used, how long the specimen was held in the killing agent, and the conditions (humidity, temperature, and light) under which the specimens were held prior to pinning. These missing data can profoundly impact DNA extraction success especially if molecular analyses are compared against the same taxa killed in a different manner (Dillon et al., 1996).

   Regardless of the specimen source, contamination of samples from previous extractions of DNA is another source that may create difficulty in sequencing DNA (Blaimer et al., 2016). Thus the loss of the specimen or parts of the specimen must be carefully weighed against the value of the intended molecular study even with the development of sophisticated amplification techniques (Dean and Ballard, 2001).

Several researchers have examined the influence of specimen age on extraction efficiency. Different researchers measure amplification success based on the number of mitochondrial COI gene fragments recovered from the PCR primer sets. Heintzman et al. (2014) characterized DNA from modern and ancient species of *Amara alpina* (Coleoptera: Carabidae). Specimens were divided into three age classes: modern (<10 years old), museum (>10 years old), and ancient (permafrost deposits). Based on the comparative protocol used, 96% of museum specimens yielded mtDNA and nuDNA, whereas 26% of the ancient specimens yielded mtDNA and nuDNA, 10% yielded either types of DNA, and 54% failed to yield any DNA.

For the age variable, Wells et al. (2001) created sequence data from the 783 bp region of the COI gene DNA from the entire thoraxes of three species of pinned adult Sarcophagidae (Diptera). The study occurred 1.0, 13.25, and 15.2 years after pinning with no differences detected among the three age categories.

Watts (2007) attempted to extract DNA from the extracted tibia of pinned damselfly (Odonata: Coenagrionidae) adults collected from 1924 to 1989. Specimens collected in 1989 had 100% PCR amplification but success decreased from 1954 to 1924, which allowed no PCR amplification. These results are counter to Goldstein and DeSalle (2003), who extracted DNA from the legs of an 80-year-old insect. The difference in results could be attributed to preservation conditions.

Strange et al. (2009) successfully generated multiloci genotypes extracted from legs of pinned specimens of bumble bees that are nearly 100 years old. However, older specimens with larger alleles had lower amplification success than newer specimens with smaller alleles, which was similar to the results of Dean and Ballard (2001).

Gilbert et al. (2007) placed whole specimens of ground beetles (Carabidae: Passinae) in a digestion buffer and through a series of steps created a pellet used to characterize DNA using PCR. Of the 14 carabids studied, 13 were sequenced—with the single failure coming from a specimen collected over 90 years ago. The carabids used in the study were treated with 100% ethanol to stop the digestion process, dried, and returned to the collection without any visible morphological changes.

The buffer used in the study entered into the body through various pathways, such as the mouth, anus, and spiracles and possibly through thin cuticular areas between sclerites, broken setae, and ectodermal glands. Another entry site might be through the right elytron and thorax where the insect pin was placed.

DNA extracted from 2-year-old museum specimens was more sheared than that of freshly assayed ones. One possible source of shearing is from cellular endonucleases. DNA-damaging nucleases remain active for longer periods of time in a moist environment (Dessauer et al., 1990; Junueira et al., 2001), and desiccating the specimen can inhibit degradation (Doyle and Dickson, 1987; Chase and Hills, 1991; Seutin et al., 1991; Lindahl, 1993; Rogstad, 1992).

For 271 pinned specimens from 2 genera and 36 species of Simuliidae (Diptera), high-quality DNA was extracted from 80% (215) specimens with most specimens over 10 years old (mean of 26 years). Of the 215 specimens, 12.5% showed full length (658 bp), 30.3% had 300 and 657 bp with 57% failed to provide sequence data (Hernandez-Triana et al., 2014). For old and degraded DNA, results are inversely proportional to the size of the largest amplicon and the selection of an effective primer is critical.

Junqueira et al. (2002) extracted DNA from the (subunit II) COII 137 bp from pinned Calliphoridae (Diptera) adults. The longer 305 bp COII was extracted from 83% of the 35 specimens tested. The calliphorids ranged from 1 to 57 years old prior to analysis.

The first molecular examination of Hemiptera was conducted by Lis et al. (2011) using five pentatomoid families consisting of 48 pinned specimens of 46 species with the oldest specimen collected in 1894. Of the 48 pinned specimens, 12S rDNA fragments of mtDNA were recovered from 10 specimens and 16S rDNA fragments were amplified from 7 of the 10 successful samples. Thus old pentatomoid specimens, some collected in 1932, can serve as a good source of amplifiable DNA.

1.17.5 Nondestructive methods

Some molecular techniques grind some parts of the specimen or the whole specimen to extract DNA, which result in both specimen loss and the ability to compare differences among similar specimens. DNA extractions from small specimens often results in the complete destruction of the specimen, making reexamination and morphological confirmation of the identification impossible. To address the issue of studying both the morphology and extracting DNA for barcoding from the same specimen, Castalanelli et al. (2010) compared five families of insects representing Coleoptera, Hemiptera, Diptera, and one family of Acarina to validate a method of extracting DNA without damaging the specimen. They used ANDE, a proprietary DNA extraction method that is stated to be an inexpensive, time-efficient, and, low toxic DNA extraction method resulting in base pairs ranging from 550 to 4000.

Philips and Simon (1995) pierced the exoskeleton of *Nemobius fasciatus* (Orthoptera: Gryllidae) with standard insect pins and then used a protocol established by Gustinich et al. (1991) for blood samples. The procedure resulted in PCR amplification and sequence of 700 bp DNA fragments of the cytochrome oxidase 2 gene while maintaining the external integrity of the exoskeleton for future taxonomic studies.

1.17.5.1 Sonication

Sonication is a novel method of extracting DNA that does not require chemicals yet maintains the morphological integrity of the study specimen suitable to be vouchered with the associated barcode. Sonication is time-efficient and inexpensive when compared with the DNeasy method of extracting DNA from the dipterans Calliphoridae and Sarcophagidae. It produced lower quality DNA but permitted inferred phylogenies similar to that obtained by the DNeasy extraction method (Stamper et al., 2017). Hunter et al. (2008) found that sonication of *Simulium posticatum* (Diptera: Simuliidae) larvae resulted in higher concentrations of DNA than unsonicated larvae.

1.18 **Vouchuring specimens**

Preserved voucher specimens are essential to open new areas of research or to clarify misidentifications. According to Rowley et al. (2007) and Mandrioli (2008), the practice of vouchering DNA sequence with the preserved specimen should be required for all studies of insect DNA. When performing DNA extraction, Rowley et al. (2007) retained some specimens that underwent the buffering process but not the final step of extraction. These controls provided a comparative material to determine any morphological changes that may have occurred during DNA extraction. If the specimen under study is extremely valuable, a leg should be removed prior to DNA extraction and placed on a card with the specimen.

Suggested readings

Molecular research
- Arensburger et al. (2004)
- Nagy (2010)
- von Hagen and Kadereit (2001)
- Wells et al. (2001)
- Wells and Sperling (2001)

Agents for killing and preserving

Agents for killing insects are discussed in Section 1.3.4. Many instances allow for insects and mites of all kinds to be killed and preserved at once in liquid agents, especially acetone. Before proceeding, however, first determine the advisability of using a liquid killing agent rather than a dry gaseous agent. Some kinds of insects are best kept dry; other kinds of insects should not be allowed to become dry. Many insects, especially larger ones, may be killed by an injection of a small amount of alcohol, and their death is nearly instantaneous, which is desirable with moths, large beetles, and large Orthoptera. Generalized directions for the treatment of different orders are provided at the end of Chapter 6, Descriptions of hexapod orders.

Preservation of insects in alcohol is a complex subject about which there is some controversy and misunderstanding. If you wish to specialize in an insect group suited to preservation in one or another kind of concentration of alcohol, consult specialists in that group or experiment to find what yields the best results.

Ethyl alcohol mixed with water (70%−80% alcohol) is usually the best general killing and preserving agent. For some kinds of insects and mites, other preservatives or higher or lower concentrations of alcohol may be better. Many histological or molecular preservation techniques call for preservation in nearly 100% (absolute) alcohol. However, absolute ethyl alcohol is often difficult to obtain; thus some collectors use isopropanol (isopropyl alcohol), with generally satisfactory results. Isopropanol does not seem to harden specimens as much as ethyl alcohol. It is readily obtained and is satisfactory in an emergency. Although controversy exists concerning the relative merits of ethyl alcohol and isopropanol, the choice of which of the two to used is not as important as the concentration to be used. This choice depends on the kind of insect or mite to be preserved.

Parasitic Hymenopterans are killed and preserved in 95% alcohol by some professional entomologists. This high concentration prevents the membranous wings from becoming twisted and folded, setae from matting, and soft body parts from shriveling. This concentration may also be desirable if large numbers of insects are to be killed in a single container, such as in the killing jar of a Malaise trap, because the insect body fluids will dilute the alcohol. On the contrary, soft-bodied insects (such as aphids and thrips, small flies, and mites) become stiff and distorted when preserved in 95% alcohol. These specimens should be preserved in alcohol of a lower concentration. Adult bees should not be collected in alcohol because their usually abundant body setae become badly matted. Adult moths, butterflies, mosquitoes, moth flies, and other groups with scales or long, fine setae on the wings or body may become severely damaged and worthless if collected in alcohol, regardless of the concentration.

Insect Collection and Identification. DOI: https://doi.org/10.1016/B978-0-12-816570-6.00002-2

> Tips on how to preserve various orders of arthropods are provided at the end of Chapter 6, Descriptions of hexapod orders.

Formalin (formaldehyde) solutions are not recommended by some workers because the specimens become excessively hardened and then become difficult to handle. However, formalin is a component of some preservatives.

Larvae of most insects should be collected in alcohol or killed in boiling water to "fix" their proteins and prevent the specimens from turning black. Larvae should be left in hot water for 1−5 minutes and, depending on the size of the specimens, transferred to 70%−80% alcohol. Large specimens or small specimens that have been crowded into one vial should be transferred to fresh alcohol within a day or two to reduce the danger of diluting the alcohol with body fluids. If the alcohol becomes too diluted, the specimens will begin to decompose. Water is not a preservative.

For some groups, preservation is better if certain substances are added to the alcohol solution. Thrips and most mites, for example, are best collected in an alcohol−glycerin−acetic acid solution. Larval insects may be collected into a kerosene−acetic acid solution. Formulas for these and other solutions are given in Appendix I.

Glycerin is an excellent preservative for many kinds of insects, and it has characteristics that should not be overlooked by collectors. Glycerin is stable at room temperature and does not evaporate as rapidly as alcohol. The periodic "topping off" or replacement of alcohol in collections is expensive and time-consuming. Glycerin-preserved material does not require the attention that alcohol-preserved material does. When specimens are preserved in glycerine, some colors are preserved better and longer, and the specimens remain supple. When specimens are removed from glycerine, they may be inspected for relatively long periods under a microscope without experiencing any shriveling or decomposing.

Some histological, cytological, or physiological studies require that specimens be preserved in special media (Walker and Boreham, 1976). For instance, mosquito eggs may be preserved in a lacto-glycerol medium (Murphy and Gisin, 1959).

Larvae and most soft-bodied adult insects and mites can be kept almost indefinitely in liquid preservatives. However, for a permanent collection, mites, aphids, thrips, whiteflies, fleas, and lice usually are mounted on microscope slides (see Section 3.3.9). Larvae are usually kept permanently in alcohol, but some specimens may be mounted by the freeze-drying technique (Dominick, 1972) or by inflation (see Section 3.1.2). Many insects collected in alcohol are later pinned for placement in a permanent collection. Hard-bodied insects such as beetles can be pinned directly after removal from alcohol, but for keeping all softer insects, such as flies and wasps, follow the process described in Section 3.2.4.

Storage of specimens

3.1 Temporary storage

After specimens have been collected, time is often not immediately available to prepare them for permanent storage. There are several ways to keep specimens in good condition until they can be prepared properly. The method chosen depends largely on the length of time that the specimens must be stored temporarily.

3.1.1 Refrigeration

Medium-sized to large specimens may be left in tightly closed bottles for several days in a refrigerator, which will still remain in good condition for pinning. Smaller specimens, such as parasitic Hymenopterans, may be left overnight in the freezer compartment of a refrigerator, without adverse effects. Some moisture must be present in the containers, so that the specimens do not become "freeze-dried," but excessive moisture will condense on the inside of the bottle as soon as it becomes chilled. Absorbent paper placed between the jar and the insects will keep them dry. When specimens are removed for further treatment, place them immediately on a piece of absorbent paper to prevent moisture from condensing on them. Insects may be placed in alcohol, as described previously, and kept for several years before they are pinned or otherwise treated.

3.1.2 Dry preservation

It is a standard practice to place many kinds of insects in small boxes, paper tubes, triangles, or envelopes for an indefinite period, allowing them to become dry. We do not advise storing soft-bodied insects by such methods because the specimens become badly shriveled and subject to breakage. Dipterans should never be dried in this manner because their head, legs, and especially the antennae can be broken very easily.

Almost any kind of container may serve for dry storage. However, tightly closed, impervious containers of metal, glass, or plastic should be avoided because mold may develop on specimens if even a small amount of moisture is present. Little can be done to restore a moldy specimen.

Dry-stored specimens must be labeled with complete collection data in or on each container (see Section 3.3). Avoid placing specimens collected at different times or places in the same container. If specimens with different collection data must be layered in the same container, include a separate data slip with each layer.

To ensure that specimens do not slip from one layer to another, cut pieces of absorbent tissue, glazed cotton, or cellucotton a little larger than the inside of the container. Place a few layers of this material in the bottom of the container, then add a few insects (do not crowd them), then add more layering material, and so on until the container is filled. If much space is left, add enough cotton to keep the insects from moving about but not enough to produce pressure that will damage them. To prevent parts of the insects from getting caught in the loose fibers, use plain cotton only for the final layer. Insect parts are very difficult to extract from plain cotton without damage.

One method of keeping layered specimens soft and pliable for several months includes the use of chlorocresol in the bottom of the layered container and a damp piece of blotting paper in the top. The container must be impermeable and sealed while stored; plastic sandwich boxes are useful containers for this method. Add about 5 mL of chlorocresol crystals to the bottom, cover with a layer of absorbent tissue, follow with the layers of specimens, then a few layers of tissue, and finally a piece of dampened blotting paper as the top layer. The cover is then put in place and sealed with masking tape. Keep boxes of layered specimens in a refrigerator (see Tindale, 1962).

Some insects, such as small beetles, should be glued to triangles (see Section 3.2.5) directly from the layers for permanent preservation. If the specimens are to be pinned or otherwise treated, they must be relaxed as described in Section 3.2.1.

3.1.3 Papering

Pinning specimens when they are fresh is preferable. However, the storage method known as *papering* has long been used successfully for larger specimens of Lepidoptera, Trichoptera, Neuroptera, Odonata, and some other groups. Papering is a traditional way of storing unmounted butterflies and is satisfactory for some moths. Unfortunately, moths often will have their relatively soft bodies flattened, legs or palpi broken, and the vesture of the body partly rubbed off. To save space in most large collections, file Odonata specimens permanently in clear cellophane envelopes instead of pinning them. Cellophane is preferred over plastic because cellophane "breathes."

Papering consists of placing specimens with the wings folded together dorsally (upper sides together) in folded triangles (Fig. 3.1) or in small rectangular envelopes of glassine paper, which are the translucent

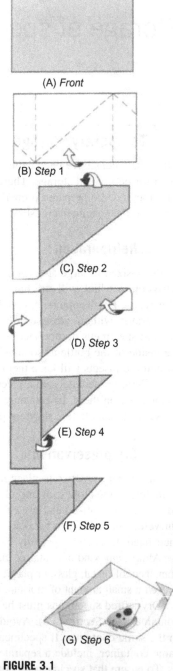

(A) *Front*

(B) *Step* 1

(C) *Step* 2

(D) *Step* 3

(E) *Step* 4

(F) *Step* 5

(G) *Step* 6

FIGURE 3.1

Steps in folding paper triangles.

envelopes familiar to stamp collectors. Glassine envelopes have become almost universally used in recent years because of the obvious advantages of transparency and ready availability. In many collections, glassine has become partly superseded by plastic. However, many collectors still prefer folded triangles of a softer, more absorbent paper, such as ordinary newsprint, and believe they are superior for preserving specimens. Specimens can become greasy after a time, and the oil is absorbed by paper such as newsprint but not by glassine. Moreover, glassine and plastic are very smooth, and specimens may slide about inside the envelopes during shipping, losing antennae, and other brittle parts. Although softer kinds of paper do not retain creases well when folded, this shortcoming may be circumvented by preparing the triangles of such material well before they are needed and pressing them with a weight for a week or so. Triangles are easy to prepare if the paper is folded, as shown in Fig. 3.1.

Some Lepidopterans are most easily papered if first placed in a relaxing box for a day or two. The wings, often reversed in field-collected butterflies, may then be folded dorsally without difficulty. Do not pack specimens together tightly before they are dried, or the bodies may be crushed. Do not store fresh specimens immediately in airtight containers or plastic envelopes, or they will mold. Write collection data on the outside of the envelopes before inserting the insects.

3.2 **Mounting specimens**

Specimens are mounted so that they may be handled and examined with the greatest convenience and with the least possible damage. Well-mounted specimens enhance the value of a collection; their value for research may depend to a great extent on how well they are prepared. Standardized methods have evolved over the past two centuries in response to the esthetic sense of collectors and to the need for high-quality research material. Although the style and technique of mounting may vary from one worker to another, the basic procedures outlined here are widely accepted. Methods of preparation are subject to improvement, but in the interest of uniformity it is best to follow currently accepted practices until the superiority of other methods has been proved.

> The value of specimens may depend on how well they are preserved.

The utility of a mounted specimen is determined by how well it is preserved, how safe it is from damage, and how much of the specimen can be examined conveniently. Utility supersedes beauty in research collections, whereas beauty should be a major consideration in nontechnical displays.

Preparation of specimens for a permanent collection is discussed here. Ideally, specimens prepared for a permanent collection should be fresh. The body tissues of fresh specimens have not hardened or dried, and thus the specimen is easy to manipulate. Specimens that have been in temporary storage must be specially treated before mounting.

Dry specimens usually must be relaxed (Weaver and White, 1980), and those preserved in liquid must be processed so that they will dry with minimal distortion or other damage. Some specimens may be kept permanently in a liquid preservative or in papers or envelopes, as discussed earlier.

3.2.1 Preparing dry specimens for mounting

Dry insects to be pinned must first be relaxed. Moistening the specimen softens the body and appendages so that the body will not break when the pin is inserted into the thorax, and parts of the relaxed specimen may be rearranged or repositioned. Insects, particularly Lepidopterans, whose wings are spread, should be relaxed even if they have been killed recently. The muscles of Lepidoptera stiffen in a matter of minutes. The thoracic flight muscles are very strong and make adjustment of the wings difficult. Treatment in a relaxing chamber, however, usually will make this procedure much easier (Fig. 3.2). In all, 8 hours in a relaxing chamber should suffice, but larger specimens may require 24–36 hours. Simply leaving specimens in a cyanide jar for a short time sometimes will relax them, but this method is not reliable. See Lane (1965) for a discussion of mounting insects of medical importance.

(A)

High humidity must be provided in a relaxing chamber for periods from several hours up to about 3 days. The actual amount of time depends upon the circumstances. Care must be exercised to ensure that the specimens do not become wet during humidification. Growth of mold should be avoided because it will destroy specimens left too long in a relaxing chamber. A chemical mold inhibitor can be added to the relaxing chamber. Parasitic Hymenoptera killed with ethyl acetate remain pliant for several hours and can be relaxed easily if they have been allowed to dry. However, some killing agents (especially chloroform, ether, and carbon tetrachloride) may harden muscles to such an extent that specimens are brittle and seemingly impervious to the humidity of the relaxing chamber. In such instances other techniques must be employed. In Korea, for example, butterflies are injected in the thoracic muscles with very hot water through a fine hypodermic needle before the wings are spread. Occasionally, some specimens cannot be relaxed satisfactorily by any method, they should be mounted soon after capture and killing.

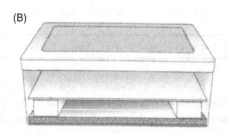

(B)

FIGURE 3.2

Insect relaxing chambers: (A) glass relaxing chamber and (B) homemade box relaxing chamber.

Glass dishes and jars with covers (low, widemouthed jars, and casserole dishes are excellent), plastic food storage containers, and earthenware crocks make excellent relaxing chambers. Glass and earthenware containers are not so immediately affected by fluctuations in temperature as are other types and thus may relax insects more evenly. Containers with a depth of 5−15 cm are most convenient; clear plastic sandwich boxes not more than 2.5 cm deep will serve for small specimens if they are not on pins. A layer of damp sand, peat, or crumpled paper toweling is placed at the bottom of the container and covered with a layer of cotton, cellulose wadding, or jeweler's cotton. This layer will not absorb water readily and will prevent direct contact between the insects and the moisture beneath.

Some workers object to the use of cotton because of the tendency for insect legs to become entangled in it and break off. If this is a problem, cover the cotton with a single piece of soft tissue paper. For very small specimens, a lining of tissue paper or some absorbent material with a smooth surface is advantageous. Heavy paper, such as blotting paper or cardboard, may be used in place of cotton, but this should be supported at least 1 cm above the moist bottom layer to avoid direct contact. Wooden or plastic strips or fine-mesh plastic screen may also be used for this purpose.

In temperate and arid regions, mold probably will not be a problem if insects are relaxed for no longer than 2 days at normal room temperature. However, relaxing chambers in regular use must be kept clean, with frequent renewal of the contents. If mold is likely to develop, as may happen with large specimens held more than 2 days, a few crystals of naphthalene, paradichlorobenzene (PDB), phenol, or chlorocresol may be sprinkled at the bottom of the relaxing chamber. A little thymol, which is more potent, may be used. Many of these chemicals may damage plastic boxes; therefore select relaxing chambers with care.

Dry insects must be carefully relaxed prior to pinning.

Insects held too long in a killing jar and those that were originally papered and pinned but not spread, or layered (placed in small boxes between pieces of soft tissue) may be relaxed by placing them in a relaxing chamber. Papered specimens will relax faster if removed from their envelopes. Beetles and other insects whose wings do not require spreading may be held less than 24 hours in a relaxing chamber. Small moths and delicate Neuroptera should also be relaxed sufficiently after 12−24 hours to allow the wings to be spread. Large moths may take 48 hours or longer if the relaxing chamber is kept at room temperature. The process can be hastened and the chance of mold developing greatly reduced if the relaxing jar is subjected to a slight raising and lowering of temperature between 18°C and 27°C. The process is greatly accelerated if the relaxing jar is placed in, or floated on, warm water for an hour or more; specimens may be relaxed within 3−6 hours in this way. If the warm-water treatment is overdone; however, then the specimens may be spoiled by the absorption of too much moisture. Some colors, especially nonmetallic greens in Lepidoptera, are unstable and may be completely lost by exposure to too much humidity. Such material requires special attention; the specimens should be left in the relaxing chamber for the shortest possible time and ideally should be pinned and spread when fresh. Experience soon enables one to judge the best procedure for the particular kind of material being prepared.

The length of time that insects may be left safely in a relaxing chamber partially depends on the temperature. At 18°C−24°C, the specimens may be left for about 3 days; beyond that time they will begin to decompose. If the relaxing chamber is placed in a refrigerator at 3°C−4°C, the specimens may be kept for 2 weeks, although they may be slightly damaged from excessive condensation over time. If relaxing chambers containing fresh specimens are placed in a deep freezer at −18°C or lower, the specimens will remain in a comparatively fresh condition for months, but not indefinitely. At extremely low temperatures, the specimens gradually desiccate and eventually will become dried. However, a freezer may be used to keep them fresh for a month or two and is a great convenience.

Even when specimens have been relaxed suitably for spreading, the wings may still seem stiff. In this instance, the wing muscles must be loosened by forcing the wings to move up and down. This may be safely done by pressing the tips of curved forceps firmly against the costal vein very near the base of the wing. The forceps should have the tips ground or honed smooth and not too sharp. Repeat this procedure separately with all four wings or they will revert gradually toward their original positions. With care, all the wings may be loosened in this way without leaving visible marks.

Occasionally, it is necessary to relax and reposition only a part of an insect, such as a leg that may conceal characters needed for identification. This is accomplished by putting a drop or two or Barber's fluid (see Appendix I) or ordinary household ammonia directly on the leg. Most household ammonia contains detergent, which helps wet and penetrate insect tissue. After a few moments, perhaps after adding a little more fluid, the body part may be pried carefully with a pin. When the part moves easily, it may be placed in the desired position and held there with a pin fixed in the same substrate that holds the pin on which the specimen is mounted. Leave the positioning pin in place until the fluid dries thoroughly.

A few methods for relaxing insects involve heat. These and other methods are summarized in a reference describing a steam-bath method (Weaver and White, 1980).

3.2.2 Degreasing

Many dried insect specimens become greasy with time. This is particularly evident in some beetles, butterflies, and moths. The grease is derived from internal fatty tissue. Traditional methods of degreasing involve soaking specimens in benzene, ethyl ether, or carbon tetrachloride for a few days (Peterson, 1964). Periodically, the fluid is replaced as it changes color. Vapor degreasing is a relatively new technique (McKillop and Preston, 1981) that employs trichloroethane as a solvent heated to vaporization in a metal tank. The apparatus is elaborate and may be available at some universities or museums. The toxicity of degreasing solvents should be kept in mind. Some curators prefer hexane because it is very clean, it does not affect pigments, and the odor is less annoying.

3.2.3 **Preparing liquid-preserved specimens**

Sabrosky (1966) discussed mounting specimens that have been preserved in alcohol. Most specimens preserved in fluid must be removed from the fluid in a way that minimizes distortion or matting of setae as the specimens dry. Only specimens with hard exoskeletons, such as beetles and some bugs (Pentatomidae, Cydnidae), may be mounted without special treatment when removed from the preserving fluid and mounted directly on pins or points. The following methods have been used routinely for removing specimens from the usual fluid preservatives. Some specimens are left in better condition than if they had been pinned while fresh. This is especially true of small Dipterans.

The following equipment is needed:

1. Two screw-top jars of about 5 cm in diameter with a cork cemented with epoxy on the top of each lid. A label on the outside should clearly identify the contents. One jar should contain about one-third Cellosolve (2-ethoxyethanol, ethylene glycol ethyl ether) and the other jar one-third xylene.
2. A small dish, such as a watch glass.
3. Absorbent tissue from which to twist small "pencils" for absorbing xylene.
4. Insect pins or double mounts (see Section 3.2.5) for mounting the specimens.
5. Adhesive in a jar with a rod in its stopper.
6. Narrow-pointed forceps.
7. A few small cards of blotting paper.

When specimens are ready for preparation, remove as many from the preservative as can be pinned or placed on triangles or card points in an hour. Experience will suggest how many specimens should be removed from the preservative. Place them on a blotting paper, then drop them from the blotter into the jar of Cellosolve, and place a label with the collection data on the pin stuck into the cork cemented to the lid of the jar. This may be done at the end of the day and the specimens left in the Cellosolve overnight. Otherwise, leave the specimens in Cellosolve for about 3 hours or longer for large specimens. They may even be left over a weekend. The same jar of Cellosolve may be used several times, up to about 10 times if the insects are small. This part of the treatment removes water and other substances from the specimens. However, Cellosolve does not evaporate readily; so it must be removed subsequently with another solvent that will evaporate readily.

> WARNING!
>   Xylene is now considered to be carcinogenic.
>   A new and widely used chemical, Histo-Clear, is a promising substitute.

The next step is to use forceps to remove the specimens carefully from the Cellosolve. Place them again briefly on blotting paper and then place them in the jar containing xylene. The identifying label on the pin in the cork must also be transferred. Small specimens should be left in the xylene for about 1 hour, larger specimens for up to 4 hour. Specimens left too long in xylene become extremely brittle and difficult to put on a pin or triangle without losing legs, antennae, or the head. As with the Cellosolve, the xylene may also be used many times until it becomes so contaminated with Cellosolve that specimens dry slowly when removed. Specimens wet with Cellosolve or xylene are somewhat pliable and appendages may be repositioned slightly.

After 1 hour in xylene, remove the smallest specimens first. Remove each with small forceps and place it in a dish. The forceps will pick up a small amount of xylene and the specimen will be left lying in it. While it is in the dish, position the specimen correctly for mounting. The wings, if present, will float out flat, sometimes requiring a little adjustment with a pin or the tip of the forceps. When the specimen is positioned correctly, take a corner of absorbent tissue and touch it to the specimen to remove the excess xylene. Larger specimens may be pinned directly in the usual manner. Just before the xylene fully dries from the surface of a small specimen, the tip of a triangle or a tiny pin called a *minuten* (already attached to its carrying insect pin) should be touched to adhesive and then be touched to the specimen, picking it up.

If a minuten is used, it may be inserted into the thorax of the specimen. A little final adjustment of position may then be made, and the specimen is ready for its label and place in the collection. Having touched the tip of the minuten to the adhesive will leave a small amount of adhesive around the place where the minuten has pierced the specimen and will keep it from working loose when fully dried. Specimens placed on standard insect pins also should have a small amount of adhesive placed around the site where the pin protrudes from the lower side of the specimen. Specimens pinned after having been in fluid preservatives do not cling as firmly to the pin as those pinned fresh.

> Specific directions for labeling and pinning insects, including double mounting and using minuten pins, follow later in this section.

This treatment will leave surface pile, setae, and bristles in a loose, unmatted, natural condition. Small insects that shrivel considerably after having been pinned fresh will usually dry in a better condition if pinned or placed on triangles after this treatment.

3.2.4 Direct pinning

This section pertains entirely to insects because mites should never be mounted on pins. Direct pinning refers to the insertion of a standard insect pin directly through the body of an insect. Only insect pins should be used; ordinary straight pins are too short and thick and also have other disadvantages. Standard insect pins are 38 mm long and range in thickness generally from size #000–#7. An integral "upset" head is made by mechanically squeezing out the end of the pin, or a small piece of metal or nylon is pressed onto the pin. A well-made upset head is considered essential by most entomologists; other kinds of heads sometimes come off, leaving a sharp point that easily can pierce a finger. Pins of #2 diameter are most useful (0.46 mm in diameter). In the past, most entomologists avoided the use of very slender pins of size 000–1, preferring to use double mounts (see Section 3.2.5), but now that soft polyethylene or plastic foam is commonly used for pinning bottoms in trays and boxes, these smaller sizes are not so impractical. Pins of larger diameter (numbers 3–7) may be needed for large insects.

FIGURE 3.3

Pinning forceps.

Standard insect pins are currently made of spring steel (which is called "black") or stainless steel with a blued or a lacquered (japanned) finish. The black pins may corrode or rust with even slight exposure to moisture or to the body contents of the insects. Stainless steel pins are more expensive than black pins, but their rustproof nature makes them desirable for use in permanent collections. However, their points are somewhat more easily turned than those of black pins when piercing an insect with a hard cuticle, and their shafts are not as rigid. For that reason, pierce an especially hard cuticle with a strong steel pin before inserting a stainless steel pin. Lacquered pins have a surface on which the insect may be less likely to become loose than it might on a bare pin.

Insect pins made of German silver or brass were once common. However, the action of the insect body contents produces a greenish verdigris on the pin inside the insect and eventually corrodes through the pin.

For handling a large number of pinned specimens, pinning or dental forceps (Fig. 3.3) may be helpful. Their curved tips permit grasping the pin below the data labels and enable the curator to set the pin firmly into the pinning-bottom material of the box or tray without bending the pin. The forceps are also invaluable when removing pins strongly corroded into cork pinning bottoms. The pin is grasped firmly above the cork and turned a little before it is lifted. Unfortunately, pinning forceps are impractical in dealing with large-winged specimens, such as many Lepidopterans.

Insects should be pinned vertically through the body with a pin of appropriate thickness. Exercise care so that the pin does not damage the legs as it passes through the body. Many insects are pinned to the right of the midline so that all the characters of at least one side will be visible. Do not attempt to pin specimens unless they are relaxed or recently killed. Inserting a pin into a dry specimen may cause it to shatter. When pinning relaxed specimens or specimens taken from Cellosolve and xylene, a little glue may be needed where the pin emerges from the specimen to prevent the dry specimen from working loose and rotating on the pin. Application of adhesive is unnecessary when mounting recently killed insects.

Standard pin placements for some of the more common groups of insects are illustrated in Fig. 3.4.

1. *Orthoptera*. Pin the specimen through the back of the thorax to the right of the midline. For display purposes, one pair of wings may be spread. Some orthopterists prefer to leave the wings folded because of limited space in most large collections (see Beatty and Beatty, 1963).
2. *Large Heteroptera*. Pin the specimen through the triangular scutellum to the right of the midline. Do not spread the wings.

3. *Large Hymenoptera and Diptera.* Pin the specimen through the back of the thorax or slightly behind the base of the forewing and to the right of the midline. Characters on the body must not be obscured. Legs should be pushed down and away from the thorax; wings should be turned upward or sideways from the body. The wings of most Dipterans and many parasitic Hymenopterans will flip upward if the specimen is laid on its back before pinning and pressure is applied simultaneously to the base of each wing with a pair of blunt forceps. Wings should be straightened if possible so that venation is clearly visible. Folded or crumpled wings sometimes can be

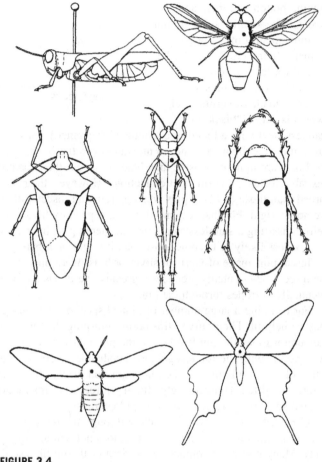

FIGURE 3.4

Proper pin positions in common insect orders; black dots indicate pin placement.

straightened by gentle brushing with a camel's hairbrush dipped in 70% alcohol. Peterson's XA mixture (equal parts of xylene and ethyl alcohol, ETOH) is recommended for treating Hymenopterans' wings.

4. *Large Coleoptera.* Pin the specimen through the right forewing (wing cover or elytron) near the base. Do not spread the wings of beetles.

5. *Large Lepidoptera and Odonata.* Pin the specimen through the middle of the thorax at its thickest point or just behind the base of the forewings. Spread the wings as described in Section 3.2.7.2.

The proper height of an insect on a pin will depend somewhat on the size of the specimen. As a guideline, enough of the pin should be exposed above the thorax to grasp the pin without the fingers touching the specimen. A specimen mounted too high on a pin probably will be damaged in handling. If a specimen is pinned too low on a pin, the specimen's legs may be broken when the pin is inserted into a tray or box. Also, insufficient space may be left for labels if a specimen is mounted too low on a pin.

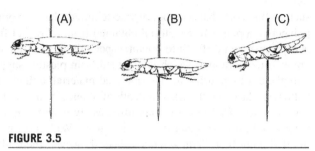

FIGURE 3.5

Pin heights: (A) correctly pinned specimen; (B) specimen mounted too low; and (C) specimen not level.

Fig. 3.5 illustrates correct and incorrect examples of pinning.

After the pin is inserted and before the specimen becomes dry, the legs, wings, and antennae should be arranged so that all parts are visible. With most insects, the legs and antennae simply can be arranged in the desired position and allowed to dry. Occasionally, the appendages must be held in place with insect pins until the specimen dries. Long-legged species or specimens with drooping abdomens can present problems. Temporarily mounting insects on a piece of styrofoam board or other soft material is recommended. Legs and abdomens may be thusly supported until dry with a series of strategically placed support pins or a piece of stiff paper pushed up on the pin from beneath. When the specimen is dry, supports can be removed. A spreading board or a block (Section 3.2.7) may be necessary for moths, butterflies, and other insects that should be mounted with their wings spread.

Some entomologists glue small insects directly to the side of a standard insect pin. This procedure is not recommended because often too much of the insect is obscured by the glue or pin and the adhesive does not bond well to the pin. A double mount is recommended for small insects.

3.2.5 Double mounts

Double mounts are intended for insects that are too small to be pinned directly on standard pins yet should be preserved dry. This term refers to insects mounted on a minuten pin or card point (triangle), which in turn is mounted or attached to a standard insect pin (Fig. 3.6). Minutens are available from supply houses in 10 and 15 mm lengths and in two or three thicknesses. They are generally made of stainless steel and are finely pointed at one end and headless on the other end. Double mounts are assembled by inserting the minuten into a small cube of soft, pithy material,

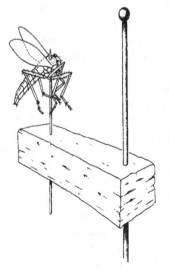

FIGURE 3.6

A properly double-mounted insect on a minuten pin.

such as fine cork, balsa wood, or fine-textured plastic. Some entomologists prefer polyporus, a pure white material obtained from a bracket fungus. Unfortunately, polyporus is expensive and difficult to obtain, especially in the United States. Many entomologists prefer silicone rubber, which is obtained from plastics suppliers. Silicone rubber is made into plaques by pouring the polymerized material, a thick creamy liquid, into a flat-bottomed plastic container to a depth of about 2.5 mm and allowing it to solidify over several hours. The rubber is then lifted easily from the mold and cut with a sharp knife or razor blade into strips and finally into cubes. With most materials, the minuten must be inserted point first. With silicone rubber, however, the dull end of the minuten is inserted first, until it strikes the surface on which the cube is lying so that it will be held firmly. Minutens should be handled with forceps; they are so thin that either end can easily pierce a finger.

Sometimes it is preferable to mount an insect on a minuten before inserting the minuten into the mounting cube. For convenience, prepare a series of minuten mounts beforehand, already attached to standard no. 3 pins. To mount extremely small insects on minutens, pick up a droplet of adhesive with the prepared minuten and touch it to the specimen between the base of the legs. When mounting an insect on a minuten, the pin needs to extend just barely through the insect. If the insect is lying on a glass surface when it is pierced with the minuten, a little extra pressure will curl the point of the minuten back into the insect and ensure that the specimen will not come off the minuten.

Many entomologists prefer to mount insects on a minuten in a vertical position in a short strip of pith or cork, with the minuten parallel to the main pin. The insect lies in an excellent position for examination under a microscope and is less liable to be damaged in handling than it would be otherwise (see Peterson et al., 1961).

Card points are slender, small triangles of stiff paper. They are pinned through the broad end (Fig. 3.7) with a no. 2 or 3 insect pin, and the insect is then glued to the point. Card points may be cut with scissors from a strip of paper; they should be no more than 12 mm long and 3 mm wide. A special punch for card points (Fig. 3.8), available from entomological supply houses, will make better, more uniform points. Many European and North American entomologists who specialize in collecting parasitic Hymenoptera prefer to use small rectangular cards for mounting specimens that are 1−10 mm long. The cards are available in various sizes and may be purchased through entomological equipment dealers. If the collector desires a specific size of card, a special card punch can be manufactured. The specimen is mounted directly onto the card, with one side of the thorax contacting the adhesive.

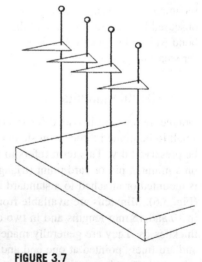

FIGURE 3.7

Proper mounting of card points.

Card points and rectangular cards should be made only from high-quality, acid-free paper. The paper should be as good as or better than that used for data labels (Section 3.3.1). Do not use "index cards" as point or rectangular-card stock. If specimens are in good condition and are well prepared, they may be kept in museum collections for a long time, perhaps for centuries. Most of the paper in common use does not have that kind

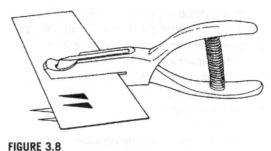

FIGURE 3.8

Point punch.

of life expectancy. Most paper typically becomes yellow and brittle with age. Paper that is made especially to last (such as that used for herbarium sheets in botanical collections) is highly recommended.

> Paper made especially to last, such as that used for herbarium sheets in botanical collections, is highly recommended.

The choice of adhesive for mounting insects on card points and rectangles is equally important. Bonding characteristics, solubility, and availability are three factors that should be considered in the selection of an adhesive. Unfortunately, the aging properties of many glues are not known and we cannot make unequivocal recommendations. The adhesive should bond to the specimen so that the specimen will not fall from the card with time. The adhesive should be soluble in a solvent that is inexpensive, nontoxic, and available. The solvent permits specimens to be remounted when necessary; expensive, toxic, or exotic solvents make remounting difficult. Ordinary white glue or carpenter's glue is readily available in the United States. It is a superior adhesive and it is soluble in water. Clear acetate cement and fingernail polish are also commonly utilized for double mounting insects, but they are less satisfactory. Fingernail polish does not bond well to some specimens and requires a special solvent to thin the polish or to remount specimens. White shellac is a possibility but is less satisfactory because it is usually too thin and flows over a specimen when applied.

For many years, some professional entomologists have used a very viscous polyvinyl acetate for double mounting and similar uses. The acetate can be obtained in granular, bead, or pellet form. A small quantity is placed in a bottle containing a glass stirring rod and covered with absolute ETOH. The acetate will dissolve in a day or two into a thick solution. If it is too thick, the acetate will "string out," and more ETOH should be added. Use the rod to stir the ETOH into the acetate. If the solution is too thin, the bottle should be left open to allow some of the ETOH to evaporate. After a period of use, the solution will thicken, and then more ETOH must be added. Specimens adhere very well to a pin or a point with this solution, and they may be removed with 95% ETOH.

Some entomologists use shellac gel (Martin, 1977) as an adhesive for double mounts. We do not recommend this material because its preparation involves boiling white shellac solution, a procedure with some danger from fire. It is hoped that a safer method of preparation may eventually be found for this useful medium.

Adhesives should not be permitted to get so thick that they form "strings." If this happens, add a little solvent to the adhesive until it attains the proper consistency. Nor should an adhesive be so thin that it flows over a specimen. Only a small amount of adhesive should be used to glue the specimen to the card point or rectangle. Excessive glue may obscure certain sutures or sclerites necessary for

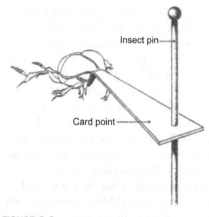

FIGURE 3.9

Double mount using card point method.

identification, just as the card point may conceal certain ventral structures if allowed to extend beyond the midline of the insect.

For most insects, the card point is attached to the *right* side of the specimen, with the left side and midventral area clear. For better adhesion with some insects, the tip of the card point may be bent downward at a slight angle to fit against the side of the specimen (Fig. 3.9). The bend increases the surface area of the point that makes contact with the specimen. Only a very small part of the point should be bent, and this is accomplished with fine-pointed forceps. With a little practice, it will be easy to judge how much of the point to bend and at what angle to fit the particular insect being mounted.

One method to ensure that the specimen is oriented properly on the point is to place it on its back with its head toward you; then with the pin held upside down, touch a bit of adhesive to the bent point and apply it to the right side of the insect. If the top of the point can be slipped between the body of the insect and an adjacent leg, a stronger mount will result. The card point should be attached to the side of the thorax, not to the wing, abdomen, or head. Some insects, such as small flies and wasps, are best mounted on unbent points.

Opinions differ on when to use direct pinning and when to use a double mount. Perhaps the decision is best determined through experience. A general rule of thumb: Do not use a double mount if you can mount a small insect on a no. 1 or 0 pin without damaging the specimen. Insects too heavy to be held on the point by adhesive yet too small to be pinned with standard pins may be attached to card points via the following method. Puncture the right side of the insect at the place where the card point usually would be placed, and insert into this puncture the tip of an unbent card point with a little glue on it. For puncturing specimens, use a needle ground and polished to make a small, sharp scalpel. Some specimens, such as moths, should never be glued to points; other specimens should never be pinned with minutens. The following suggestions will serve as a guide.

1. Small moths, caddis flies, and neuropteroids.

   Mount the specimen on a minuten inserted through the center of the thorax with the abdomen positioned toward the insect pin. The mount must be sufficiently low so that the head of the pin can be grasped easily with fingers or pinning forceps. Do not glue moths to points. Ideally, such specimens should be spread in the conventional manner despite their small size.

2. Mosquitoes and small flies (recently killed).

   Pin the specimen with a minuten through the thorax with the left side of the specimen positioned toward the main pin. Note that the minuten is vertical. This is more advantageous than if it were horizontal because the specimen is less liable to come into contact with fingers or pinning forceps. Placing a small amount of glue on the tip of the minuten before piercing the specimen will help hold soft-bodied insects (see Schlee, 1966).

3. Small wasps and flies (not recently killed).

   Mount the specimen on an unbent card point, with the point inserted between the coxae on the right side of the insect (keeping clear of the midline), or glue the tip of the point to the mesopleuron. Alternatively, microhymenoptera can be mounted directly on rectangular cards.

4. Small beetles, bugs, leafhoppers, and most other small insects.

   Glue the card point with its tip bent down to the right side of the specimen.

3.2.6 Pinning blocks

Pinning blocks (Fig. 3.10) allow for insects and labels to be easily mounted at uniform heights on the pin. Specimens should be mounted on the pin such that the pin can be grasped between the thumb and index finger without touching the specimen and yet keeping the specimen above the bottom of the box or tray.

After piercing the insect through the appropriate spot on the specimen, insert the pin into the deepest hole in the pinning block until the pin goes no further. This pushes each specimen to the same height on each pin. Repeat this process for positioning label heights on pins in subsequent holes (Fig. 3.11). Pinning blocks improve the general appearance of a collection and help preserve specimens from breakage.

Double mounts should conform to the same rule as for direct pinning. Do not place a double mount too high or too low on the pin. Double-mount cubes or points may be adjusted at any

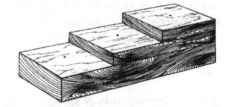

FIGURE 3.10

Three-step wooden pinning block.

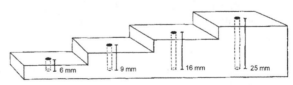

6 mm   9 mm   16 mm   25 mm

FIGURE 3.11

Plastic pinning block illustrating hole depths.

time; a directly pinned insect is virtually impossible to move without damage after it has dried. If points become loose on the main pin, place a little adhesive at the connection. Consult Borgmeier (1964) for additional details.

3.2.7 Mounting and spreading boards and blocks

The appearance of an insect collection can be enhanced if specimens are pinned in a lifelike position. Most entomologists find that a mounting board made of styrofoam, cork, or cardboard is convenient. Pins on which fresh specimens have been mounted should be inserted into the mounting board deep enough that the specimen rests slightly above the flat surface. The legs, wings, abdomen, and antennae can then be conformed into the desired position by strategically placing holding pins. Leave the specimen in place until dry (sometimes 1–2 weeks). Once dry, the specimen will hold this shape indefinitely.

In advanced collection, mounting boards allow an entomologist to prominently display certain specific characters used in identification.

Insects to be preserved with their wings spread uniformly are generally set and dried in this position on spreading boards or blocks. Spreading boards are more commonly used than spreading blocks. These pinning aids vary greatly in design, but the same basic principle is inherent in all of them: a smooth surface on which the wings are spread and positioned horizontally; a central, longitudinal groove for the body of the insect; and a layer of soft material into which the pin bearing the insect is inserted to hold the specimen at the proper height. An active collector will need several spreading boards because the insects must dry for a considerable time (about 2 weeks for most specimens, even very small ones) before being removed from the boards. Spreading boards may be purchased from biological supply houses or may easily be made as described here if the proper materials can be obtained. When purchasing spreading boards, avoid (1) too hard or too soft a material for the pinning medium under the central groove, (2) too hard an upper pinning surface, and (3) top pieces without the same thickness at the center (an especially common fault in beveled boards). This last defect may be corrected by sanding down the higher side; evenness is especially critical when working with small specimens.

3.2.7.1 Construction of spreading boards

A spreading board of simple design (Fig. 3.12) requires the following materials:

1. Two top pieces, 9 mm by 4.8 cm by 38 cm, preferably of seasoned basswood (a fine-grained, durable wood from trees of the genus *Tilia*). Holes made in basswood by insect pins tend to close after they are removed. If the surface of the board is lightly sanded after use, especially when working with small specimens, its smooth, even quality can be maintained through many years. Basswood is sometimes known as "whitewood." Wood from trees of the genus *Liriodendron* is also sold under this name. If basswood cannot be obtained, well-seasoned white pine selected for softness and "pinnability" is serviceable. A third choice is 20 cm (8 in.) beveled redwood siding. Beveled top pieces are desirable because the beveling, sloping inward, compensates for the tendency of the wings of spread specimens to drop slightly after the specimens are

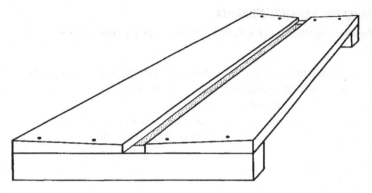

FIGURE 3.12

Spreading board.

removed from the board. The 20 cm siding is actually 19.1 cm wide, and a 0.9 m piece of it will provide two pairs of top pieces. One pair cut slightly more than 4.8 cm wide from the wide side of the board and planed to exact width will make a pair 11 mm thick at the narrow side, and another pair cut from the same side of the board will provide a second pair about 8 mm thick at the narrow side. Redwood is stiff and fine-grained; it splits and splinters easily.

2. Two end pieces of any good, fine-grained wood, 2 cm^2 × 10 cm.
3. One strip of entomological or gasket cork or foam, 6 mm × 3 cm × 34 cm.
4. One base of plywood or any fine-grained wood, 6 mm × 10 cm × 38 cm.

These materials are for a spreading board with a 6 mm wide central groove. Boards with grooves of several widths will be needed. For the larger Lepidoptera (macrolepidoptera or "macros"), the most useful widths are 3, 6, and 9 mm. For very large moths, a width of 17 mm is required; the board will also have to be as much as 15 cm in total width, with a groove depth of 16 mm. For small moths (microlepidoptera or "micros"), special boards with groove widths of 1.5–2 mm will be needed; with the groove shallow enough for minutens, the width and thickness of the top pieces (1) must be altered accordingly. The pinning medium (3) could be of polyethylene foam, but to give specimens firm support, the entire depth between the top pieces and the base would have to be filled with the material. A dense, finely textured plastic foam known as Plastazote is better than polyethylene for entomological applications. A strip of model maker's balsa wood, selected for pinning softness, may also be satisfactory.

The end pieces (2) should be glued with epoxy or another good adhesive and nailed to the top pieces (1), with the proper groove width maintained. Then the pinning strip (3) should be firmly glued to the underside of the top pieces (1), on the same side on which the end pieces (2) were fastened, and should cover the central groove. Finally, after the adhesive has set, the base (4) should be attached. It is affixed to the end pieces (2) with two flat-headed wood screws (no. 5, 19 mm) countersunk into the base piece and screwed into each end piece. The base may be removed easily later if replacement of the pinning strip is necessary.

3.2.7.2 Using the spreading boards

Before spreading specimens, the spreading boards and the following materials should be at hand.

1. Pins (called setting pins) of size 00 or 000 for bringing wings into position. Setting pins used by some lepidopterists are made by inserting a minuten into the end of a round matchstick and securing it with a drop of glue.
2. Strips of glassine or tracing paper (the translucent, smooth paper used for tracing, not what a draftsman calls tracing paper). Cellophane, plastic film, or waxed paper should not be used. Their disadvantages include expanding with moisture, becoming electrostatically charged, and containing a substance that pulls scales off the wings. The strips of tracing paper should be long enough to extend from the base of the wings to a little beyond the end of the wings of the specimens being spread. Strips about 25 mm long are convenient for spreading most Lepidopterans. Short strips are appropriate when spreading specimens that have been relaxed from a dried condition; strips long enough to cover several specimens in a row on the board are commonly utilized for recently captured insects. The strips are often applied with a narrow fold alongside the body of the specimen with the fold upward; this provides a rounded edge that reduces the likelihood that a sharp edge will displace a row of scales. This fold may be made by holding the strip on a spreading board with 2–5 mm of it overhanging the edge of the board, running a finger along the overhang to bend it down, and then firmly folding it back.
3. Glass-headed pins at least 2.5 cm long for holding the strips in place. Ordinary no. 2 or 3 insect pins with nylon heads may also be used, but some lepidopterists find them hard on the fingers.

With this equipment ready, the collector is prepared to mount and spread the specimens. The specimens must be properly relaxed (see Section 3.2.1), even the recently collected ones, before any attempt is made to spread the wings. Insert an insect pin of appropriate size through the middle of the thorax, leaving at least 7 mm of pin above the specimen. The pin should pass through the body as nearly vertically as possible to avoid having the wings higher on one side than on the other. Pin the specimen into the central groove of the spreading board so that the wings are exactly level with the surface of the board. Carefully draw each wing forward, with the point of a setting pin inserted near the base and immediately behind the strong veins that lie near the front of the wings (Fig. 3.13). If care is taken not to tear the wing, the fine setting pins should leave holes so small that they are barely visible. The hind margin of the forewings should be at right angles to the groove in the board. Bring the hind wing into a natural position, with its base slightly under the forewing. The setting pins will hold the wings in position until they can be secured with the paper strips.

The strip is placed near the body of the insect, with its fold upward and toward the insect. A glass-headed pin is inserted in the middle of the folded part of the strip just outside the margin of the forewing. The pin may be tilted slightly away from the wing to keep the strip down against the wing. The tip is then carefully stretched backward and another pin is placed just behind the hind wing. A third pin in the notch where the

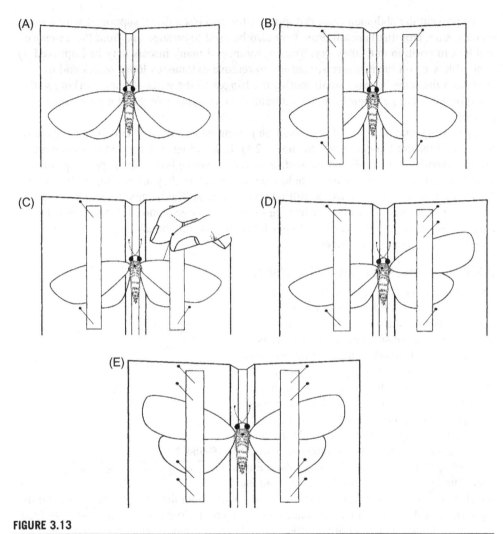

FIGURE 3.13

Steps (A to E) to follow when using the spreading board.

forewings and hind wings meet is usually enough. None of the three glass-headed pins on each side of the specimen should pass through the wings. After the paper strips are in place, the setting pins may be removed. Twisting the setting pins a little as they are removed will prevent a possible bent tip from hooking onto a wing vein and pulling the whole wing out of place.

Fresh specimens may be arranged closely on the board in series of five or even more before the paper strips are applied to cover all the wings. Relaxed specimens should be treated individually because they dry so quickly that antennae may break or the wings curl if the spreading is not completed promptly.

To prevent the abdomen from drooping as the specimen dries, support it with a pin on each side, crossing beneath. Pins may also be used to arrange and hold the antennae and legs in position until they dry. The appearance of many insects may be improved by gently blowing on them before spreading, to remove extraneous loose scales and to straighten the setae or, with small moths, the fringes of the wings. When working with small insects, a large magnifying lens mounted on an adjustable stand may be very helpful.

Specimens relaxed from a dried condition present additional problems. The wings may be stiff and require loosening (see Section 3.2.1). If the wings of a relaxed specimen are turned upward and do not lie on the surface of the spreading board, the paper strips may need to be pinned over the wings to hold them down before they are positioned. Because the wings of relaxed specimens are still relatively stiff, skillful manipulation is needed to spread the wings without tearing or leaving excessively large pinholes. If the wings do not move readily under gentle pressure, do not force and possibly break them. Return the specimen to the relaxing chamber.

3.2.7.3 Construction of spreading blocks

The spreading block, a modification of the spreading board, is designed to accommodate only one specimen at a time. Blocks often are preferred by specialists in microlepidopterans but can be also used for other insects. The design is simple (Fig. 3.14), consisting of a wooden cube about 2 cm on a side for most insects, with a groove across the middle of one face. The width and depth of the groove vary to suit the size of the insect to be spread, usually 1.5−2 mm in width and deep enough to accommodate a strip of fine cork, polyporus, or similar pithy material into which the pin or minutens are lodged to hold the

FIGURE 3.14

Spreading block for small insects.

insects being spread. The groove should be cut parallel with the grain of the wood, and the top surface of the block should be sanded very smooth. Before the pinning strip is wedged or cemented into the bottom of the groove, a hole about 1 mm in diameter should be drilled squarely in its middle. The pin can extend into this hole when the insect becomes level with the spreading surface. A few gashes, made by pressing the blade of a thin knife in the upper corners of the block near each end of the groove, should be made to catch the thread that will hold the wings of the insect.

Suggested reading

Spreading boards and blocks
- Edwards (1963)
- Lewis (1965)
- Tagestad (1974, 1977)

The insect to be spread, mounted either on a standard insect pin or on a minuten, is pinned into the groove as with a spreading board, and the wings are manipulated by gentle blowing and using a setting pin. A piece of fine silk or nylon thread is then caught in one of the knife gashes and brought over the wings, and, if necessary, once around the block and again over the wing at another point, and finally caught again in the knife gash. A small piece of tracing paper may be placed on the wings before passing the thread over them, but if special scale tufts are found on the wings, it is better to omit the paper and leave the tufts in a natural position.

Specimens on spreading boards or blocks should be placed in a warm, well-ventilated place to dry for at least 2 weeks. Even very small moths require that long to dry completely and become stiff. If they are placed in a low-temperature oven, such as is used for drying plant specimens, 2 days may suffice. Specimens relaxed from a dry condition, as already noted, dry quickest, but even they should be left for several days. Fresh specimens, even large ones, may be dried in 2 days or less with heat. Where humidity is low and there is ample sunshine, the spreading boards or blocks may be placed in cardboard cartons painted black and left out in full sunlight for about 2 days. Occasionally, specimens may become greasy, but otherwise no harm results. The spreading boards or blocks must be kept where they are safe from mice, bats, dermestid beetles, lizards, psocids (book lice and bark lice), and ants, especially in tropical climates. One preventive measure that is sometimes advisable involves placing the boards or blocks on bricks set in pans of water. If they are hung from the ceiling, then a mosquito net around them may be necessary.

Again, we must emphasize that the collection data are as valuable as the specimen. Always keep temporary data labels with specimens on spreading boards or associated with them in some way to ensure that there is no confusion or loss of data when they are removed from the boards. Specimens are scientifically worthless without accurate collection information.

Spreading is a highly individualistic skill, subject to wide variation. Nearly everyone, with practice, evolves his or her own technique, so that two workers may appear to follow different procedures and yet produce equally good results. There is no single standardized technique with respect to the fine points of spreading.

3.2.7.4 Riker mounts

Sometimes, it is desirable to prepare specimens for exhibition in such a way that they may be handled for close examination without risk of damage. Riker mounts have long been used for this purpose. They may be purchased from entomological supply houses, but similar cases may be easily constructed. The Riker mount (Fig. 3.15) is a flat cardboard box about 3 cm deep, filled with cotton, and with a pane of glass or plastic set into the cover. Unpinned specimens are placed upside down on the glass of the cover, spread into position with some cotton held in place by small weights, and allowed to dry thoroughly (about 2 weeks). Then the weights are removed, enough cotton is added to hold the specimens firmly in place, a little fumigant is added to kill any pests or their eggs that might have been laid in the box, and the bottom part of the box is put in place. When the box is closed, it should be sealed completely to prevent access by pests. Plant material or

FIGURE 3.15

Riker mounts.

associated insect frass may also be dried in place with the specimens to make valuable reference collections.

Riker mounts are practical only for relatively large insects, such as butterflies, larger moths, beetles, and dragonflies, which are suitable for such display. Although Riker-mounted specimens are useful for classroom instruction and general display, they are not used for storage of insects in a scientific collection, where specimens must be available for examination from all angles under magnification. Riker mounts should be inspected periodically for pests and kept out of sunlight, which will cause fading of colors and general deterioration. It is sometimes desirable to put pinned specimens into Riker mounts. To do so, remove the pin or cut it flush with the surface of the insect (Holbrook, 1927).

3.2.7.5 Inflation of larvae

A common practice in the 19th and early 20th centuries was to preserve larvae, mainly caterpillars, by inflation. That practice has largely been abandoned in favor of alcoholic preservation or freeze-drying. These latter methods permit more thorough examination of all parts of the specimens, even internal organs, which must be removed before inflation. Some of the colors of the larger larvae are better preserved in inflated specimens than in alcohol, but color photography has made preservation of the larval colors less essential. However, the technique is still potentially useful and, if done well, is not to be discouraged. For instructions on how to inflate larvae, consult Banks (1909, pp. 69–70), Hammond (1960), and Martin (1977).

3.2.7.6 Artificial drying

The disadvantages of the conventional methods of drying are numerous.

1. Time is involved in transferring the specimens or the liquids.
2. Often the appendages must be teased away from the body.
3. The body sometimes shrivels.
4. Specimens become hard or brittle.
5. Compounds such as xylene and acetone are toxic.
6. Modern techniques for artificially drying specimens preserved in fluid involve freeze-drying and critical-point drying (CPD).

Many soft-bodied insects and other arthropods may be preserved by freeze-drying in a very lifelike manner and with no loss of color. Freeze-drying is a great improvement over traditional methods of preservation in fluids or by inflation, especially with lepidopterous larvae, in which important characters of color and pattern are largely lost in all the traditional methods. Unfortunately, the cost of freeze-drying equipment is high, amounting to several hundred dollars (Blum and Woodring, 1963; Roe and Clifford, 1976).

Briefly, the procedure consists of killing the insect by first freezing it in a natural position and then dehydrating it under vacuum in a desiccator jar kept inside a freezer at $-4°C$ to $-7°C$. With a vacuum of 0.1 μm at $-7°C$, a medium-sized caterpillar will lose about 90% of its moisture and about 75% of its weight in 48 hours. The frozen condition prevents distortion while drying. The time required to complete drying is variable, at least a few days with small specimens and more than a week with larger specimens. When dry, specimens can be brought to room temperature and pressure and permanently stored in a collection. Like all well-dried insect specimens, they are rather brittle and must be handled carefully. Freeze-drying has also yielded excellent specimens of plant galls formed by insects. Freeze-drying can distort or destroy ultrastructure by differential thermal expansion or the formation of ice crystals from unbound water. Thus the technique is not appropriate for tissues.

CPD (Gordh and Hall, 1979) is a procedure that was developed to prepare specimens for study under the scanning electron microscope. Like freeze-drying, CPD is relatively expensive and sophisticated because special equipment and chemicals are required. Nevertheless, the results are ideally suited for specimens collected into or preserved in a fluid. The technique involves passing specimens through an intermediate fluid (acetone, ETOH, and Freon) and then into a transitional fluid (CO_2, Freon, and nitrous oxide), and then subjecting them to a critical temperature and pressure.

The CPD procedures are simple and relatively rapid. Specimens in alcohol are placed in a small mesh-screen basket with a lid. Specimens then are dehydrated through a series of increasing concentrations of ETOH, ending in a solution of 100% ETOH. The specimens within the basket are passed through two washes of 100% ETOH. Specimens that have been preserved in 70% ETOH can be taken through the alcohol series without rehydration. Leave the basket of specimens in concentrations of alcohol for 20−30 minutes. Recently, collected specimens preserved in alcohol or small Lepidoptera larvae may require 1−2 hours in each concentration. Specimens can be left for longer periods in concentrations of alcohol above 50%. After removal from the last wash of 100% ETOH, the specimens in their basket are placed within the chamber of the critical-point drier and are processed according to drier instructions. Several critical-point driers are manufactured. Liquefied CO_2 (research grade) is the ideal transitional fluid because it is easy to use, comparatively inexpensive, and less noxious than others.

CPD has distinct advantages over conventional methods of specimen preparation. Many specimens can be handled simultaneously, which saves considerable time. Pigmentary colors remain lifelike. Specimens do not collapse or shrivel. Appendages can be manipulated and are more supple than in air-dried specimens. CPD is most effective for small, soft-bodied insects, such as parasitic Hymenopterans and Nematocerans. As freeze-drying and CPD become more popular, we may see excellently prepared dry specimens of immature insects in entomological collections alongside pinned adults.

3.2.7.7 Embedding

Preservation of various kinds of biological specimens in blocks of polymerized transparent plastics was popularized some years ago and is still of some interest. The process is rather complicated and laborious, but if carefully done it will yield useful preparations, especially for exhibits and teaching. Directions for embedding insect specimens may be found in the following references and in directions furnished by suppliers of the materials (Foote, 1948; Fessenden, 1949; Hocking, 1953).

3.2.8 Mounting specimens for microscopic examination

Insects are mounted on pins to allow them to be moved, sorted, or viewed under a microscope when necessary. Many of the external morphological features of insects, which are important for identification, can be seen under low-power microscopy, such as with a dissecting scope. Pinned specimens can be handled most easily and with the least amount of damage if they are transferred from the permanent storage box to a small observation base (Fig. 3.16) constructed of cork and a small amount of modeling clay or Plasticine mounted on a small wooden or cardboard base. The base should be small enough to sit directly on the stage of the dissecting microscope, allowing enough room for it to be turned as needed. The pinned insect can then be inserted into either the clay or the cork at nearly any angle such that observations can be made of any side of the insect. Commercially available

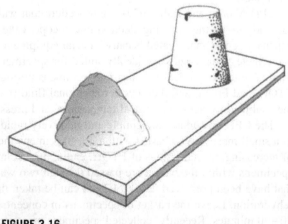

FIGURE 3.16

Observation block.

observation stages (Fig. 3.17) can be
purchased that offer more flexibility
under a microscope.

> The techniques and materials used to prepare
> specimens for high-power microscopic
> examination vary considerably and are
> influenced by the kind of insect or mite as well
> as the researcher's preferences and objectives.

Many insects and mites must be
examined at relatively high magnification
with a dissecting microscope (80–250 ×)
or a compound microscope (320–970 ×).
This is particularly true of small mites,
thrips, whiteflies, aphids, scale insects,
fleas, lice, and some parasitic Hymenoptera.
The microscope is also necessary to clearly
see minute details of larger insects, such as
the genitalia. Small specimens (≈ 1–4 mm
long) and parts of specimens must be
specially prepared and placed temporarily
or permanently on microscope slides. We
call this process *whole-mount preparation.*

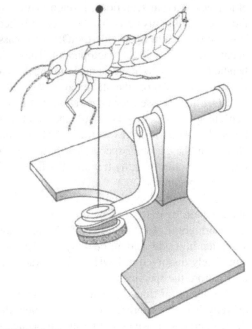

FIGURE 3.17

Commercially available observation stage.

Whole-mounts are distinct from slide preparation of tissues for histological purposes, although
many of the techniques are similar. Slide mounting is sometimes not advisable for large, thick,
or complex structures that must be examined from several angles. Complex or large structures
may be examined in a liquid (such as glycerin) and preserved in microvials. Whichever course
is adopted, the preliminary treatment of specimens is the same.

The techniques and materials used to prepare specimens for high-power microscopic
examination vary considerably and are influenced by the kind of insect or mite as well as the
researcher's preferences and objectives. The information given here will provide the reader
with basic concepts and principles involved in preparing specimens for such study. For more
specific instructions, consult the references listed in the section "Suggested reading" at the
end of Section 3.2.8.6.3. For reagents and media formulas mentioned here, see Appendix I.

Microscopy methods vary considerably according to the condition of the specimen, the
objectives of the study, and the mounting medium employed. However, certain features are
common to all processes. The following procedures are given for whole-mounting
specimens to be examined microscopically and do not apply to histological preparations.

3.2.8.1 Maceration

Maceration is the chemical removal of muscles and other soft tissues while leaving the
sclerotized or chitinized parts of insects that are needed for identification. Maceration

eliminates external secretions, foreign matter, some organs, muscles, and fat body without damage to chitinous parts. This is accomplished by immersing the specimen in a 5%–10% solution of sodium hydroxide (NaOH) or potassium hydroxide (KOH). Other macerating agents are lactic acid and lactophenol. These chemicals are strongly caustic and must be handled carefully to avoid damage to the skin and eyes. If spattered on the skin, immediately wash off the chemical with water.

KOH may be used for maceration, but this chemical must be used cautiously because delicate specimens may be easily and quickly damaged or completely destroyed in it. Often, NaOH performs as well as KOH or even better. Delicate specimens should be cleared in NaOH; other specimens should be cleared in KOH. Fine ducts lost with KOH remain even after lengthy treatment with NaOH, which will damage only teneral or newly emerged specimens.

NaOH supplied by chemical firms in pellet form is most convenient; three pellets in about 10 mL of water may be used for a day. The solution, however, becomes useless if left overnight. Even if NaOH boils dry on a hotplate set a little above its boiling point, a specimen in it will seldom be damaged. Under similar conditions, a specimen may be completely dissolved in KOH. Adding water to a specimen boiled dry in NaOH solution may restore the specimen.

The length of time the specimen is subjected to maceration is determined by the size and physical characteristics of the specimen, the concentration of the chemical agent, and the temperature at which the treatment occurs. Overnight treatment at room temperature is generally adequate for most specimens; more refractory specimens may require longer periods at higher temperatures. The cuticle of relatively large, strongly sclerotized insects sometimes must be punctured with a fine needle to allow the agent to penetrate the body. Puncture the specimen along the intersegmental membrane between the sclerites but not through them. Heating accelerates the action of the macerating agent, but care must be taken to avoid damage by excessive action, especially if the specimen is teneral. Do not heat the genitalia of microlepidoptera because heat distorts the sclerotized parts. Immersion of the genitalia in cold 10%–20% KOH solution overnight is recommended. Boiling for 1 min in a 10% solution of NaOH will clear most other small genitalia. For directions on how to macerate insect genitalia, see Section 3.2.8.7.

3.2.8.2 Washing

When the specimen is adequately macerated, remove the caustic macerating agent from the specimen by washing it in distilled water, when available, or tap water. Remove the specimen from the macerating container and place the specimen in a small dish, a watch glass, or a small plastic bottle cover. If the specimen is placed in water for several minutes for manipulation, dissection, or examination, it will then be ready for further treatment. Adding a drop of acetic acid (white vinegar) will guarantee that no caustic substance remains.

3.2.8.3 Dehydration

Specimen dehydration is usually necessary before the specimen is mounted on a microscope slide. Dehydration is essential if the mounting medium has a resin base.

Dehydration is accomplished by immersing the specimen in a graded series of ETOH. After maceration, wash the specimen in distilled water and place the specimen sequentially in 50% ETOH, 70% ETOH, 80% ETOH, 90% ETOH, 95% ETOH, and two washes of absolute ethyl alcohol. Each wash extracts water from the tissues such that the final treatment with absolute ethyl alcohol leaves the specimen dehydrated. The amount of time in each wash varies with the size and nature of the specimen. Some specimens must gradually dehydrate over a period of several hours to avoid distortion.

3.2.8.4 Staining

Specimen staining is sometimes necessary with insects and mites because their immersion in the mounting medium may make colorless and transparent tissues virtually invisible if the medium has a refractive index close to that of the tissues of the specimen (Stein et al., 1968). If a phase-contrast microscope is available, then staining (even with colorless specimens) is not necessary. Staining specimens can be a complicated matter, and histological texts often devote considerable attention to the process. Consult a good histological text for kinds of stains and protocols for specific results (Humason, 1979). Bleaching, usually accomplished with hydrogen peroxide, may also be required in very dark-colored specimens.

Several kinds of stains are available from biological supply houses. Acid fuchsin is generally used for aphids, lice, and scale insects. Thrips and fleas should never be stained; most acarologists do not stain mites if they are to be mounted in Hoyer's medium. A simple stain for the exoskeleton of insects is made by dissolving a few mercurochrome crystals in water. Specimens may be immersed in the stain solution for a few minutes, depending on the degree of staining needed, and then briefly rinsed in water. Consult Gier (1949) and Carayon (1969) for information pertaining to cuticular stains.

3.2.8.5 Bleaching

If specimens are too dark to reveal sufficient detail after maceration, they may be bleached in a mixture of one part strong ammonia solution to six parts hydrogen peroxide solution. The length of time the specimen is left in the ammonia-peroxide solution depends on the amount of bleaching needed. Specimens should be examined frequently and care must be exercised because specimens are irrevocably damaged by excessive treatment with bleach.

> Small specimens ($\approx 1-4$ mm long) and parts of specimens must be specially prepared and placed temporarily or permanently on microscope slides.
> We call this process whole-mount preparation.

3.2.8.6 Mounting medium

Selection of a mounting medium is very important. Mountants are substances in which a specimen is placed for observations, usually beneath a coverslip for microscopic observation. Mountants are classified as temporary or permanent. Mountants may be solid (plastic), semisolid (gelatin), liquid (glycerin), or gas (air). Mountants miscible in water include Hoyer's medium, lactic acid, polyvinyl alcohol (PVA), and glycerin jellies. Mountants miscible in alcohol include resins, gum mastic, Venice Turpentine, and Gum Sandarac. Mountants miscible in aromatic hydrocarbons include natural resins (Canada balsam) synthetic resins (Permount, Euparol, Histoclad, Piccolyte, Hyrax), and mineral oils (Nujol).

3.2.8.6.1 Mounting

Specific mounting techniques depend on the purpose of the preparation. Some workers feel that preparations may be needed only temporarily in routine work and may be discarded after examination. We recommend permanent microscope slide mounts whenever possible or maintaining the preparation permanently in glycerin in a microvial. If the preparation is kept in a microvial, see Section 3.2.8.7. If the specimen is mounted on a microscope slide, then further treatment depends on the mounting medium.

3.2.8.6.2 Temporary mounting

A temporary mount can be made with lactic acid, glycerin, or other medium on a 2.5×7.5 cm^2 cavity slide. The specimen is placed near the edge of the cavity and wedged into position by manipulating a coverslip over the cavity and the specimen. A fine needle will help bring the specimen into a desired position before the coverslip is centered over the cavity. After the specimen is positioned and the coverslip is centered, a commercial ringing compound, nail polish, or quick-drying cement is used to seal around the edge of the coverslip.

Temporary slide mounts may be kept for a year or more without damage to the specimens. However, these mounts occupy more space in a collection than other slide mounts and are more expensive than conventional slides. Thus the specimens usually are placed in vials of alcohol for storage. Temporary mounts are advantageous, in that the specimen can be turned easily and viewed from many angles. However, because of the thickness of the mounts, a vertical illuminator operated through a microscope or some alternate method of direct lighting is generally needed. Genitalia and other body parts may be examined. Drawings can be made with the aid of an ocular grid in the microscope, while the parts are lying in water in the dish in which they were dissected and extended. Water gives contrast to the structure, which may be difficult to see in glycerin. The water should be "dead," or boiled to drive out gases that may form bubbles in or on the object. The object may be held in place with a minuten bent at a 90-degree angle and laid over the object or by piercing it at a convenient place.

Hoyer's medium and PVA are aqueous-mounting media. Slides made with aqueous mounting media are not permanent. Their impermanence creates several problems that must be evaluated by preparators before these mounting media are selected. Advocates of Hoyer's medium note that the US National Collection at the Smithsonian Institution contains some 40-year-old slides of mites mounted in Hoyer's medium that are still in good condition. However, 40-year-old mounts are still young in terms of slide preparations, and such mounts may eventually deteriorate.

In spite of its popularity, this medium contains chloral hydrate, a controlled substance that cannot be obtained without a permit. Furthermore, chloral hydrate progressively clears specimens. This means that over a period of years, the specimen becomes transparent and sometimes invisible. Remember that Hoyer's medium is a water-based medium and thus impermanent. Some workers ring Hoyer's mounts with glyptol or fingernail polish. These ringing agents fail in time, the mountant dries, and the specimen must be removed, or it is ruined.

Microscope slide preparations made with Hoyer's medium or PVA, particularly of large or thick specimens, tend to crystallize with age and eventually require remounting. Remounting is a relatively simple yet time-consuming process that exposes specimens potentially to damage or destruction each time they are remounted. To remount a specimen, soak the slide in water until the cover glass can be removed, then lift the specimen carefully, and transfer it to a new slide. Specimens permanently mounted require no additional curatorial time, and specimens are not periodically exposed to damage or destruction. Considering the amount of time and effort devoted to whole-mount preparations, a permanent mounting medium is preferred.

Some preparators find specimens easier to arrange if the Hoyer's medium is diluted with water. However, in the process the mounts may collapse as the excess water evaporates. If Hoyer's medium is selected, then we strongly recommend that the medium be prepared exactly as directed (see Appendix I) and used undiluted. Aqueous media are affected by ambient moisture; mounts made in very humid conditions may not dry satisfactorily. These problems notwithstanding, Hoyer's medium is preferred by some acarologists because its refractive index is excellent for use with mites, and specimens can be mounted directly from the collection fluid without clearing or fixing. The specimens are cleared after mounting by heating the slides briefly on a hotplate set at 65°C until the medium barely begins to bubble. Do not allow Hoyer's medium to boil, or the specimens may be damaged beyond use. Such mounts can be prepared quickly for immediate study but should be placed in an oven for curing.

3.2.8.6.3 Permanent mounting media

The standard medium for permanence is Canada balsam. Specimens must be dehydrated through a graded series of ethyl alcohol concentrations before mounting them in Canada balsam. Some preparators are reluctant to use Canada balsam because it has some drawbacks. It yellows with age, which sometimes complicates photography. Inexperienced workers find it difficult to manipulate delicate specimens in balsam. Mites require special treatment, primarily because their cuticle differs from that of insects. These difficulties notwithstanding, Canada balsam is selected as a mountant for many groups of insects, including scale insects, thrips, and parasitic Hymenoptera.

Euparal is a satisfactory mounting medium for most insects (other than scale insects and thrips). This synthetic material has been used for many decades, with favorable results. When it was unobtainable, especially during World Wars I and II, an inferior compound was used. Euparal may be obtained from medical or entomological supply houses and other sources, all of which import the same from Germany. Euparal's formula is a proprietary secret. Euparal does not require dehydration of specimens before mounting. Good preparations may be obtained with specimens taken directly from 80% ETOH and with specimens immersed in 95% ETOH for only a minute. The medium is water-white and remains so indefinitely. For remounting, in case of breakage, specimens may be removed by soaking in absolute ethyl alcohol.

Euparal has a decided advantage over other media, in that small air bubbles trapped in slide preparations are absorbed by the medium during drying. The process sometimes requires several days. Euparal shrinks considerably in drying, which is also a disadvantage. In moderately thick preparations, this results in shrinkage away from the edges of the coverslip. Shrinkage may be countered by adding additional Euparal until there is no further shrinkage. Alternatively, a large coverslip, 2.2 cm in diameter, will pull down around the edges when drying instead of allowing the medium to draw inward. Euparal is relatively fast drying. Allow the slide to remain overnight in an oven set at about 35°C or in the open at room temperature for a few days.

To place specimens in the medium, put one or more drops of the medium in the center of a clean 2.5×7.5 cm^2 glass slide. Determining the precise amount of medium to use requires some experience. Sufficient mountant is needed to run under the entire coverslip. Because Euparal shrinks, a little more mountant is required than with other media. An excess of any medium around the edge of the coverslip is undesirable. Place the cleared and washed (also stained or bleached, if necessary) specimen in the medium on the slide. Make certain that the specimen is completely immersed and that air bubbles are absent. Arrange the specimen in the desired position with a fine needle. If the specimen is thick, place a few pieces of broken coverslip or plastic sheet around it to prevent crushing when the coverslip is applied. With some preparations (such as ovipositors of tephritid flies), a considerable amount of pressure during drying is desirable to obtain maximally flattened and comparable preparations. Then gently lower a coverslip onto the specimen with forceps, holding the coverslip at a slight angle so that it touches the medium first at one side to prevent air entrapment. Apply gentle pressure with the forceps to fix the position of the specimen. Prepare specimens in more than one position (e.g., dorsal side up and dorsal side down). Do not mount parts of more than one specimen on one slide, because all individuals in the series may not be taxonomically identical.

Suggested reading

Mounting specimens

- Hood (1947)
- Mitchell and Cook (1952)
- Baker (1958)
- Murphy and Gisin (1959)
- Hazeltine (1962)
- McClung (1964)
- Richards (1964)
- Strenzke (1966)
- Singer (1967)
- Wirth and Marston (1968)
- Willey (1971)
- Freeman (1972)
- Grimstone (1972)
- Burrells (1978)
- Dioni (2014)
- Chick (2016)

Coverslips are an important consideration in slide preparations and can be purchased from many biological supply houses. Coverslips are manufactured in many sizes and grades. Outline shapes include rectangles, squares, and circles. The size of coverslip selected should be influenced by the size of the specimen. The coverslip should not be so large that the specimen occupies a minute portion of the area covered. The coverslip should not be so small that all of the specimens are not adequately covered. The specimen should be centered under the coverslip and in the center of the slide. This may be accomplished by taping a microscope slide on the bottom and left edges of a card stock (Fig. 3.18). Place a microscope slide on the taped bottom slide and mark the two top corners. Remove the slide and draw diagonal lines from each dot to the top corners of the taped slide; where the two lines intersect, center a round coverslip. Trace the outline of the coverslip, which serves as a target for positioning the specimen within the mountant. With use of this template, all specimens in a slide collection are placed in the same area of the slide. This placement consistency is helpful when examining many specimens under the microscope and very helpful when the microscope is fitted with a mechanical stage.

3.2.8.6.4 Ringing

Canada balsam, Euparal, and other permanent mounts do not require ringing. However, special compounds should be applied around the edge of the coverslip and the adjacent area of the slide to seal temporary mounts, such as aqueous media and media that do not harden as they dry. Transparent fingernail polish is used by some workers, but this is not effective for more than a year. Glyptol is a product utilized as a ringing compound by some taxonomists interested in preserving specimens in a temporary mountant. Gyptol lasts for several years, but its life expectancy for slide ringing is not known. A second application of Glyptol after the first has dried is likely more effective than one application.

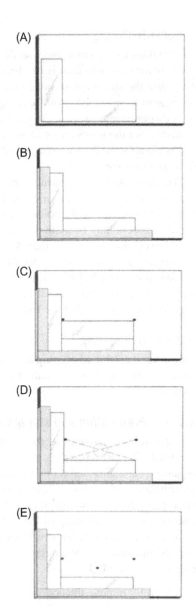

FIGURE 3.18

Steps in making a slide template.

3.2.8.6.5 Curing

Allow slides to dry completely while they remain in a horizontal position. Do not handle slides before the mountant is dry because movement, gravity, and touching of the coverslip can cause the specimen to move in the mountant. In the case of small specimens, setae can be sheared from the body or appendage. Avoid storing slides in a slide box, mailing them, or allowing people to use them until the mountant has dried completely. If an oven is available, set the temperature at 40°C−45°C and leave the slides in it for at least 24 hours and preferably for 3−4 weeks. Avoid excessive heat because this can cause bubbles to form and ruin the mount.

The final stage in preparing permanent mounts is thorough drying or hardening of the mounting medium. Special microscope slide warming trays and ovens may be purchased through biological supply houses. However, special warming equipment is not necessary, and the process may be done in any clean environment or in an oven under gentle heat. Avoid excessive heat (more than 40°C), which can cause bubbles in the mountant or movement of the specimen under the coverslip. The microscope slides should be carefully labeled, preferably before drying. If more than one mount is being made at a time, some recognition mark or code must be applied to reagent containers and anything associated with the specimen so that the final mount may be correctly labeled.

3.2.8.6.6 Labeling

Collection data should accompany specimens at all times during preparation. As with pinned specimens, the information on a microscope slide label is as important as the specimen.

3.2.8.7 Preparation and storage of genitalia

The structures at the apex of the insect abdomen are called the *postabdomen, terminalia,* or *genitalia*. The last term is more restrictive and refers morphologically only to certain organs of the eighth and ninth abdominal segments. These structures are very important in identification. Many insects cannot be identified to species without critical examination of these parts. In some insects, these parts are seen easily without special preparation; in other insects, special positioning of the genitalia when the insects are pinned is sufficient. In many insect species, these structures are withdrawn, folded, or convoluted. For these species, the abdomen or a large part of it must be removed and the genitalia specially prepared as follows:

1. Carefully remove the abdomen by grasping it with forceps near the thorax. Bending the abdomen slightly upward and then downward will usually break it free of the specimen. Perform this operation over a small dissecting dish containing water or 70% ETOH into which the part can fall. If the specimen is in a fluid and soft, then the abdomen may be severed with fine scissors.
2. Place the severed abdomen in a small beaker or crucible containing three pellets of NaOH in about 10 mL of water. Then set the container on a hotplate at a temperature a little above that needed to boil the solution. Place a cover loosely over the container to prevent the specimen from being thrown out and lost if the solution "bumps" when

heating. Use a copper mesh screen between the hotplate and crucible to eliminate "bumping." Allow the solution with the specimen in it to boil for a minute. Great care must be taken to avoid getting NaOH on any part of the skin. If that should happen, wash it off immediately with plenty of water.

3. Remove the specimen with forceps and return it to the dissecting dish. Clean the tips of the forceps with water to remove caustic material. Examine the abdomen to see that muscles and most internal organs have been dissolved. If they have not completely dissolved, then return the specimen to the solution and heat it a little longer.

4. When the specimen is well macerated, use a pair of no. 1 stainless steel insect pins (glued in wooden handles with a drop of epoxy) to pull the genitalia into an extended position. Remove unwanted parts or debris. If much unwanted material is present, transfer the specimen to a clean dish of 70% ETOH. Water or dilute alcohol is better than glycerin in which to examine small, colorless specimens because fine structures are more clearly visible. Tap water may be used, but it should be boiled to remove dissolved gases that may collect on and in the specimen. These gases may be very difficult to remove. The specimen then may be examined and identified; it may be placed in an aqueous solution of a few grains of dry Mercurochrome for staining (see Section 3.2.8.4). The specimen may be held at various angles with a bent minuten for study or sketching. If the specimen is preserved for permanent reference, then it may be stored in a microvial or mounted on a microscope slide.

When several specimens must be identified simultaneously, they may be placed in a "spot," "well," or "depression" plate. The plate is a white or black ceramic dish, generally with 12 wells on the surface. Place one specimen in each well. With more than one specimen per well, be absolutely certain that the abdomens are correctly associated with the specimens. To ensure this, place the abdomens in the wells in the same order or configuration as the specimens are arranged in their holding container, and mark one side of the plate to indicate its orientation.

Add water and one pellet of NaOH to each well. After the NaOH has dissolved, place one abdomen or part of it in each depression and warm the plate gently under an incandescent bulb for about 1 hour. After some of the water has evaporated, replace it with fresh distilled water. Also at this time, examine the abdomens and expel any large air bubbles trapped within that might prevent penetration of the caustic. Reposition the plate under the bulb. A thin stream of macerated tissues soon will issue from the abdomens into the fresh solution.

After an hour or so, depending on the degree of maceration desired, transfer the abdomens for a few minutes to the wells of a second plate that are filled with 70%− 80% ETOH. Add a small amount of acetic acid to neutralize the caustic. While in these wells, the abdomens may be gently manipulated to remove any remaining tissues. Wash and dry the first plate, place two drops of glycerin in each well, top with 70%− 80% alcohol, and transfer the abdomens to this plate. Exercise care to keep them in the proper order.

The abdomens may now be examined or left in a clean place for several days if necessary. Avoid the accumulation of dust by covering the container. The glycerin will not evaporate. If the genitalia are to be permanently preserved, place the parts in a microvial (Fig. 3.19). Mount the specimen on a microscope slide only if the specimen is relatively flat and all important characters can be seen in the final position. For example, the ovipositors of fruit flies (Tephritidae) are flat and may be fully extended; the ovipositor and sheath, including spermathecae, can be mounted on a slide with all necessary characters well displayed. The postabdomens of male tephritids are not suited to such treatment because they are nearly as thick as wide and must be examined in profile, ventral, and posterior views.

5. If the specimen is mounted on a microscope slide, place the specimen in a small dish of 95% ETOH briefly (1 minute is usually sufficient). Place a few drops of Euparal in the center of a clean microscope slide. Remove the specimen from the ETOH and immediately place it in the Euparal. Position the specimen and break any large bubbles present before carefully lowering the coverslip. If insufficient Euparal is present to run to the entire circumference of the coverslip, add a little mountant at the edge of the coverslip. This process can be messy. However, the process can be neat if forceps are used to collect the Euparal from the bottle. With pressure from the thumb and forefinger, nearly close the tips of the forceps and immerse them in the Euparal. This action collects fluid between the tips and shafts of the forceps. Next, touch the tips of the forceps to the microscope slide at the edge of the coverslip. When contact is established between the slide, coverslip, and forceps, release the closing pressure on the forceps and allow the tips to expand. Capillary action will conduct the Euparal beneath the coverslip and increase the volume of mountant beneath the coverslip. Repeat the process until the mountant fills the area beneath the coverslip.

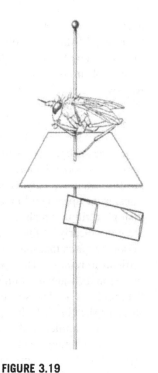

FIGURE 3.19

Specimen mounted with microvial containing genitalia.

Label the slide and allow the mountant to dry overnight in a warm oven or for a few days in a clean, dust-free environment (see Section 3.2.8.6.5). Small bubbles will disappear, and the specimen will become a little more transparent.

The specimen or its parts may be kept in a microvial if the preparation is not suitable for mounting on a slide. The best microvials are made of transparent plastic with neoprene stoppers. Microvials with an inner lip are particularly desirable. Earlier workers used glass microvials with cork stoppers, but the tannin in the cork is injurious to the specimen. Tannin in solution also affects the glycerin in which the specimen is kept and the pin on which the preparation is held. Whatever kind of microvial is used, before placing the specimen in the vial, add just enough glycerin to the bottom to cover the specimen completely. A throwaway injection syringe is excellent for this purpose. It may be kept filled with enough glycerin for many preparations. A small container of squeezable plastic with a fine tubular nozzle is made for model makers to dispense plastic cement. This is an excellent glycerin dispenser. After placing the specimen in the vial, add the stopper. A dull-pointed pin inserted between the stopper and vial allows air to escape as the stopper is pressed into the vial. When the stopper is completely seated in the vial, the pinpoint is extracted. The insect pin holding the specimen and label is then thrust through the stopper, thereby preserving the specimen and its parts together. The parts may be removed from the microvial and reexamined in water or ETOH at any time and then replaced.

With the aid of an ocular grid in the microscope, the genitalia may be examined and sketched when lying in a small dish of water. (Water provides more contrast than glycerin to delicate structures that may be difficult to see.) Lay a bent minuten over the object to hold it in place, or pierce the object at a convenient place. The specimen can be held in place by a dab of petroleum jelly, but the jelly must be dissolved before the specimen is replaced in a microvial or mounted on a slide.

Note: Dissecting, staining, and mounting Lepidopteran genitalia are highly specialized procedures that are not included here. See Clarke (1941) for a description of mounting genitalia of Lepidoptera on slides.

3.2.8.8 Mounting wings

Wings of many kinds of insects can be mounted on microslides for detailed study or photography. Wings covered with scales, such as wings of Lepidoptera and mosquitoes, must first have the scales removed or at least bleached for study of the venation.

Wings are bleached by immersion in an ordinary laundry bleach (sodium hypochlorite) solution. Wet the wings with ethyl alcohol to activate the bleach. Immersion in the bleach for 1−3 minute is usually sufficient. When the veins become visible, remove the specimen or part from the bleach and wash it in water. Sometimes the scales must be removed under water by brushing the wings carefully with a fine brush or with the tip of a small feather. The descaled wing may then be stained, if desired, in eosin-Y or in an aqueous solution of mercurochrome crystals for a few to several hours and then washed again. The wing is then ready for mounting as described here or it may be dried on a slide, placed under a coverslip, and the coverslip ringed with a ringing compound to hold it in place.

Wings not needing descaling may be removed from a fresh specimen. Wings on a dried specimen should be treated with a drop of household ammonia (containing detergent) placed at the base of the wing and allowed to stand in a closed receptacle for about an hour. The wing may be removed with fine-pointed forceps by piercing the body cuticle surrounding the wing base and then pulling the wing loose. This method yields the complete wing, including the basal sclerites. Then wet the wing with 70% ETOH and place it in water for about 10 minute to soften it. If the wing is from a dried specimen, place the wing in water and then heat the water carefully until it just begins to boil. This will aid in removing air from the larger veins. While the wing is in the water, carefully remove dirt that may be present with a fine brush, but avoid removing setae. Also remove unwanted parts of thoracic cuticle and muscles at the base of the wing.

Next, place the wing in 95% ETOH for about a half a minute and add a few drops of Euparal to a slide. Remove the wing from the alcohol and immediately place it in the Euparal on the slide. Position the wing as desired and make sure that its basal part is well stretched. Alternatively, especially with very delicate wings, arrange the wet wing on the bare slide first and then pour the mounting medium on top. Carefully apply a coverslip, touching it to one side of the Euparal first at a slight angle from horizontal to avoid entrapping bubbles. Press the coverslip down on the wing carefully to expand it as much as possible and to force bubbles out of the basal veins and elsewhere. Then cure the slide in a warm oven overnight or in open, clean air for a few days.

3.2.8.9 Mounting larvae of Diptera, Coleoptera, Lepidoptera, and other groups

The study of the immature stages of many insects is important for identification, especially of economically important species. Special techniques are often needed to identify immature insects because their cuticle is soft. CPD and freeze-drying leave the immature stages of most insects well preserved. Unfortunately, this technology is not always available. A method is given here for preparing dipterous larvae that may also be used for the immature stages of other groups. Dipterous larvae, especially those of the higher Diptera, have mouthparts (a cephalopharyngeal skeleton), anterior and posterior spiracular structures, anal plates, cuticular spicules, and other features that are taxonomically important. These parts usually must be examined at high magnification and require special treatment. The larvae of Diptera, Coleoptera, Lepidoptera, and many other groups are best killed in boiling water because it leaves them in prime condition for study.

For cursory examination of the internal cephalopharyngeal skeleton, place the larva in a dissecting dish with a few drops of water. Pierce the cuticle in a few places near the anterior end of the larva and apply a few drops of pure liquid phenol. Do not get phenol on your skin; if you do, wash with water immediately. In a short time the tissues will become transparent. The larva may be returned to 75% ETOH after examination, and the tissues will again become opaque.

For more detailed study and permanent larval preparation, place the specimen in water in a dissecting dish and cut the cuticle with fine dissecting scissors along one side. Start near the anterior end, pass below the spiracle, and continue almost to the posterior end. Then place the larva in an NaOH solution and boil as described in Section 3.2.8.1. When the larva is well macerated, remove the body contents, almost separate the posterior spiracular area from the remainder of the integument, and pull the cephalopharyngeal skeleton a short way out of the body. Place the integument in 95% ETOH while adding a few drops of Euparal on a slide. Then put the integument in the Euparal, opened outward so that the cephalopharyngeal skeleton with the mouth hooks lies away from the integument and the posterior spiracular area lies with both spiracles upward. Apply the coverslip and carefully press it into place. This should give a clear view under high magnification of the cephalopharyngeal skeleton in lateral view, the anterior spiracles, all structures of dorsum and venter of one side, anal plates, and posterior spiracles. The last, often somewhat domed or conical protuberances, may be distorted; the sunray setae and relationships of one spiracle to the other should be easily observed.

As with the genitalia, the larval integument is sometimes best preserved in glycerin in a microvial. Other parts of the insect body, such as antennae, legs, and palpi, may be mounted on slides in Euparal in the same manner as described for the genitalia, wings, and larvae.

References by Clarke (1941), Hardwick (1950), and Zimmerman (1978) concern specialized procedures for making slide mounts of lepidopterous genitalia. Richards (1964) and Wirth and Marston (1968) describe methods using Canada balsam for mounting aphids, scale insects, and other small insects. For methods to use with mites (Acari), see the references at the end of Section 3.2.8.6.3. Further procedures are also given in Appendix II.

3.3 Labeling

To have scientific value, specimens must be accompanied by a label or labels giving, at the very minimum, information about where and when the specimen was collected, who collected it, and the host or food plant. During preparation and mounting, specimens should bear temporary labels with this information. Any time a sample is subdivided, the label must be copied so that every specimen continues to be accompanied by the data. Many collectors keep a field notebook to record more detailed information, such as general ecological aspects of the area, abundance and behavior of the specimens, and other observations noted at the time of collection. Byers (1959) describes a rapid method for making temporary labels in the field. A system of code numbers may be used to associate field notes with the specimens collected. However, the code number should appear in addition to the basic data on the label with the specimen. *A code number by itself is never a valid permanent label for a specimen.*

Samples taken for forensic evidence must be sealed and fully labeled. Chain of custody must also be documented, including the name of the person in charge of specimen collection, dates of collection, and similar information, each time the samples are transferred from one person to the next, verified by signatures.

3.3.1 Paper

The selection of an appropriate paper for labels is very important. Ordinary paper contains acids that weaken the cellulose fibers and subsequently break the paper down over time. Most paper manufactured today is produced from wood pulp that is bleached. Chlorine residue from the bleaching process increases acidity. Other sources of acidity include rosin and alum that are used to prevent ink from running. Paper manufactured from rags is superior because cotton fibers in the paper have superior strength and hold up over time and in solution (alcohol). The best paper has 100% rag content. The paper used for making labels should be heavy so that the labels remain flat and do not rotate loosely on the pin. The surface of the paper should be smooth enough to write on with a fine pen. Lined ledger paper (100% rag and 36 lb weight) is best. Smooth-calendered, two-ply Bristol board is also good; it is usually available at art supply stores. Also desirable is a heavy, high-rag-content paper that is used for professional-grade herbarium sheets; it may be obtained from biological supply houses. Labels made from poor-quality paper become yellow and brittle with age, tend to curl, disintegrate in liquid preservatives, and are generally unsatisfactory.

> The best paper has 100% rag content.

3.3.2 Pens

Handwritten labels are made with "rapidograph" or technical drawing pens. Several brands (Staedtler, Koh-i-nor, Rotring) are available from art stores, craft shops, and bookstores. Technical drawing pens come in several sizes that produce lines of different widths. For most label information, 0.25 mm (no. 000) to 0.30 mm (no. 00) are suitable sizes for achieving small, legible writing. Proper care and maintenance of technical writing pens is necessary for optimal performance.

3.3.3 Ink

The ink should be a superior grade of India ink that is permanent and will not "run" if the labels are placed in jars or vials of liquid preservative. Be sure the ink is completely dry before placing the label in the liquid. When possible, use a waterproofing spray (artist's fixative) on the labels after they are dry. India ink is not always available or convenient to use when collecting in the field. However, labels written with a firm hand and with a moderately soft lead pencil are satisfactory. Labels created using a ballpoint pen or hard-lead pencil soon fade and become illegible, especially when placed in liquids.

3.3.4 Lettered and printed labels

Labels may be lettered carefully by hand with a fine-pointed pen, such as a crow quill or Hunt's 104 drawing pen point. Personal computers and printers also can provide a rapid, inexpensive, and professional label for museum-quality specimens (Inouye, 1991). Printed labels, with four-point type, are preferable and are advisable if more than a few labels of one kind are needed. They may be printed in advance, with blank spaces left for the date, or the full data may be printed. Alternatively, computer-generated labels with the proper font and size printed on sufficiently heavy, good-quality paper can be printed by the sheet. Then permanent labels can be cut from the sheets. Photo-offset methods can also produce satisfactory labels from typewritten copy, but the proper kind of paper must be specified. Print on only one side of a label. A satisfactory alternate approach is to order printed labels from a commercial firm.

> The maximum size of a label should be about 7×18 mm.

3.3.5 Size of labels

The size of the label is important. Labels must be neither too large nor too small. In determining the size of labels, a relationship must be established between the size of the insect on a pin and the amount of data a label will hold. Most insects are small and the collection data can occupy more label space than the planar surface area occupied by the individual insects. Try to make labels of a certain maximum size, and use multiple labels for each specimen to accommodate all of the information that must be presented. The maximum size of a label should be about 7×18 mm^2. In four-point type this translates into five lines of 5 pica length, or about 13 capital letters. Commercially printed labels can be much smaller. Use several labels (if needed) for each specimen to accommodate all of the information that must be presented. Large beetles and butterflies need larger labels, but avoid excessively large "barn door" labels because they do not hold well on a pin. Even with very small insects, do not skimp on the amount of data just to make a small label. One advantage of moderately large labels for small insects is that if a pin with such a label is accidentally dropped, the label will often keep the insect from being damaged. However, large labels may damage nearby insects in a box when the pin holding the label is removed from the box. If capital and lowercase letters are used, then spaces between words are not necessary. If there is any chance of ambiguity, then use full spellings. With only one line of data, the label should be wide enough so that when the pin is inserted, all data are legible.

3.3.6 Label data

The information provided on the data label is as important as the specimen on the pin. The information must be concise, accurate, and unambiguous. Indispensable data must answer the questions of where and when the specimen was collected and include the name of the collector. These data, usually known as locality, date, and collector data, should be given as follows.

3.3.6.1 Locality

The collection locality should be given in such a manner that it can be found on any good map. If the place is not an officially named locality, then it should be given in terms of approximate direction and distance from such a locality. Alternatively, the coordinates of latitude and longitude may be given. The Smithsonian Institution (US National Museum) recommends that for localities in the United States and Canada, the name of the state or province be spelled in capital letters, such as ALBERTA, UTAH, KANSAS, INDIANA, and NO. DAKOTA. This method should also be used for foreign countries, such as NETHERLANDS, CHINA, EQUADOR, and HONDURAS. The next subordinate region should be cited in capitals and lowercase letters, such as counties and parishes in the United States and Canada and provinces elsewhere. Here are a few examples, with virgule (/) indicating the end of a line: ARIZONA:Cochise Co./15kmNEPearce (=15 km northeast of Pearce); NEWFOUNDLAND:Hermitage Dist./12kmWStAlbans; EGYPT: Cairo/Suez Road 38kmW/Suez; EGYPT:Mud.-Al-Tahrir/22 km/SWAbulMatamir; or EGYPT:Mud.-Al-Tahrir/30°05E,30°15N. Current two-letter abbreviations for American states and zip codes should not be used because they are not self-explanatory and may not be permanent.

3.3.6.2 Date

Avoid ambiguity in providing the date of collection. Cite the day, month, and year in that order. Use the international convention of writing day and year in Arabic numerals and the month in Roman numerals without a line over and under the numerals. (Americans tend to put the month before the day; Europeans tend to put the day before the month. A Roman numeral indicating the month avoids confusion involving dates such as 4-8-78 and 8-4-78). Place a period or hyphen between each number: 4. VIII. 78 (=August 4, 1978), or 16. V.86 (=May 16, 1986). If a few consecutive days have been spent collecting in one locality but not more than a week, then the extreme days may be cited. For example, 1−6. IX. 1989 means that specimens were collected from September 1 to 6, 1989. If three consecutive nights of light trapping were at one place, the median day may be cited, such as 19. VIII. 1980, for trapping on the nights of August 18−20, 1980. For reared specimens, the dates of collection of the immature stages and of adult emergence should be cited. "Pupa 10-XI-1981, emer. 14-VIII-1982" indicates that the pupa was collected on November 10, 1981, and that the adult emerged on August 14, 1982.

3.3.6.3 Collector

The name of the collector is regarded as invaluable information on data labels. Spell out the last name of the collector or collectors, using initials for given names. If the last name is a common one, such as Smith, Jones, or Williams, always include middle-name initial. If a collection party consists of more than three collectors, then use the leader's name followed by et al.: T. J. Gibb et al.

3.3.6.4 Other data

Many kinds of information may be important but not relevant or available for all insects. For instance, cite the hosts of parasitic insects and the host plant for phytophagous insects when this information is known. Details of the habitat (elevation, ecological type, and conditions of collection) are important and are usually put on a label in addition to the primary data. Such data may include "swept from Salsola kali." "McPhail trap in orange grove," "at light." "3200 m" "sandy beach," and "under bark dead Populus deltoides." Do not use vernacular names of hosts unless the host is common and widespread, such as orange or horse. If the specific name of a host is not known, at least give the genus. *Vaccinium* sp. is better than no name or *huckleberry*. Even the family name of the host is helpful if no more specific name is available. The presumed nature of the association between insect and plant should be clearly indicated, for example, *Resting on flowers of Vaccinium* sp. The word "ex" (Latin for "out of") should mean that the insect was observed feeding on or in or was bred from the mentioned plant, for example, "Ex seed *Abutilon theophrasti.*"

Forensic evidence should also include additional data, such as agency of collector, case number, victim or subject's name, a brief description of the item, and name and signature of the person sealing the evidence.

As mentioned earlier, it is a good idea for collectors to keep a notebook in which details of locality, habitat, and other important data are recorded. Each individual locality may be assigned a notebook or code number with which the collection jars and vials are marked until the specimens can be prepared. However, citation of such a number on permanent labels is virtually worthless. Even when large files of such data are kept, they are seldom available many years later when a researcher needs to know what a cited notebook number means. Cryptic codes should be avoided. If information is maintained on a computer, then the files should be copied onto separate storage disks and stored in a safe place away from the computer.

3.3.7 Placing the labels

For double-mounted insects, insert the pin through the center of the right side of the label (Fig. 3.20), with the long axis of the label oriented in the same direction as the card point. Take care that the pin is not inserted through the writing on the label. For specimens mounted by direct pinning, the label is centered under the specimen, with the long axis of the label coinciding with the long axis of the specimen. The left margin of the label is toward the head of the insect. An exception to this is when specimens have the wings spread, such as Lepidoptera. In these cases, the label always should be aligned transversely, at right angles to the axis of the body, with the upper margin toward the head. Labels may be moved up the pin to the desired height using a staging or pinning block

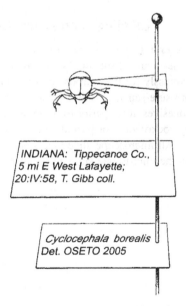

FIGURE 3.20

Proper labeling of double mount.

(Fig. 3.20). The middle step of the block will give about the right height if only one label is used. With more than one label, space the labels on the pin beneath the specimen so that the information on the labels can be read without moving any of them.

3.3.8 Labeling vials

Specimens preserved in fluid should be accompanied by one rather large label that includes all collection data. Do not fold the label because small specimens may be damaged or lost when the label is removed. Multiple labels or small labels that float in the vial may also damage specimens. Furthermore, when two labels lie face to face, they cannot be read.

> Vial labels should be written with a moderately soft lead pencil or with India ink.

Always place labels inside the vial. Labels attached to vials may become defaced, destroyed, or detached, regardless of the method or substance used to affix them to the vial. The label should be written with a pencil or in ink. If a pencil is used, then select a moderately soft lead pencil because hard-lead pencil writing becomes illegible in liquid. If a pen is chosen, then select India ink and ensure that the ink is well dried so that it will not dissolve or run when immersed in the liquid. Never use a ballpoint or felt-tip pen for labeling vials.

3.3.9 Labeling microscope slides

Preserving specimens on microscope slides is time-consuming and tedious. The labels for slide-mounted specimens should be given equal care and consideration. Labels made expressly for this purpose can be obtained from biological supply houses. Labels with pressure-sensitive adhesive are now available that seem durable with time. Modern adhesives are superior to the older glues, which often failed with time or were consumed by cockroaches or psocids. Labels should *not* be affixed to slides with transparent tape.

All microscope slide labels should be square (Fig. 3.21). Specimens should be centered on the microscope slide and a label placed on either side of the specimen. Never put labels

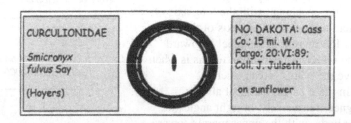

FIGURE 3.21

A properly labeled, microscope slide—mounted insect.

on the underside of a microscope slide. The label on the left side of the specimen should contain taxonomic information. The label on the right side of the specimen should include collection information. The sequence of information should also be standardized. Put as much data on the label as possible, including the kind of mounting medium in which the specimen is preserved. This kind of information is important when specimen remounting is required (Waltz and McCafferty, 1984).

3.3.10 Identification labels

When specimens are sent to an expert for identification, they should be accompanied by permanent collection labels giving all essential data. If associated field notes are available, copies of these should accompany the specimens. When the identification has been made, the scientific name of the specimen and the name of the identifier should be printed on a label associated with the specimen. On pinned specimens, this information is always printed on a separate label placed below the collection label or labels on the same pin. When a series of specimens consists of the same species, the identification label is often placed only on the first specimen in the series, with the understanding that all other specimens to the right in that row and in following rows belong to the same species. The series ends with another specimen bearing an identification label. Identifications for specimens preserved in alcohol or on slides may be written on the same label as the collection data or on a separate label, depending on the preference of the collector or person making the identification.

3.4 Care of the collection

If care is taken and a few basic precautions are followed, a collection of insects or mites can be maintained indefinitely. The information given here is general; institutions and individuals will want to adapt materials and procedures to fit their own needs and resources.

3.4.1 Housing the collection

The serious collector should consider using standard equipment for housing a collection because this ensures uniformity of container size and style when additions are necessary. Homemade boxes, drawers, and cabinets may serve a useful purpose, but they cannot be easily incorporated into museum collections. Standard equipment is available from many entomological and biological supply houses.

Material preserved in liquid usually needs little attention beyond the occasional replacement of preservative and stoppers. Small vials may be stored in racks so that the stoppers are not in contact with the liquid (Fig. 3.22). The use of storage racks for vials expedites rearrangement and examination of the material. Alcohol evaporates in closed containers and vials should be examined periodically to ensure that the specimens do not become dry. Specimens that have become dry in vials, because of the evaporation of preservative, become ruined. If the vials cannot be inspected frequently, then vials containing larvae or large insects should have their stoppers replaced by cotton plugs. Several cotton-plugged vials can be placed upside down in a large jar filled with preservative. Use of cotton plugs is not recommended for very small or delicate specimens because they may become entangled in the cotton fibers. Jars with screw tops or clamping lids, as are used in home canning, are ideal for this purpose. Jars specifically designed for museum use can be obtained from biological supply houses. Stoppers of neoprene or other synthetic materials generally are superior to cork stoppers, but good-quality cork stoppers are usually preferred to plastic screw tops, which are easily broken. Some rubber or rubber-like stoppers in contact with 80% ETOH over long periods of time become bleached and swollen. When specimens stored in such containers are critical-point dried, an undesirable residue forms on the specimens. Many of the newer hanged plastic stoppers are excellent.

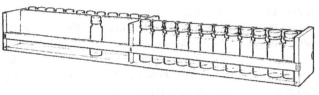

FIGURE 3.22

Vial rack.

(A)

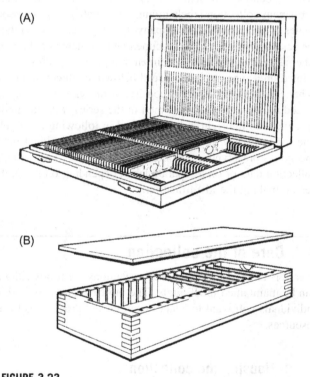

(B)

FIGURE 3.23

Storage containers for microscope slides: (A) plastic slide storage box and (B) wooden slide storage box.

Microscope slides are usually stored in wooden or plastic boxes obtainable from biological supply houses (Fig. 3.23). The inner sides of the boxes are slotted to hold the slides vertically and to separate them from one another. Slide boxes are available in sizes made to hold from 50 to 100 or more slides. If slides are stored vertically, then they must be thoroughly dried before storage or the coverslips may move. Some workers store the slide boxes so that the slides rest horizontally. This is especially desirable if the slides are made with Hoyer's medium, which may become soft under very humid conditions. Several slide-filing systems are available from suppliers, but whatever system is used, care should be taken to ensure that additional similar equipment will be available in the future for expansion of the collection.

Small plastic slide boxes, usually made to hold five slides, are convenient for keeping slides in a unit-tray system along with pinned specimens. This is particularly advantageous when genitalia are mounted on slides, because the preparations are readily apparent to visiting researchers examining the pinned specimens.

Pinned specimens should be kept in any standard, commercially available insect drawer (Fig. 3.24). These standards are named US National Museum, California

(A)

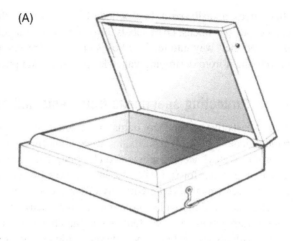

(B)

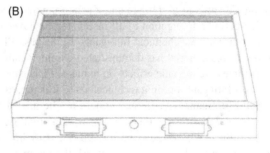

IGURE 3.24

F Storage containers for pinned insects: (A) Schmitt box and (B) Cornell drawer.

Academy of Sciences, Cornell, or Schmitt sizes. Larger North American collections typically use the unit-tray system, with various sizes of unit trays made to fit into a drawer. The pinning bottoms of the unit trays and boxes are now generally made of polyethylene foam. The older standard was pressed cork, but that was extremely variable in quality and usually contained enough tannin to corrode pins and eventually to cement the pins firmly into the pinning material. Polyethylene foam is available in large sheets that are cut to the desired size and cemented into boxes or unit trays.

A serviceable substitute for polyethylene is 6 mm thick balsa wood boards, obtainable from model-maker supply houses. These boards should be individually selected for softness because they are frequently excessively hard. Another good substitute, especially for temporary storage of pinned specimens, is double-thickness corrugated board, which is often used to separate layers or rows of cans in cartons. Single-thickness corrugated board will not hold an insect pin firmly, and the harder board used for making cartons is not suitable.

Any box used to store insect specimens must be nearly airtight to keep out museum pests. The most commonly encountered pests include dermestid beetles, psocids (book lice), and silverfish. In a period of a few months, these insects (and certain other pests) can devour specimens, chew labels, or otherwise ruin a collection made over many years. These pests find their way into the best boxes or insect drawers. Constant vigilance is necessary to prevent them from destroying valuable specimens and priceless scientific information.

3.4.2 Protecting specimens from pests and mold

To kill pests that are actively damaging a collection, apply a liquid fumigant, which acts more rapidly than solid fumigants. Examples of liquid fumigants include carbon tetrachloride, chloroform, ethyl acetate, ethylene dichloride, methyl bromide, and sulfuryl fluoride. These may be registered for restricted use in museums by certified exterminators only. Exercise extreme care when working with any of these compounds. Liquid fumigants volatilize rapidly, may be flammable, and are toxic to humans. Work outdoors if possible and use some kind of fumigation chamber. One day in the chamber usually is sufficient to kill the pests.

Individuals interested in pest damage, identification, and control in museums should consult Edwards et al. (1981) and Zycherman and Schrock (1988).

The periodic fumigation of all insect storage boxes is necessary. The best-made insect drawers provide for chemical fumigants. In the past, PDB was used in museums as a fumigant. However, this material has demonstrated chronic health effects and is under EPA investigation for possible carcinogenic effects on humans. PDB should be avoided. Naphthalene is widely used as a fumigant, but in a technical sense it is regarded as a repellent.

Solid fumigants (repellents) should not be placed loose in a box of pinned specimens. If crystals or flakes of naphthalene are used, a small quantity should be placed in a cloth bag or in a pillbox whose top is perforated with tiny holes. This container is pinned firmly into one corner of the box of specimens. Mothballs may be pinned in a box by attaching the mothball to the head of an ordinary pin. This is done by heating the pin and forcing its head into the mothball. When moving boxes, be careful that the mothballs and fumigant containers do not come loose and damage the specimens. To keep pests out of Riker mounts and other display cases, sprinkle naphthalene flakes on the cotton when the mount is prepared.

Another useful method to kill pests in a collection is to cut strips of dichlorvos (738 Vapona strips, No-pest Strips, Vaponite, and Nuvan) into small pieces, and secure them in the insect drawers. Dichlorvos is registered for nonrestricted use, but only pest strips should be used. This method gives a fairly rapid kill while avoiding the hazards of using flammable liquids. Under conditions of high humidity, Vapona strips corrode metals and dissolve polystyrene plastic.

Mold is another serious problem with insect collections, especially in moist, warm climates. Mold is a kind of fungus that readily attacks and grows on insect specimens. Once a specimen has become moldy, it is extremely difficult to restore it. If only a few filaments or hyphae of mold are present on a specimen, they may be removed carefully with forceps or with a fine brush and the specimen dried well in a warm oven and then returned to the collection. Only keeping the collection in a dry place will prevent mold. In humid climates, it is sometimes necessary to keep insect and other kinds of collections in rooms with artificial dehydration. Some microscope slide mounting media are also subject to molding.

3.5 Packaging and shipping specimens

In mailing insects and mites, there is always a risk of damaging or losing specimens. By following the recommendations given here, the risk can be reduced (Sabrosky, 1971; Stein, 1976; Hoffard, 1980).

3.5.1 Packing materials

Often specimens must be mailed to colleagues or museums. Specimens are at high risk of damage when they are in the mail service. Thus mailing specimens is a very important process that requires considerable thought and attention. Mailing cartons should be constructed of strong corrugated board or other stiff material: wood is advisable for overseas shipments. Do not use cardboard boxes that have been sent through the mail several times. They tend to be damaged. Screw-top mailing tubes are ideal for small items, particularly slide-mounted specimens and specimens in alcohol. All containers must be sufficiently large to include ample packing material to minimize the effects of jarring (Fig. 3.25). A rule of thumb suggests a minimum of 5 cm on all sides between the specimens and the inner surface of the mailing carton. The packing material is intended to absorb shock. Containers holding specimens should not be packed so tightly that this objective is negated. Suitable packing materials include excelsior, shaved wood, crumpled newspapers, styrofoam chips, or plastic bits. Clear plastic sheet material with a regular pattern of bubbles (bubble wrap or blister pack) is superior. This material is very lightweight and has excellent shock-deadening properties.

3.5.2 Pinned specimens

Pinned specimens should always be placed as described here in a small box with a pinning bottom. The box should be well wrapped and placed in a larger carton with at least 5 cm of lightly packed packing material between it and the carton on all sides.

1. Select a sturdy pinning box with a good pinning bottom. The pinning bottom should not be very hard, because hard substrates tend to transmit shock. The pinning substrate should be at least 6 mm thick and cemented securely to the bottom of the box. The box should have a tight-fitting lid or a flap-type lid held in place with a strip of masking tape. Do not mail specimens in an open-top museum tray.
2. Pin the specimens firmly into the pinning bottom. Leave sufficient

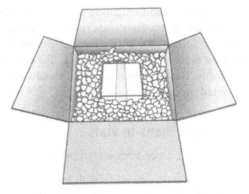

FIGURE 3.25

Packaging specimens for shipment.

space between pins so that the specimens can be removed easily. Specimens pinned too closely together sometimes are damaged by adjacent specimens or their labels when removed. Place bracing pins on both sides of heavy or long-bodied specimens to prevent them from rotating (cartwheeling) on their pins and damaging adjacent specimens. Bracing pins may be put on the sides of data labels or card-mounted specimens to prevent cartwheeling. Microvials should have an additional pin at the end of the vial to keep it from coming off its stopper. Pin a loose ball of cotton in the corner of the box. This captures antennae and legs that might fall from specimens in transit. If loose appendages are not captured, they can damage specimens. Vials other than microvials should be wrapped in a box separate from pinned specimens.

3. The box in which the specimens are pinned should be shallow enough so that the heads of the pins almost touch the lid. If the box is too deep, a piece of firm cardboard should be cut to fit into the box and lie on top of the pins. If there are only a few specimens in the box, a few extra pins should be added near all of the corners to keep the cardboard level. Attach a tab made of a piece of adhesive tape folded double, with the ends left free to attach to the top of the inserted cardboard. The insert may be lifted out by the tab. The space between the insert and the lid of the box should be filled with packing material to keep the insert pressed lightly against the tops of the pins when the lid is in place. This prevents the pins from working loose and damaging specimens in transit. Use cotton batting, sponge, or bubble wrap between the lid and insert. Do not use excelsior or any loose material because this can work its way beneath the insert and damage specimens.

4. If only one or two specimens are being shipped, they may be placed in a straight-sided plastic vial with a press-on or screw-on top. The vial should be of sufficient diameter to hold the labels in a normal position. A cork stopper cut to such a length that its larger end is a little greater in diameter than that of the inside of the vial is pressed tightly into the bottom of the vial. This will provide a good pinning bottom into which one or two pinned specimens may be firmly pressed. Attach the cover of the vial, wrap the vial in enough packing material to hold it firmly in a mailing tube, and attach the cover of the mailing tube, and it is ready to ship.

We recommend fumigating boxes before they are shipped. This eliminates the transmission of museum pests (psocids, silverfish, and dermestid beetles) to other collectors or institutions. Do not leave loose fumigant in the box with the specimen or any fumigant balls on pins in containers. They are especially prone to work loose and damage specimens. Boxes that are received should also be fumigated to prevent the acquisition of unwanted museum pests.

3.5.3 Specimens in vials

The following procedures are recommended:

1. Fill each vial completely with liquid preservative. Air bubbles in vials promote movement of specimens and labels. Stopper the vial tightly by holding a pin or piece of wire between the inner surface of the vial lip and the stopper. Press the stopper into place and permit air or excess fluid to escape; then remove the pin or wire. Cork

stoppers should not be used because they crumble and often leak. If they are used, make certain they do not have defects that will allow leakage while specimens are in transit. Screw-top vials should be firmly closed and sealed with a turn and a half of plastic adhesive tape around the lower edge of the cap and part of the vial. Some workers seal vials with paraffin. In our experience this is unnecessary because the paraffin often breaks loose and will not prevent leakage.

2. Wrap each vial with cotton, tissue, foam padding, paper toweling, or similar material. Allow no piece of glass to come into contact with another piece of glass. Several individually wrapped vials may be bound or held with tape or rubber bands as a unit, or they may be placed in a small cardboard box with enough packing material to ensure that they are not shaken around. Placing all of the vials in a self-sealing plastic bag prevents fluid from damaging pinned material in the event that a vial breaks.

3.5.4 Loading cartons

After pinned specimens or specimens in vials, or both have been prepared properly, they should be placed in a strong carton large enough to hold at least 5 cm of packing material around all sides, including the top and bottom. Use ample packing material to prevent the contents from shifting within the carton, but do not pack the material tightly. Packing material should be resilient, to absorb shocks and prevent damage to the contents being shipped. One or a few vials may be shipped in a mailing tube, as previously described. When shipping more than one box or packet of vials, tie or wrap them together as a unit before placing them in the larger carton. Individual boxes or vials may easily be overlooked and lost when unpacked. For this reason, a packing list should be included in the carton. Vials of specimens in fluid are much heavier than boxes of pinned specimens. Cartons containing many vials may be packed somewhat tighter in the carton than cartons containing only pinned specimens. Pinned and liquid-preserved specimens may be shipped in the same carton. When many vials of liquid-preserved specimens are involved, ship them separately.

3.5.5 Shipping microscope slides

Make certain that slides are thoroughly dried and cured. Slides may be shipped in holders made expressly for that purpose (Fig. 3.26). Slide mailers are available from biological supply houses. Even in these holders, wrap a little soft tissue around each end of each slide so that the coverslip does not come into contact with anything. The slides may also be shipped in standard storage boxes. Place soft tissue around each end of each slide and between the slides and the box lid to prevent movement. The box should then be wrapped

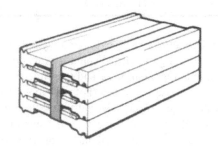

FIGURE 3.26

Microscope slide shipper.

to hold the lid firmly. The slide box may be treated as described here for pinned and liquid-preserved specimens and placed in a carton with them. If slide holders are not available, a few slides, each wrapped with tissue, may be tied together at each end with tape, rubber bands, or string and wrapped in strong paper. Place the wrapped slides in a mailing tube or carton with packing material.

3.5.6 Shipping live specimens

Most insects and mites intended for a collection or submitted to experts for identification should not be shipped alive. To protect American agriculture, federal law prohibits the importation and movement of live pests, pathogens, vectors, and articles that might harbor these organisms unless the shipments are authorized by the US Department of Agriculture. Shipments of live material must comply with all federal, state, and local regulations (see sample form in Appendix IV). Shipments of live insect material without valid permits may be seized and destroyed by Animal and Plant Health Inspection Service (APHIS) quarantine inspectors. In addition to meeting federal laws, the shipment of some species must be approved by state officials. To obtain federal authorization, write well in advance of shipment to: APHIS, USDA, Federal Building, Hyattsville, MD 20782. Each request is evaluated according to its individual merits.

Suggested reading

Collections
- VanCleave and Ross (1947) *(artificial drying)*
- Woodring and Blum (1963)
- Harris (1964)
- Flaschka and Floyd (1969)
- Dominick (1972)
- Berte (1979)
- Gordh and Hall (1979)
- Hower (1979)
- Kissinger (1982) *(lettered and printed labels)*
- Ellis et al. (1985)
- Clark and Gregg (1986) *(care of the collection)*
- Darling and Plowright (1990)

Immature stages (egg, larva, nymph, and pupa) shipped for rearing elsewhere should be placed in tightly closed containers without vent holes. These insect stages require only a minimum amount of air and therefore the insects will not suffocate. Pupae should be packed loosely in moist (not wet) moss or a similar material. Larvae should be packed with adequate food material to last until their arrival. Most beetle larvae and some caterpillars (especially cutworms) should be isolated because they are cannibalistic. To prevent excessive accumulation of frass (fecal material) and moisture, do not overload containers. Plant material held without ventilation becomes moldy, especially when kept in plastic bags. For this reason, pieces of the host plant bearing such insects as scale insects

(Coccoidea) should be partially or completely dried before being placed in a container. If this is not possible, they should be packed in a container (such as a paper bag) that will permit drying to continue after closure. Live Heteropterans and other small, active insects are killed easily by excessive moisture in the container. Therefore provide several tiny vent holes or place a fine-mesh screen over one end of the container when shipping such insects.

Some containers designed to hold living insects are strong and can be shipped without additional packing. Generally, the containers should be enclosed in a second carton, with enough packing material to prevent damage to the inner carton. In all cases, affix a permit for shipping live insects in a conspicuous place on the outside of the shipping container. Some express mail service companies will not accept living insects for shipment. Check with company agents before taking the material in for shipment. If specimens are sent overseas, be certain that importation permits of the destination country are obtained, and use airfreight to send the parcel. Check with international carriers for special requirements.

There are no restrictions on mailing dead specimens within the United States or to other countries. If the specimens are sent out of the country, US Customs requires that the contents and value of the package be listed. The statement "Dead Insects for Scientific Study, of No Commercial Value" will suffice. All packages should be marked "FRAGILE," and a complete return address should be included on the outside of each container.

> To protect American agriculture, federal law prohibits the importation and movement of live pests, pathogens, vectors, and articles that might harbor these organisms unless the shipments are authorized by the US Department of Agriculture.

Classification of insects and mites

Introduction 126
Chapter 4: Classification of insects and mites 129
Chapter 5: Synopsis of insect orders 147
Chapter 6: Descriptions of hexapod orders 187
Summary 235

Introduction

The number of insects and mites known today staggers the imagination. More than 1.5 million species of insects have been described and estimated of the number of undescribed species range in the millions. There are more than 50,000 described species of mites, and many times more than that number remains undescribed. Associated with this enormity of numbers, we find a biological diversity not seen among other animal groups. Insects and mites can be found in almost every conceivable environment. Mites are found living on plants, molds, and stored foods, in the nares of sea snakes, in the cloacae of turtles, beneath the elytra of beetles, and in the quills of birds. Insects can be collected from such diverse and seemingly inhospitable habitats as in petroleum pools or saltwater marshes, within the canopy of tropical trees, or deep inside solid wood, at high altitudes in the atmosphere or deep inside caves and from every continent, including Antarctica. The habits of insects and mites vary from parasitic to free-living and from beneficial to highly destructive. Many insects and mites have developed intimate associations with humans. Some are extremely valuable and provide pollination, food, pest control, and other benefits to humans. Other destructive forms can cause enormous losses to agricultural crops, food, clothing, and other materials of value to humans and can injure people or transmit devastating diseases.

Because of the potential destructiveness of some species of insects and mites, outbreaks of pests must be detected rapidly, and their population sizes must be estimated accurately. Correct identification of a newly detected pest or newly collected or discovered insect or mite is of paramount importance, because its scientific name is the key to all known information about the insect, its habits, its behavior, and its potential threat or benefit to human welfare. Control recommendations, where necessary and prudent, hinge directly on an accurate identification of the pest. Land grant universities also include cooperative extension specialists who can serve as professional resources for arthropod information and identification. A list of state land grant universities, together with street and e-mail addresses and telephone numbers, is provided in Appendix III. Requests for specific identification can also be submitted through the Systematic Entomology Laboratory in Beltsville, MD. Directions for submitting samples are provided in Appendix IV.

Insects and mites can and should be observed in their natural environments, but most of them, especially the many small ones, must be collected and properly preserved before they can be identified. Correct identification seldom is easy. To identify specimens correctly, they must be properly labeled and preserved in the best condition possible. The identification of a particular insect or mite usually requires examination of minute details of its anatomy with the aid of a hand lens or microscope. If the microscopic details of a specimen's anatomy are concealed, damaged, or missing because of improper handling or preservation, then identification is sometimes impossible, and information about the species to which it belongs cannot be made available. Therefore adequate preservation and proper labeling of specimens are essential to their identification.

The reasons for identifying insects and mites are many and varied. Insects and mites may be collected to obtain general study material (as for a school course) or for personal interests (as in a hobby). Professional entomologists may survey for the presence or abundance (relative or absolute numbers) of pests. Scientists may acquire material for biological, physiological, ecological, and other types of studies. The pest control industry often requires specimens to be collected and retained to document past history as well as to justify the need for pesticide applications. It is essential that forensic entomologists and crime scene investigators collect, preserve, and retain specimens that may become important evidence in legal cases. The purpose for making an insect or mite collection may demand various special collecting methods.

What to collect depends on the purpose for which the collection is intended. When important pest insects and mites are needed for identification, specimens should be collected in large numbers. A sample of 20 specimens should be considered the minimum; larger numbers are desirable. Remember, excess specimens can be discarded or exchanged, but additional specimens cannot always be collected when needed. All available life stages should be collected. Frequently, insects and mites cannot be identified accurately from immature stages, and they must be reared to the adult stage for correct identification. Photographers should collect the specimens they photograph when positive identification is desired. Minute or critical diagnostic characters often are not discernible in photographs. If specimens are destined for display cases, it may be important to collect and identify a sample of the host plant for the display as well. Collections that include gall-producing insects and mites must also display the gall. Reference collections often are made more valuable if they contain samples of associated insect damage or frass, which can help identify similar problems later (Kosztarab, 1966). For example, the architecture of ground nests for bees, wasps, mole crickets, and termites can be preserved with wax or fiberglass resins (Chapman et al., 1990; Brandenburg and Villani, 1995).

Biology students in high school and college are often required to collect and identify specimens from as many orders or groups as possible. Other beginning enthusiasts attempt to collect every specimen they find. The experience and knowledge gained in making a general collection are of value in helping the collector decide on a specialty. However, with so many different kinds of insects from which to choose— over 80,000 described species in North America alone—most students eventually concentrate on one or two of the major insect or mite groups. Specimens other than *those in a chosen group may be collected for exchange with other collectors or for deposition in an institutional collection.*

Suggested reading

- Seber (1973)
- Lee et al. (1982)

Part 2 presents basic identification techniques and descriptions of the different arthropod groups. References listed in the text and in Suggested readings *are intended for those who desire to more specific information about the topic.*

Classification of insects and mites

4

Classification is essential in any type of biological work. Identification is an element of classification. Identification to the species or genus level is necessary to determine the best collecting methods and techniques for preservation and how best to control pests. Identification to the family or order level is necessary for the most elementary study of insects and mites. The keys and diagnostic information presented here are designed to enable a novice collector to recognize the most common orders. Students must remember that the number of insect and mite species is immense. We cannot provide a complete classification here, and none is available in any other single reference. Keys to aid in the diagnosis to family levels may be found in general entomology textbooks such as Borror et al. (1989) and Gullan and Cranston (2004). A glossary of basic entomological terms is often helpful in understanding complex terminology associated with identification and classification. A basic glossary is provided at the end of this work. Nichols and Schuh (1989) is a more extensive English-language glossary that may be helpful in entomological work.

Classification involves the arrangement of organisms in groups based on the set of "characters" common to each group. Traditionally, the "characters" have been based on anatomical study. However, another method of classification is molecular taxonomy. In molecular taxonomy, proteins and genes are used to determine evolutionary relationships. The percentage of DNA that is similar between the species is taken into consideration. Signature sequences, such as those derived from ribosomal RNA, also serve to determine the separation between different arthropods. Once a species has been carefully examined, specific genes or proteins can be used as molecular clocks that determine the divergence from a common ancestor.

Insect Collection and Identification. DOI: https://doi.org/10.1016/B978-0-12-816570-6.00004-6

> *Kingdom*
> *Phylum*
> *Subphylum*
> *Class*
> *Subclass*
> *Order*
> *Suborder*
> *Superfamily*
> *Family*
> *Subfamily*
> *Tribe*
> *Genus*
> *Subgenus*
> *Species*
> *Subspecies*

Categories commonly recognized are arranged in order of rank from highest to lowest.

Groups are arranged in a taxonomic hierarchy that proceeds stepwise from the highest to the lowest level: kingdom, phylum, class, order, family, genus, and species. In many cases, there are intermediate super divisions and subdivisions as well as the less used groups of tribe, cohort, and phalanx. This stepped arrangement is known as a hierarchical system and was developed by the Swedish naturalist and taxonomist Carl Linnaeus (1707–78). The kingdom Animalia is divided into several phyla. The phylum Arthropoda contains insects, mites, and all other animals with an external skeleton and segmented appendages. This phylum is divided into classes as shown in the following key. Additional keys are given to orders and in a few cases to suborders. For further identification, the reader is directed to the vast taxonomic literature.

The keys given here are in a format often found in North American publications. Each couplet is composed of two statements, labeled **a** and **b**. Start at couplet **1a**. Read the two choices and select the one that best describes your specimen. At the end of each couplet is a named taxon or a number directing you to another numbered couplet from which you must again choose. Follow the numbers, comparing alternatives in each couplet, until a name is reached. A similar procedure may be followed with other keys in the literature to which you may wish to refer. The use of keys is far more satisfactory than comparison with pictures or even with identified specimens because the most important diagnostic characteristics are specifically stated. The characters used in taxonomic keys often involve seemingly minute differences that might not be suspected by an inexperienced person.

We urge students to refer to an elementary textbook on entomology and to become familiar with the basic elements of insect anatomy. Even words such as "face" and "leg" do not mean quite the same when referring to an insect as they do in reference to a human. It is also necessary to distinguish between an immature and an adult. This kind of information is found in general textbooks.

Scientific names are used throughout the world in many kinds of publications. Each species has one scientific name, and all the biological information regarding a species is tied to the scientific name. For instance, the scientific name *Homo sapiens* refers to humans and not other primates. In essence, the scientific name is a shorthand method of conveying a considerable amount of information. The scientific name of a species is composed of two words: the genus (written with an initial capital: *Homo*) and the species (a specific epithet or trivial name written in lowercase: *sapiens*). This combination, known as a *binomen*, must be written in a typeface different from the text in which the name appears. Scientific names are usually italicized and are often followed by the name of the person (the author) who described the species (e.g., the honeybee: *Apis mellifera* Linnaeus). When italics are not possible, scientific names are underlined, which conveys the same meaning as italics. The author's name is never italicized or underlined. Each genus name may be used only once in the animal kingdom; the same species may be used in different genera but not more than once in a genus. The application of scientific names is regulated by the International Commission on Zoological Nomenclature, which publishes the *International Code of Zoological Nomenclature*. Similarly plant names for algae, fungi, and plants are governed by the International Code of Nomenclature.

Relatively few people are qualified to identify organisms, including insects, to species, and therefore scientific names cannot always be applied with confidence. Humans have applied common names to insects in all languages, but sometimes these names do not translate. Also, one common name may be applied to several species. Further, one species may have several common names even in different regions of the same country. This imprecision in the application of names creates problems. In an attempt to solve part of this problem, within the United States, certain insects and mites of recognized pest status have common names selected by a committee of the Entomological Society of America. The standardized list is published in *Common Names of Insects and Related Organisms*. This document was last revised in 1997 (Bosik, 1997). Authors' names are not used with common names.

Common insect names ending in words applied to most insects of an order (such as *horse fly* and *house fly* of the order Diptera) are written as two words. Names such as *butterfly* and *caddisfly* for insects belonging to other orders (i.e., not true flies) are written as one word.

The names of groups above the rank of genus are single words, capitalized but not printed in a different typeface from that of text in which they appear. Only the names of families and subfamilies are regulated by the International Code, which specifies that family names must end in "-idae" and subfamily names in "-inae." These endings are added to the stem of the genus name that is considered typical of the higher taxon, and the generic name is called the *basonym* of the group name. All group names are plural.

Names above the rank of family are not regulated by the International Code and are therefore subject to considerable difference in usage. They are sometimes formed, as are those of families and subfamilies, on a basonym, but many are formed on a word descriptive of a diagnostic character (e.g., the name of the order Diptera means "two-winged" in classical Greek). For these reasons, alternative names of some classes and orders are cited here.

The name of the person who first proposed a scientific name is often appended to the name as a reference; sometimes the year of the proposal is also added after a comma. These references, when added to a binomen, are placed in parentheses if the species name was originally combined with a different genus name. This name and year reference is not a necessary part of the binomen, but it is of value to taxonomists because it sometimes helps verify that different researchers are discussing the same taxon.

4.1 **Key to classes of Arthropoda**

This key includes all living classes of Arthropoda except Pentastomida (or Linguatulida) and Tardigrada. Pentastomids and tardigrades are numerically small groups that are characterized by small to minute body size and lack antennae. These groups are rare and not likely to be confused with other Arthropoda. In fact, some authorities exclude them from the Arthropoda.

1a. Antennae absent (in immature stages and in some adult insects, the antennae are considerably reduced or absent; such forms will not key well here) 2

1b. Antennae present 4

2a. Body usually with seven pairs of appendages, including five pairs of legs; abdomen rudimentary (sea spiders; not further considered here) ... . **PYCNOGONIDA**

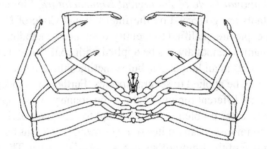

2b. Body with six pairs of appendages (rarely fewer) including four pairs of legs (rarely five); abdomen usually well developed but sometimes fused with cephalothorax 3

3a. Abdomen with book-like gills on ventral surface; large animals up to 50 cm long with hard, expanded shell and long, spine-like tail (horseshoe crabs; not further considered here) Class MEROSTOMATA, order or subclass .... **XIPHOSURA**

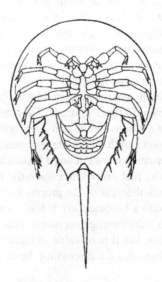

3b. Abdomen without book-like gills; smaller forms rarely over 7 cm long, body not as above (spiders, scorpions, mites, etc.; see Section 4.3.1) .... **ARACHNIDA**

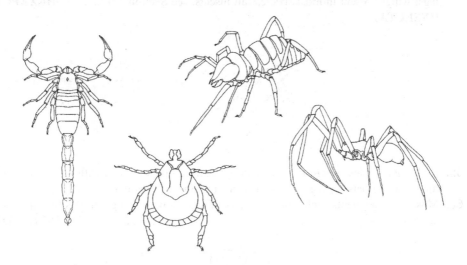

4a. With two pairs of antennae (one pair may be rudimentary in sow bugs); head and thorax fused to form cephalothorax; breathing by gills (crabs, lobsters, shrimps, sow bugs, etc.; see Section 4.5) .... **CRUSTACEA**

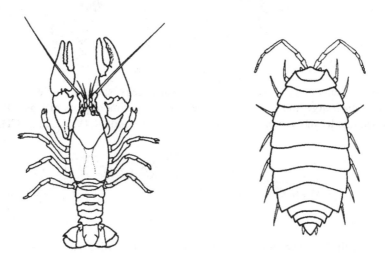

4b. With one pair of antennae; head and thorax separate; breathing by tracheae 5

5a. Body with head, thorax, and abdomen; thorax with three pairs of legs at some stage in the life cycle; abdomen sometimes with appendages that resemble thoracic legs; wings present in most species (all insects; see Section 4.6) ... . **HEXAPODA (INSECTA)**

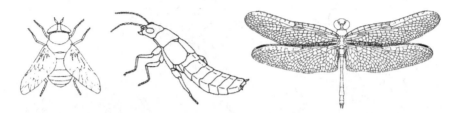

5b. Body more or less worm like; most body segments behind head with legs; body with nine or more pairs of legs; wings absent (myriapodan classes) 6

6a. Most body segments each with two pairs of legs; slow-moving animals (millipedes; see Section 4.4) .... **DIPLOPODA**

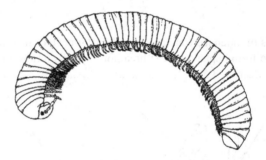

6b. Most segments of body with at most one pair of legs; speed variable 7

7a. Body rather large, more or less flattened, with 15 or more pairs of legs; often reddish brown, rapidly moving animals (centipedes; see Section 4.4) .... **CHILOPODA**

7b. Body small to minute (not over 8 mm long), usually cylindrical, with 9—12 pairs of legs; whitish or pale colored 8

8a. Antennae branched; nine pairs of legs; body minute (1–1.5 mm long); found in leaf litter etc. (pauropods; not further considered here) ... . **PAUROPODA**

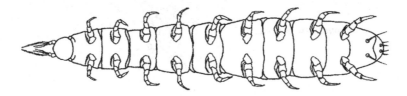

8b. Antennae not branched; 10–12 pairs of legs; body to 8 mm long, cylindrical, centipede-like; found in moist habitats (symphylans; see Section 4.4) ... **SYMPHYLA**

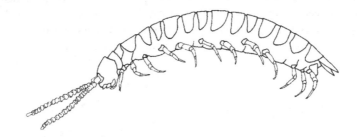

4.2 Class Arachnida

The Arachnida are the second largest class of arthropods in terms of number of species and species important to agriculture. Most arachnids, including all spiders, are predaceous. Some arachnids are parasitic on animals including humans. Many mites (subclass Acari) are important pests of plants. All arachnids should be collected and preserved in 70%–80% alcohol. Mites should be treated specially (see Section 3.2.8).

4.2.1 Key to orders of Arachnida

1a. Abdomen not segmented or if segmented (rarely) then with distinct sclerites (as in Asiatic family Liphistiidae); spinning organs on ventral side of the abdomen 2

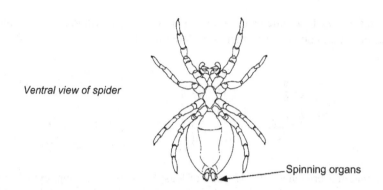

Ventral view of spider

Spinning organs

1b. Abdomen distinctly segmented; without silk-spinning organs on abdomen 3

2a. Abdomen joined to cephalothorax by narrow, short stalk; abdomen usually soft and weakly sclerotized; abdomen with spinning organs (spiders; not further considered here) ... . **ARANEIDA**

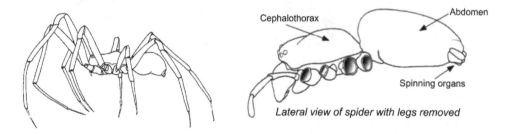

Cephalothorax   Abdomen

Spinning organs

Lateral view of spider with legs removed

2b. Abdomen broadly joined to cephalothorax; abdomen comparatively tough or comparatively strongly sclerotized; abdomen without spinning organs (some mites spin silk from palpi or mouthparts) (mites and ticks; see Section 4.2) ... . **Subclass ACARI**

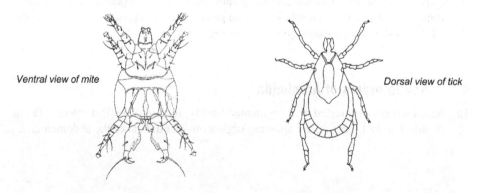

Ventral view of mite   *Dorsal view of tick*

3a. Abdomen with posterior segments forming a long, tail-like projection 4

3b. Abdomen without a tail-like projection 6

4a. Tail-like projection of abdomen six-segmented, ending in bulbous, claw-like sting; abdomen broadly joined to nonsegmented cephalothorax; venter of second abdominal segment with a pair of comb-like organs, the pectines (scorpions; widespread in warm, dry areas; not further considered here) ... . **SCORPIONIDA**

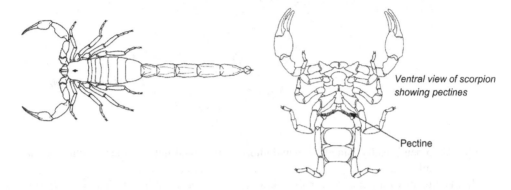

Ventral view of scorpion showing pectines

Pectine

4b. Tail-like projection of abdomen very slender, many segmented but not ending in sting; abdomen narrowed to base without ventral comb-like organs 5

5a. Anteriormost leg-like appendages (pedipalpi) slender; body minute, not more than 2 mm long; found in warm areas (not further considered here) ... . **PALPIGRADI (MICROTHELYPHONIDA)**

Lateral view of Palpigradi

5b. Anteriormost leg-like appendages (pedipalpi) very stout, contrasting with very long first pair of legs; body 2−65 mm long (whip scorpions, vinegaroons; not further considered here) .... **UROPYGI** (THELYPHONIDA; PEDIPALPIDA in part)

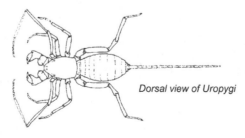

Dorsal view of Uropygi

6a. Abdomen constricted at base; front legs very long with long tarsi; pedipalpi clawed at apex; tropical animals 4–45 mm long (tail absent, whip scorpions; not further considered here) .... **AMBLYPYGI** (SCHIZOMIDA; PEDIPALPIDA in part)

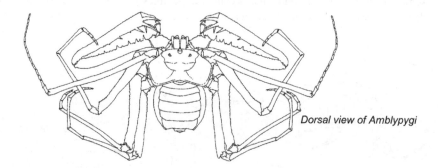

Dorsal view of Amblypygi

6b. Abdomen broadly joined to cephalothorax; front tarsi not lengthened; other characters variable 7

7a. Pedipalpi with large, pincer-like claws (pseudoscorpions; not further considered here) .... **PSEUDOSCORPIONIDA** (CHELONETHIDA)

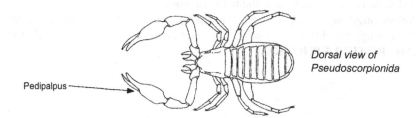

Pedipalpus ⟶

Dorsal view of Pseudoscorpionida

7b. Pedipalpi without pincer-like claws 8

8a. Head distinct from three-segmented thorax; chelicerae large and powerful with pincers moving up and down; pale-colored, nocturnal, large animals (up to 7 cm long); occur in Florida and the southwest (wind scorpions; not further considered here) .... **SOLPUGIDA**

Chelicera

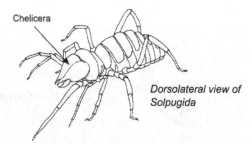

Dorsolateral view of Solpugida

8b. Cephalothorax present, not divided into head and three segments; chelicerae usually smaller, pincers not moving up and down; other characters variable 9

9a. Abdomen apparently four-segmented with lateral and dorsal sclerites, and small, several-segmented endpiece; eyes absent; heavy-bodied animals 5−10 mm long with moderately long legs; tropical to Texas (ricinuleids; not further considered here) .... **RICINULEI**

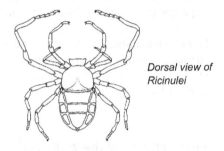

*Dorsal view of
Ricinulei*

9b. Abdomen usually appearing seven-segmented from above without separate lateral sclerites; body 5−10 mm long with legs very long; distribution widespread (daddy long legs, harvestmen; not further considered here) ... . **OPILIONES (PHALANGIDA)**

*Lateral view of
Opiliones*

4.3 Subclass Acari

The acarines are so varied in form that their anatomical terminology has developed along considerably different lines than that of the insects. Most acarines are very small and require special techniques for preservation and examination. The small size, morphological plasticity, and complex terminology make acarines considerably more difficult to identify than most insects. Only recently has their higher classification attained a measure of stability.

Past acarologists considered the Acarina as an order equivalent to the other orders of the class Arachnida (such as Araneida and Scorpionida). Current students of the mites regard Acarina as the subclass Acari without reference to the status of other orders. The classification of the subclass Acari is variously constructed. Wolley (1988) divides the sublcass Acari into the orders Astimata, Gamasida, Opilioacarda, Ixodida, Holothyrida, Oribatida (Sarcoptiformes, Oribatei, Cruyptostigmata, and Oribatoidca), and Actinedida (Trombidiformes and Prostigmata). Evans (1992) lists the orders Notostigmata, Holothyrida, Ixodida, Mesostigmata, Prostigmata, Astigmata, and Oribatida. The group names used here are left without rank designation until problems with higher classification are resolved. Specific procedures for mounting mites may also be found in Appendix II.

A few definitions for the more unfamiliar words may be helpful in the following key. *Sejugal furrow* is a line of demarcation that separates the podosoma and opisthosoma; *hypostome* is the anteroventral region of the gnathosoma (foremost part of the mouthparts); *Haller's organ* is a sensory organ found on tarsus I of ticks.

4.3.1 Key to some primary groups of the subclass Acari

1a. Without visible stigmata (breathing pores) posterior of coxae II; coxae not free, often fused with ventral body wall forming coxosternal regions delimited by epimera but lacking sternum; sejugal furrow or interval present causing legs III to be farther from legs II than the latter are from legs l; number of legs sometimes reduced ... . **Order ACARIFORMES, including PROSTIGMATA, ... ASTIGMATA, and CRYPTOSTIGMATA**

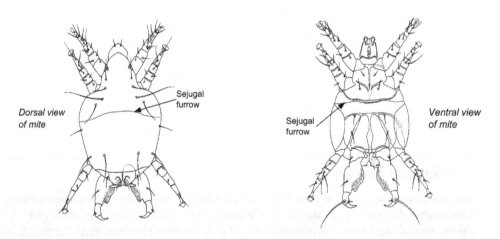

1b. With one to four pairs of dorsolateral or ventrolateral stigmata posterior of coxa II; coxae free, distinct; sternum nearly always present (lacking in Ixodida); sejugal furrow or interval lacking; distance between legs II and III not greater than between I and II and III and IV, all of which are present .... **Order PARASITIFORMES** 2

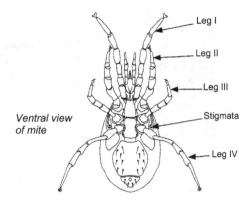

Ventral view of mite

Leg I
Leg II
Leg III
Stigmata
Leg IV

2a. Pedipalpal tarsus without claws; hypostome modified into piercing organ with backward-directed teeth; stigmata present behind coxa IV or lateral above coxal intervals II to III, each surrounded by stigmal plate; sternum absent; Haller's organ present on upper side of tarsus I (large, blood-sucking acarines called ticks; length usually well over 2 mm) .... **Suborder IXODIDA**

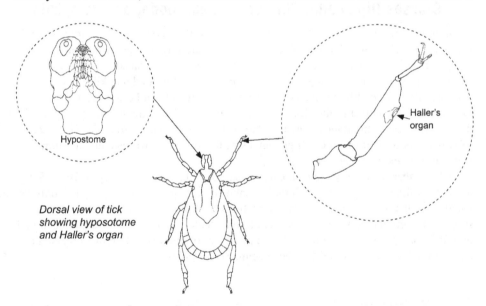

Hypostome

Haller's organ

Dorsal view of tick showing hyposotome and Haller's organ

2b. Pedipalpal tarsus with terminal, subterminal, or basal claw (simple or tined); hypostome serving only as floor of gnathosoma, without teeth; Haller's organ lacking (usually smaller mites) ... ... **Suborders MESOSTIGMATA, HOLOTHYRINA, and OPILIOACARIDA**

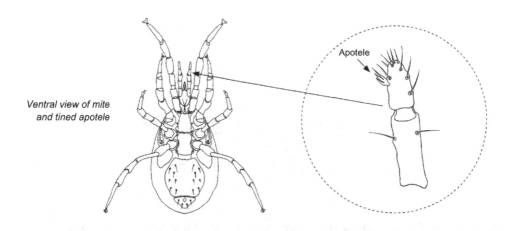

*Ventral view of mite
and tined apotele*

Apotele

4.4 **Classes Diplopoda, Chilopoda, Pauropoda, and Symphyla**

These four classes were once combined as one class, the Myriapoda, and are still known as
the myriapodan classes. All millipedes (Diplopoda) are phytophagous, but most species do
not seriously injure plants. The centipedes (Chilopoda) are predaceous and possess a pair
of strong, poison-supplied fangs called *toxicognaths* used to hold and kill their prey.
Toxicognaths are modified legs, and the venom they inject can be painful. Centipedes are
fast moving, in contrast to the slow-moving millipedes. The minute Pauropoda are of little
direct agricultural importance. One member of the Symphyla, *Scutigerella immaculata*
(Newport), is sometimes a pest in greenhouses; the species is whitish, about 8 mm long,
and may become very abundant.

   All members of the myriapodan classes may be killed and preserved in 70%–80%
alcohol. Millipedes have the discouraging habit of curling up when dying. The only simple
solution to this problem is to remove them from the killing fluid before they stiffen,
straighten them out, and then return them to the alcohol in a narrow, straight-sided vial
until they become rigid. Millipedes killed in vapors of ethyl acetate remain relaxed and
easily straightened before placing the specimens in alcohol.

4.5 **Class Crustacea**

The Crustacea are a large class divided into 8 subclasses and about 30 orders. Most
species of Crustacea are aquatic and are found especially in salt water. Many
crustaceans, including crayfish, lobsters, and prawns, serve as important human food.
Sow bugs including pill bugs belong to the subclass Malacostraca anti the order Isopoda.
Sow bugs are terrestrial isopods that are the only Crustacea of agricultural importance.

Sow bugs are found under stones, logs, and debris on the ground. They feed on vegetable matter and may become pests of tender plants. Sow bugs are distinguished by a depressed body, seven pairs of walking legs, and a shield-shaped head; the other body segments extend to the side. The first pair of antennae (antennules) are vestigial; the conspicuous antennae represent the second pair. Eggs are carried by the female in a brood sac on the underside of the body. Breathing involves paired gills on the lower hind part of the body. Because the gills must remain moist, sow bugs cannot withstand drying. Pill bugs are capable of rolling themselves into a tight ball for protective purposes. Crustaceans may be best killed and preserved in 70%–80% ethyl alcohol.

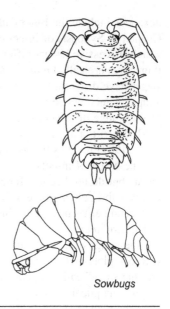

Sowbugs

4.6 **Class Hexapoda (Insecta)**

All Arthropoda including insects are characterized by hardened or sclerotized parts of the integument. After the arthropod emerges from the egg stage, the animal must molt or shed the hardened cuticle to grow and increase in size. These hardened parts called *sclerites* restrict growth. The process of molting must take place several times during the insect's period of growth, the nymphal or the larval stage. The number of molts varies among species of insects. In some instances, the number of molts is species specific, and in other instances, the number of molts may be influenced by sex of the individual, nutrition, or environmental conditions.

The term *instar* refers to the immature insect between molts. The term *stadium* refers to the amount of time between molts. During the molt, each instar of the growth stage increases in size over the preceding instar. During the molt, the animal may also change its form. The amount of change in form can range from insignificant to profound. The transformation in shape or form is called *metamorphosis*. The primitive

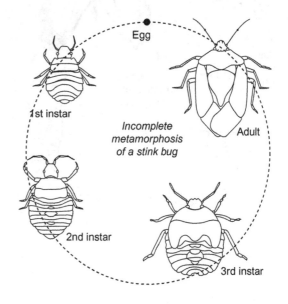

hexapods include the Entognatha and the division Exopterygota of the Pterygota (see Chapter 5: Synopsis of insect orders). These groups undergo gradual metamorphosis in which successive instars differ relatively little in form. Some aquatic exopterygote immatures are called *naiads* and may be entirely different from the adult because the two stages have different life habits. The immature stages of terrestrial insects with gradual simple or incomplete metamorphosis are called *nymphs*. The term *larva* is applied to the feeding stage of insects with complete metamorphosis, although some authors (McCafferty, 1981; Merritt et al., 1984) use *larva* to refer to the immature stages of aquatic insects. Stehr (1987) reviews the use of *larva*, *nymph*, and *naiad*.

The wings of insects with "incomplete" metamorphosis develop externally in the later instars but do not become functional until the adult stage. An exception to this rule involves the Ephemeroptera, which experience a molt after the wings become functional. The technical term for adult insects is *imago*; the first imago adult of the Ephemeroptera is called a *subimago*.

The wings of insects with complete metamorphosis typically develop internally. Insects with complete metamorphosis change radically in form from one stage to the next. A "resting stage" called a *pupa* (pl., pupae) is interpolated between the larva and the adult stages. The pupal exoskeleton protects the insect during the virtual reorganization of its entire body. The stages in the development of insects with complete metamorphosis include the egg, several larval instars, sometimes a prepupal stage, the pupa, occasionally a subimago, and the imago or adult.

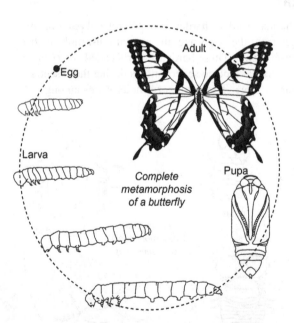

Complete metamorphosis of a butterfly

Some insects are wingless (either primarily or secondarily) in the adult stage and thus are difficult to distinguish from immature stages. Both the immature and the adult insects are treated in the key to the orders. Considerable knowledge about the various groups of insects is needed to determine whether a specimen is an immature or an adult. If functional wings are present, the specimen is certainly an adult (or a subimaginal mayfly). Compound eyes are not found in immature endopterygote insects.

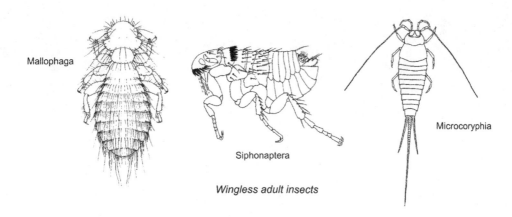

Mallophaga

Siphonaptera

Microcoryphia

Wingless adult insects

The classification of the insects into orders (ordinal classification) adopted here is a conservative one used in some textbooks. Recent studies on various aspects of insect morphology and physiology have led to the proposal of a considerably different classification, but only time and further study will determine its acceptance or rejection.

A fundamental dichotomy, based on the presence or absence of wings, divides six-legged arthropods (hexapods) into the subclasses Apterygota and Pterygota. The Apterygota of earlier authors included four orders: Protura, Collembola, Diplura, and Thysanura (Richards and Davies, 1977). In recent classifications, the term Apterygota may be restricted to the Microcoryphia (= Archaeognatha) and Thysanura (Watson and Smith, 1991). Other classifications have abandoned the traditional dichotomy of Apterygota—Pterygota and adopted a neutral viewpoint. In the opinion of some phylogeneticists, a more fundamental feature of evoludonary importance involves the ability to extend and retract the mouthparts. Hexapods that maintain their mandibles concealed when not in use are called Entognatha; hexapods whose mandibles remain exposed and not extensible are called Ectognatha. Entognathous Hexapoda includes the orders Protura, Collembola, and Diplura; ectognathous Hexapoda includes apterygotes of the orders Microcoryphia and Thysanura (Boudreaux, 1979; Borror et al., 1989) and the pterygotes.

The following synopsis of the classification includes some synonyms and most of the common English vernacular names for members of the various orders, along with a key to these groups.

Synopsis of insect orders

5

Disagreement exists over the higher classification of insects. Some suborders may be considered as orders, while some orders may be considered as suborders. There is value in understanding the basis for grouping the various orders of insects, because each grouping relates to the development or morphology of the insects in question. Here, we use the term *hexapod* to refer to wingless arthropods with six legs and three body tagmata. Hexapods can be further classified into two groups, Entognatha and Ectognatha, depending on whether the mouthparts are concealed within the head or exposed. In the following list, alternative ordinal names are given in parentheses and common names are given in brackets. The primitive insects with concealed mouthparts (Entognatha) include the following orders.

5.1 Subclass Entognatha: primitive wingless hexapods

1. Protura (Myrientomata) [proturans]
2. Diplura (Aptera) [diplurans]
3. Collembola [springtails]

All Entognathans should be collected and preserved in a similar manner. They may be collected with a Berlese funnel (Christiansen, 1990; Copeland and Imadate, 1990; Ferguson, 1990), an aspirator, or a small brush dipped in alcohol. To collect the species of Collembola that are often found on the surface of water, use a small dipper (Davidson and Swan, 1933). Specimens should be killed and preserved in 80% alcohol. All Entognathans are delicate. To minimize the risk of breakage, completely fill their vials with fluid. Microscope slide preparations are necessary for critical work with most Entognatha.

5.2 Subclass Ectognatha: primitive wingless hexapods

4. Microcoryphia (Archaeognatha) [bristletails]
5. Thysanura (Zygentoma) [silverfish, firebrats]

The subclass Insecta is restricted to the so-called Pterygota, which displays six legs, three body tagmata, and one pair of antennae and wings. The main difference between the hexapods and the insects is that the hexapods are primitively wingless

Insect Collection and Identification. DOI: https://doi.org/10.1016/B978-0-12-816570-6.00005-8

(Apterygota = without wings), while the Insecta possess wings during at least sometime in their life (Pterygota = having wings).

5.3 Subclass Pterygota (Insecta): winged and secondarily wingless insects

Insects that possess wings belong to the Pterygota. However, insects such as lice and fleas do not possess wings and still are classified among the Pterygota. This seeming contradiction is simply explained. Wingless pterygote insects have been shown to be closely related to winged insects, and the ancestors of these wingless pterygotes probably had wings. The wingless condition is common among parasitic insects, insects that live in caves or other subterranean habitats, and insects that live on islands. Some social insects, including termites and ants, display wings only in sexual forms. The workers and soldiers are wingless. Sexuals shed their wings after a nuptial flight. The subclass Insecta contains two divisions, Exopterygota and Endopterygota, depending on whether the wings are developed internally or externally. Furthermore, the Exopterygota includes insects with wings that develop externally and have simple metamorphosis. [Note: In some classification systems, Pterygotan orders are further subdivided into Paleoptera and Neoptera, based on the ability to flex wings over the abdomen. Members of the paleopteran orders (Ephemeroptera, Odonata) lack a wing-flexion mechanism and are unable to move wings posteriorly over the abdomen.] Orders of Exopterygota include the following.

5.3.1 Exopterygota

6. Ephemeroptera (Ephemerida, Plecoptera) [mayflies]
7. Odonata [dragonflies, damselflies]
8. Orthoptera [grasshoppers, locusts, katydids, crickets]

   Metamorphosis (Greek, meta = change of; morphosis = form) is another feature of pterygote insects that is important to understand, but difficult to define in a way satisfactory to everyone. Entomologists typically use the term in a restrictive sense to mean the morphological and physiological changes from the immature to the adult insect. Anatomical change is achieved through molting. Metamorphosis may be gradual, as during the transition of structure within a stage, such as the nymphs of a bug. Metamorphosis may be radical, as seen between stages, such as the transition from a caterpillar to a butterfly. Molting serves at least two purposes: growth in size and change in shape.

   When viewing all insects and their close relatives, we can see trends in the expression of metamorphosis. For instance, *ametabolous* (Greek, a = without; metabole = change) development involves organisms that lack metamorphosis. Specifically, ametaboly is applied to the Apterygota in which the adult closely resembles the immature stages. Ametaboly is viewed as the primitive condition among hexapods. *Paurometabolous* (Greek, pauros = little; metabole = change) development refers to metamorphosis in which the changes of form between immature and adult are gradual and inconspicuous. Paurometabolous insects are terrestrial. Examples of paurometabolous insects include Dermaptera, Orthoptera, Embiidina, Isoptera, Zoraptera, Psocoptera, Phthiraptera, and most Hemiptera. The situation is not straightforward, as noted below.

Hemimetabolous (Greek, hemi = half; metabole = change) development refers to metamorphosis in which form changes between immature and adult are made in one radical move. The anatomical changes between stages are conspicuous. Hemimetabolous development involves aquatic immatures because most hemimetabolous insects are aquatic as immatures. Examples of hemimetabolous insects include Plecoptera, Ephemeroptera, and Odonata. *Holometabolous* (Greek, holos = entire; metabole = change) development involves the interpolation of a pupal stage between immature and adult. This is the most advanced type of metamorphosis found among the Insecta. Examples include Neuroptera, Coleoptera, Strepsiptera, Trichoptera, Lepidoptera, Mecoptera, Siphonaptera, Diptera, and Hymenoptera. Fitting all of the Insecta into one metamorphic category or another is difficult because exceptions occur.

Until recently, the Blattodea and Mantodea were considered suborders of an order called the Dictyoptera. The following orders were included with the Orthoptera for many years:

9. Blattodea [cockroaches]
10. Mantodea [mantids]

There is value in understanding the basis for grouping the various orders of insects, because each grouping relates to the development or morphology of the insects in question. The ever-changing science of insect taxonomy and systematics makes insect collection, identification, and preservation techniques not only that much more critical, but also both challenging and rewarding. For example, older entomologists will quickly notice that termites (previously recognized as belonging to their own order, Isoptera) are now joined with their close relatives into the single order Blattodea. Sucking and chewing lice are no longer separate orders (Anoplura and Mallophaga), but rather belong to the common order Phthiraptera. The long embraced, two-order classification for all "true bugs" (Hemiptera and Homoptera) has been recently reduced into a single order called Hemiptera. Changes at lower levels (family, genus, and species) occur more frequently. For instance, the family Lygaeidae recently has been split to form 10 separate families, whereas insects previously known as Cicindelidae are now part of the family Carabidae, Lyctidae is now included in Bostrichidae, Buchidae is considered part of Chrysomelidae, Nymphalidae includes Danaidae, as well as Satyridae and Apidae now include both Anthophoridae and Bombidae. These are but a few of the recent family changes to contemporary insect classification. Genus and species level taxonomy is even more fluid.

11. Phasmatodea (Phasmatoptera, Phasmoidea, Phasmida) [leaf insects, walkingsticks]
12. Gryllobattodea (Notoptera) [rock crawlers]

13. Dermaptera (Euplexoptera) [earwigs]
14. Isoptera [termites]
15. Embiidina (Embioptera) [webspinners]
16. Plecoptera [stoneflies]
17. Psocoptera (Corrodentia) [psocids, booklice]
18. Zoraptera [zorapterans]

    The following two orders are often combined into the order Phthiraptera. This name is a tongue-twister that is derived from the Greek (phthir = lice; aptera = wingless) and is used to include all lice. This name has an element of redundancy because all lice are wingless. In some classifications, the Phthiraptera are regarded as the order embracing all lice, including Mallophaga and Anoplura. In other classifications, the Mallophaga and Anoplura are regarded as distinct orders. Triplehorn and Johnson (2005) recognize Phthiraptera as an order, and place the Anoplura as a suborder, and subdivide the Mallophaga into two suborders or divisions: the Amblycera and Ischnocera. Here, we follow the classification used by Triplehorn and Johnson (2005) with the proviso that other classifications exist.

19. Phthiraptera (Mallophaga [chewing lice] and Anoplura [sucking lice])
20. Thysanoptera [thrips]

    The higher classification of the two orders, Hemiptera and Homoptera, remains unresolved to the satisfaction of all entomologists. Some workers recognize these as distinct orders. Under such a scheme, the Hemiptera include several suborders (Stys and Kerzhner, 1975). An alternative classification (Waterhouse, 1971) considers the Hemiptera as an order that includes the suborders Heteroptera and Homoptera. Here, we adopt the view that the Hemiptera s. str. and Homoptera s. str. should be combined into a single Hemiptera with Heteroptera representing Hemiptera s. str. and Sternorrhyncha and Auchenorrhyncha representing Homoptera s. str. No matter what we call them or how we classify them, the two are similar in several respects. In both groups, the mouthparts are adapted for piercing/sucking, and maxillary and labial palpi are not evident. In Hemipteras. str. the mandibles and maxillae have become stylets that are surrounded by a "beak" that arises from the front part of the head. Members of the Homoptera s. str. show sucking mouthparts that arise from the posterior-ventral part of the head. Both groups usually have two pairs of wings. Adults of the Hemiptera s. str. show the basal part of the forewings thickened and leathery and the apical part membranous. The hind wings are entirely membranous and are covered by the forewings except during flight. When at rest, the wings are held flat over the back, with the apical parts of the forewings overlapping.

21. Hemiptera (Heteroptera ["true bugs"], Sternorrhyncha [cicadas, treehoppers, spittlebugs, etc.] and Auchenorrhyncha [whiteflies, scales, etc.])

5.3.2 Endopterygota

The next group of orders includes those insects whose wings develop internally. These insects are further classified as the holometabolous insects. Holometabolous insects undergo complete development characterized by egg, larval, pupal, and adult stages.

All holometabolous insects display a pupal stage. The term *pupa* (Latin, puppet) originally was used for the chrysalis of a butterfly. Subsequently, entomologists have applied the term to all holometabolous insects. The pupal stage is the last major event in the anatomical maturation of insects through the process of metamorphosis. The pupa represents a phase during which larval anatomical features are destroyed and adult features are constructed. The degree of reconstruction differs among groups. During this process, the pupa is relatively immobile. In holometabolous insects, the pupal stage occurs in time between the larva and the adult.

The division Endopterygota includes insects with complete metamorphosis (=Holometabola) and wings that develop internally. These insects are the most abundant pterygotes in terms of number of species. Orders of endopterygotes include the following:

22. Coleoptera [wood-boring beetles, weevils, leaf beetles, ground beetles, bark beetles]
23. Strepsiptera (sometimes included in Coleoptera) [twisted-winged parasites]
24. Mecoptera [scorpionflies, hangingflies]
25. Neuroptera [snakeflies, antlions, lacewings, mantidflies, dobsonflies, hellgrammites]
26. Trichoptera [caddisflies]
27. Lepidoptera [butterflies, moths]
28. Diptera ("True" flies) [house flies, deer flies, mosquitoes, gnats, midges]
29. Siphonaptera [fleas]
30. Hymenoptera [sawflies, ichneumons, ants, wasps, bees]

Disagreement exists over the higher classification of insects. Some suborders may be considered as orders, while some orders may be considered as suborders. Here, we use the term *hexapod* to refer to two groups of arthropods: the Entognatha and the Ectognatha, with both having six legs and three body tagmata. The term *insect* is restricted to the so-called Ectognatha, which have mouthparts freely exposed, six legs, three body tagmata, and one pair of antennae and some with wings.

5.4 **Key to orders of hexapoda (Insecta)**

The following key includes suborders sometimes considered as orders. Any comprehensive survey of the larger orders of insects will include exceptions and aberrant forms that do not fit well in a brief key. Treating all aberrant forms is impractical, particularly when including known immature forms. Some orders appear more than once in the key. With specimens that do not key satisfactorily here, the reader should consult other references or an experienced systematic entomologist. This is especially true of pupal forms of Trichoptera, Mecoptera, and some Neuroptera. Interested students should obtain a copy of Triplehorn and Johnson (2005) for comprehensive keys to the North American insect fauna.

1a.   Wings present and well developed..2

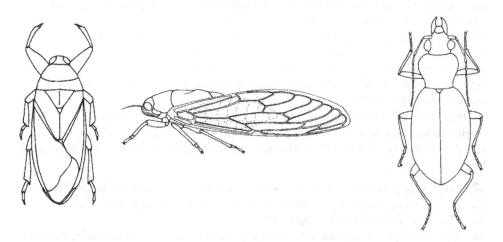

1b.   Wings absent or not suitable for flight (wingless adults and immature
stages) 34

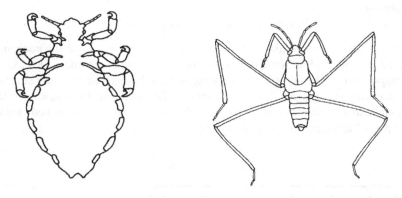

2a.   Forewings at least partly horny, leathery, or strongly differing from completely
membranous hind wings; hind wings sometimes absent 3

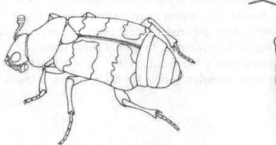
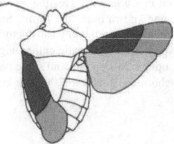

2b.   Forewings completely membranous or membranous at base; hind wings variable, usually present and membranous, sometimes modified 11

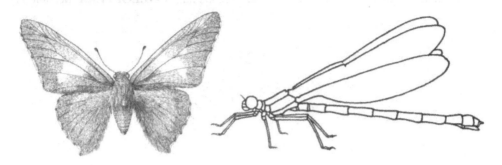

3a.   Forewings (wing covers, elytra) uniformly horny or leathery and without apparent veins; hind wings, if present, folded lengthwise and crosswise, concealed beneath forewings when at rest; mouthparts including mandibles 4

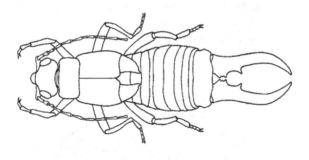

3b.   Forewings (hemelytra, tegmina) with variable texture and with veins; hind wings not folded crosswise; mouthparts variable... 5

4a.   Apex of abdomen with heavy, forceps-like cerci; wings short and most of abdomen exposed; hind wings delicate, almost circular, radially folded ....
....**DERMAPTERA**, Section 6.13

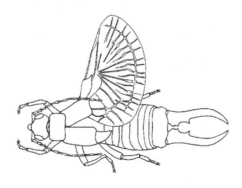

4b.   Apex of abdomen without heavy, forceps-like cerci; wings usually long or covering most of abdomen; if forewings short, then hind wings elongate or absent ....**COLEOPTERA**, Section 6.22

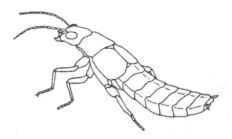

5a.   Mouthparts adapted for sucking, forming a jointed or segmented beak; forewings variable; hardened or leathery basally and membranous apically, usually held flat on abdomen with apices overlapping................**HEMIPTERA**, Suborder HETEROPTERA, Section 6.21 or forewings completely or predominantly membranous, held more or less rooflike over abdomen. Beak arising from anterior part of head and usually capable of movement forward of head or beak arising from posterior-ventral part of head and projecting downward and rearward between forelegs .... **HEMIPTERA**, Suborders .... AUCHENORRHYNCHA and STERNORRHYNCHA, Section 6.21

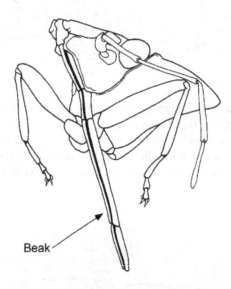

Beak

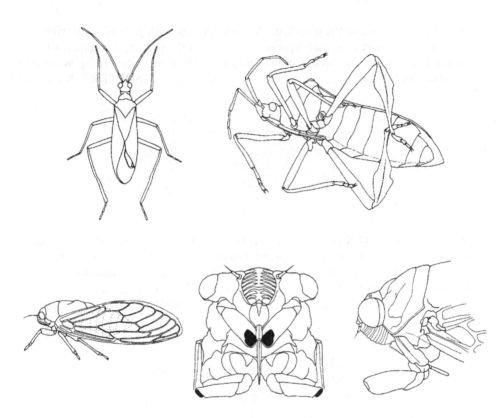

5b. Mouthparts adapted for chewing, forming mandibles that move laterally 6

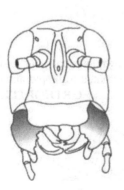 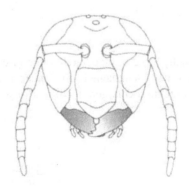

6a.   Hind wings not folded at rest, similar to forewings; both pairs of wings with
thickened, very short basal part separated from rest of the wing by a suture, easily
detached at suture; body soft and usually pale-colored; social insects living in colonies
(see also couplet 32)... **ISOPTERA**, Section 6.14

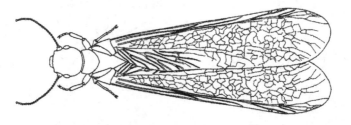

6b.   Hind wings folded fanwise at repose, broader than forewings; both pairs of wings
without basal suture for wing detachment; body hardness and coloration variable 7
7a.   Minute insects, usually less than 6 mm long; forewings small, clublike;
antennae short, with few segments; parasites of other insects...
.... **STREPSIPTERA** males, Section 6.23

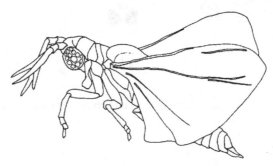

7b.   Usually large or moderately large insects; forewings usually flat and long;
antennae usually lengthened and slender, many-segmented (orthopteran orders).......... 8
8a.   Hind femora enlarged, modified for jumping .... **ORTHOPTERA**, Section 6.8

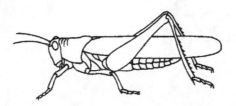

8b.   Hind femora not enlarged, similar in size and shape to femora of other legs or not apparently adapted for jumping 9
9a.   Cerci short, not segmented; body usually elongate and slender, sticklike................
.... **PHASMATODEA**, Section 6.11

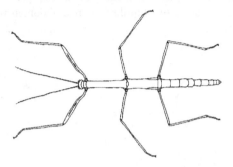

9b.   Cerci long or short but segmented; body usually not elongate or sticklike 10
10a.   Forelegs raptorial with tibial and femoral spines modified for grasping prey; middle and hind legs adapted for walking ....**MANTODEA**, Section 6.10

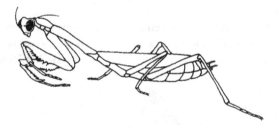

10b.   All legs similar in size and shape, without spines adapted for grasping prey; all legs adapted for walking .... **BLATTODEA**, Section 6.9

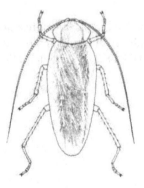

11a.   Two well-developed wings present: forewings obviously used in flight; hind wings modified, sometimes small and clublike 12
11b.   Four wings present: forewings obviously used in flight, hind wings sometimes small but flat or straplike and not clublike 14
12a.   Mouthparts forming a piercing-sucking or lapping proboscis, rarely rudimentary or absent; hind wings replaced by clublike halteres; abdomen without tail filaments....... .
.... **DIPTERA**, Section 6.28

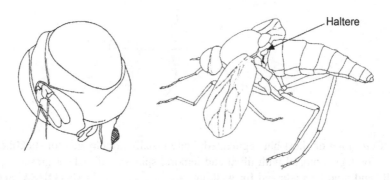

12b.   Mouthparts not functional; hind wings not formed into clublike halteres; abdomen with tail filaments 13
13a.   Hind wings not halterelike; antennae inconspicuous, with small scape and pedicel, flagellum bristlelike; forewings with numerous crossveins (few mayflies) ....
.... **EPHEMEROPTERA**, Section 6.6

13b.   Hind wings reduced to halterelike structures; antennae conspicuous, flagellum not bristlelike; forewings with venation apparently reduced to one forked vein (male scale insects) .... **HEMIPTERA**, Suborder **STERNORRHYNCHA**, Section 6.21

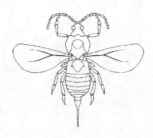

14a.　Wings long, narrow, almost veinless, with long marginal fringe; tarsi one- or two-segmented, with apex swollen; mouthparts conical, adapted for piercing and sucking plant tissues (minute insects) ....**THYSANOPTERA**, Section 6.20

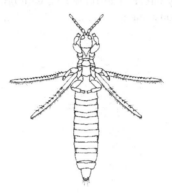

14b.　Wings relatively broad, veins usually conspicuous and at least one crossvein present, marginal fringe absent or not longer than width of wing; tarsi with more than two segments and apex not swollen; mouthparts variable 15

15a.　Wings, legs, and body at least partially covered with elongate, flattened scales (setae) and often with hairlike setae; wings hyaline (transparent) under color pattern formed by scales; mouthparts tonguelike (rarely rudimentary), forming a helically coiled tube; small mandibles present only in a few families of small moths with wingspread not over 12 mm .... **LEPIDOPTERA**, Section 6.27

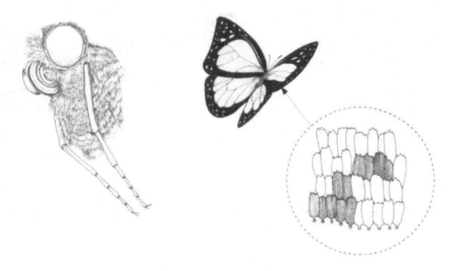

15b. Wings, legs, and body not covered with flattened scales, although a few scales sometimes present; color pattern of wing involving wing membrane or hairlike setae or both; mandibles typically present 16

16a. Hind wings usually larger than forewings, with broad anal area, plaited when wings folded; antennae conspicuous 17

16b. Hind wings usually not larger than forewings, without plaited anal area; antennae often inconspicuous, bristlelike 19

17a. Tarsi three-segmented; cerci well developed, usually long and many-segmented .... **PLECOPTERA**, Section 6.16

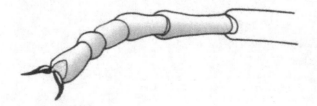

17b. Tarsi five-segmented: cerci not well developed 18

18a.   Wings with several subcostal crossveins, surface without hairlike setae or scales
.... **NEUROPTERA**, Suborder **MEGALOPTERA**, Section 6.25

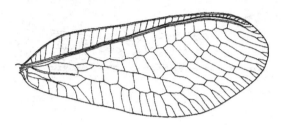

18b.   Wings without subcostal crossveins, surface with hairlike setae or scales ....
.... **TRICHOPTERA**, Section 6.26

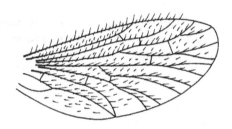

19a.   Antennae short, bristlelike; wings with numerous crossveins forming a network;
mouthparts with mandibles near eyes 20
19b.   Antennae large or wings with a few crossveins or mouthparts near beak 21
20a.   Hind wings much smaller than forewings; abdomen with long tail filaments ....
....**EPHEMEROPTERA**, Section 6.6

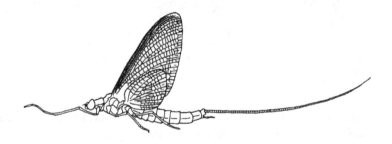

20b.   Hind wings about as large as forewings; abdomen without long tail filaments .....
.... **ODONATA**, Section 6.7

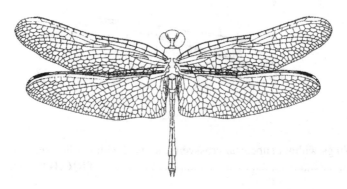

21a.   Head beneath eyes beaklike with mandibles at apex; hind wings not folded; wings usually with color pattern and numerous crossveins; male genitalia usually swollen, turned forward, and with strong pair of forceps ....
.... **MECOPTERA**, Section 6.24

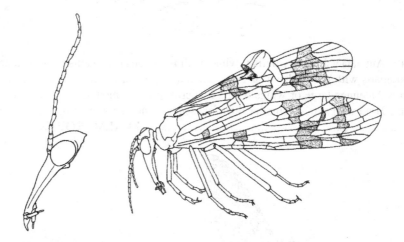

21b.   Head beneath eyes not beaklike or formed into a conical tube; wing characters variable; male genitalia without conspicuous forceps 22

22a.   Mouthparts sometimes absent, when present consisting of proboscis without chewing mandibles; cerci absent; wings with few crossveins 23

22b.   Mouthparts with mandibles adapted for chewing; cerci sometimes present; wing venation variable 25

23a.   Wings covered with scales that form color pattern; antennae with many
segments; mouthparts (when present) consisting of helically coiled haustellum
(tongue) .... **LEPIDOPTERA**, Section 6.27

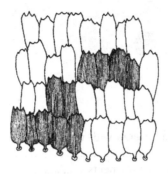

23b.   Wings not covered with scales; antennae with few segments; mouthparts forming
a segmented piercing beak 24
24a.   Beak arising from anterior part of head .... **HEMIPTERA**,
Suborder .... **HETEROPTERA** Section 6.21

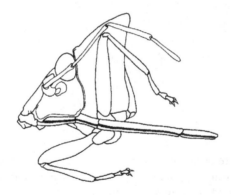

24b.   Beak arising from posterior part of head, extended downward between forelegs
.... **HEMIPTERA, Suborder AUCHENORRHYNCHA**
and .... **STERNORRHYNCHA** Section 6.21

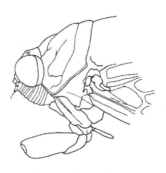

25a.   Body and wings covered with whitish powder; wings bordered anteriorly by very narrow cell without row of crossveins; insects less than 5 mm long ....
NEUROPTERA, Suborder PLANIPENNIA (Coniopterygidae), ... Section 6.25

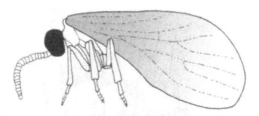

25b.   Body and wings not covered with whitish powder, other characters differing ... 26
26a.   Tarsi five-segmented 27
26b.   Tarsi with four or fewer segments 30
27a.   Prothorax typically fused with mesothorax; forewings with fewer than 20 cells; hind wings smaller than forewings; abdomen usually constricted at base, forming a petiole; sting or appendicular ovipositor present .... **HYMENOPTERA**, Section 6.30

27b.   Prothorax more or less free, sometimes long; forewings with more than 20 cells; forewings and hind wings approximately equal in size; abdomen not constricted to form petiole; sting or appendicular ovipositor not present 28
28a.   Prothorax cylindrical, much longer than head; forelegs similar to other legs, not enlarged .... **NEUROPTERA**, Suborder **RAPHIDIODEA**, Section 6.25

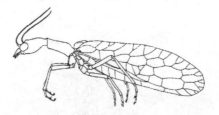

28b.   Prothorax not longer than head; if longer, then forelegs enlarged and adapted for grasping prey 29

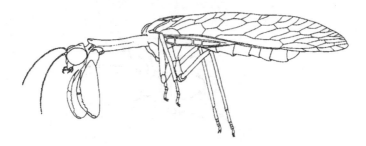

29a.   Costal cell with many crossveins ....
.... **NEUROPTRA, Suborder PLANIPENNIA**, Section 6.25
29b.   Costal cell without many crossveins .... **MECOPTERA**, Section 6.24
30a.   Wings equal in size or hind wings rarely larger; tarsi three- or four-segmented 31
30b.   Hind wings smaller than forewings; tarsi two- or three-segmented 32
31a.   Forebasitarsi not swollen; wings dehiscent (see also couplet 7)..............................
.... **ISOPTERA**, Section 6.14

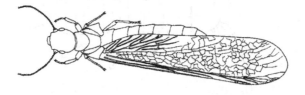

31b.   Forebasitarsi swollen .... **EMBIIDINA**, Section 6.15

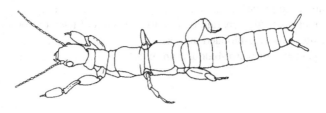

32a. Cerci absent; wings remain attached to body; antennae slender, with 13 or more segments (insects commonly collected) .... **PSOCOPTERA**, Section 6.17

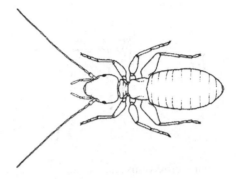

32b. Cerci present, although short, ending in bristle; wings shed; antennae with nine beadlike segments (insects seldom collected) .... **ZORAPTERA**, Section 6.18

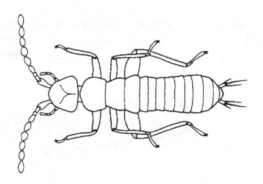

33a. Body with more or less distinct head, thorax, and abdomen; legs jointed, enabling animal to move 34
33b. Without distinctly separate body parts, or without legs, or not able to move ... 79
34a. Parasites of warm-blooded animals 35
34b. Not parasites of warm-blooded animals 39

35a.   Adult body strongly compressed laterally; mouth forming a short, sharp, downward-projecting beak; powerful jumping insects; adults found on vertebrate hosts; immatures found in nest of host .... **SIPHONAPTERA**, Section 6.29

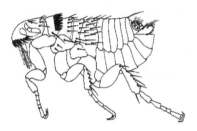

35b.   Adult body not compressed laterally; mouthparts variable; not powerful jumping insects; habitats variable 36

36a.   Mouthparts with mandibles adapted for chewing, directed forward; insects generally oval in outline with more or less triangular head; head wider than width of prothorax; parasites of birds and mammals ....**PTHIRAPTERA**, Suborders **AMBLYCERA** and **ISCHNOCERA** ....Section 6.19

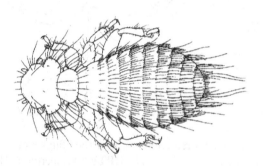

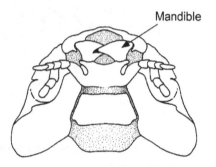

Mandible

Or

Mouthparts piercing-sucking with stylets retracted into head; head narrower than width of the prothorax; body shape variable; feeding habits diverse ....**PHTHIRAPTERA**, Suborder **ANOPLURA** ....Section 6.19

36b.   Mouthparts forming a beak or otherwise modified; body shape variable; feeding habits .... diverse 37

37a. Antennae inserted in pits and not visible when viewed from above (also maggot-shaped larvae without antennae) .... **DIPTERA** (a few families), Section 6.28

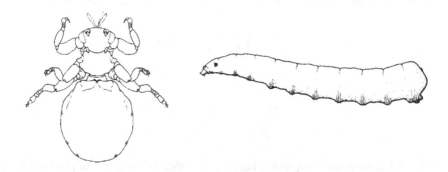

37b. Antennae present, short but not in pits 38
38a. Beak not jointed; tarsi forming a hook for grasping hairs of host; parasites remaining on host **PHTHIRAPTERA**, Suborder **ANOPLURA** .... Section 6.19

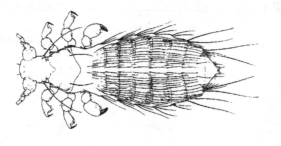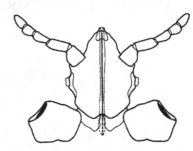

38b. Beak jointed; tarsi not hooked; parasites not remaining on host (bed bugs and related insects) .... **HEMIPTERA**, Section 6.21
39a. Aquatic insects, usually breathing by gills; larval and some pupal forms 40
39b. Terrestrial insects, breathing by spiracles or rarely without breathing organs .. 48
40a. Mouth forming a strong, pointed, downward-curved beak Immature....................
....**HEMIPTERA**, Section 6.21

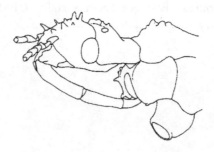

40b.   Mouth with mandibles 41

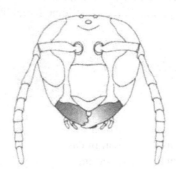

41a.   Mandibles extending straight forward, united with maxillae to form piercing jaws Some larval .... **NEUROPTERA.** Section 6.25

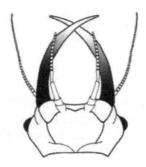

41b   Mandibles moving laterally, forming biting jaws 42

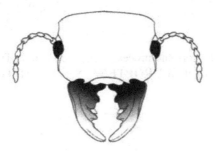

42a.   Immature insects living within cases formed of sand, pebbles, leaves, twigs, etc.: usually with external tracheae serving as gills .... Some larval **TRICHOPTERA.** Section 6.26

42b.   Immature insects not living within cases 43
43a.   Abdomen with lateral organs serving as gills (a few larval Trichoptera and Coleoptera also key here) 44

Lateral organ

43b.   Abdomen without external gills (some larval Trichoptera also key here) 45
44a.   Abdomen with two or three long tail filaments ....
.... Immature **EPHEMEROPTERA**, Section 6.6

44b.   Abdomen with short end processes (larvae of some Trichoptera key here) .... Larval **NEUROPTERA**, Suborder **MEGALOPTERA**, Section 6.25

45a.   Lower lip (labium) folded backward, extensible, and furnished with a pair of jawlike hooks .... Immature **ODONATA**, Section 6.7

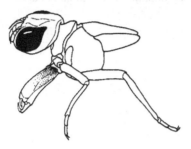

45b.   Labium not so constructed 46

46a.   Abdomen with nonjointed false legs (pseudopods) arranged in pairs on several segments .... Few larval **LEPIDOPTERA**, Section 6.27

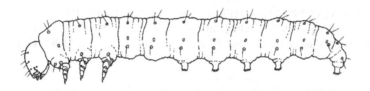

46b.   Abdomen without pseudopods 47

47a.   Thorax in three loosely united divisions; antennae and tail filaments long and slender .... Larval **PLECOPTERA**, Section 6.16

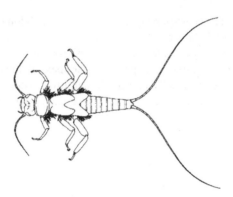

47b.   Thoracic divisions without constrictions; antennae and tail filaments short (larvae of some aquatic Diptera and Trichoptera also key here) ....
.... Larval **COLEOPTERA**, Section 6.22

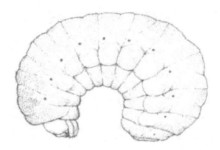

48a.   Mouthparts retracted into head and difficult to see; antennae sometimes absent; venter of abdomen with appendages; very delicate, small-to-minute animals 49
48b.   Mouthparts external, conspicuous; antennae always present; venter of abdomen rarely with appendages; body typically larger than a few millimeters 51
49a.   Head pear-shaped; antennae absent; abdomen without long cerci, pincers, jumping apparatus, or basal ventral "sucker" .... **PROTURA**, Section 6.1

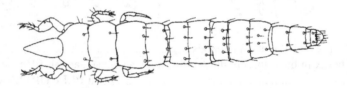

49b.   Head usually not pear-shaped, antennae present; abdomen with long cerci, pincers, or basal ventral "sucker" 50
50a.   Abdomen with six or fewer segments, with forked "sucker" at base below and usually with conspicuous jumping apparatus near apex; abdomen lacking conspicuous long cerci or pincers; eyes usually present though often reduced in size.........................
.... **COLLEMBOLA**, Section 6.3

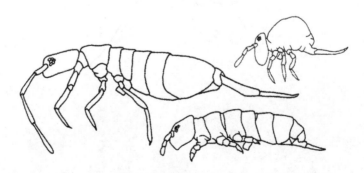

50b.   Abdomen with more than eight evident segments but lacking a "sucker" at base; abdomen ending in long, many-segmented cerci or strong pincers; eyes and ocelli absent .... **DIPLURA**, Section 6.2

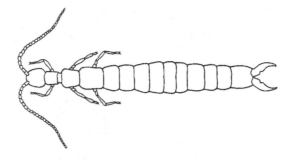

51a.   Mouthparts with mandibles adapted for chewing 52
51b.   Mouthparts in form of proboscis adapted for sucking 74
52a.   Body usually covered with scales; abdomen with three prominent tail filaments and at least two pairs of ventral appendages (styli) 53

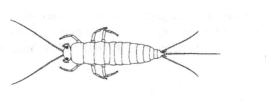 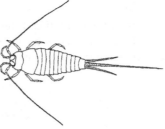

52b.   Body not covered with scales; abdomen without three tail filaments or ventral styli 54
53a.   Compound eyes small and widely separated; body flat ....
....**THYSANURA**, Section 6.5
53b.   Compound eyes large, nearly touching; thorax humpbacked ....
....**MICROCORYPHIA**, Section 6.4

54a.   Abdomen bearing ventral pairs of false, nonjointed legs (pseudopods) that differ from true legs on thorax; thorax and abdomen not distinctly separated; body caterpillar-like; larval forms 55

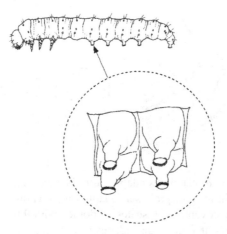

54b.   Ventral surface of abdomen without legs or pseudopods; other characters different 56

55a.   Abdomen with five or fewer pairs of pseudopods, none on first, second, or seventh segments; pseudopods tipped with many tiny hooklets and rarely present on second and seventh segments .... Most larval **LEPIDOPTERA**, Section 6.27

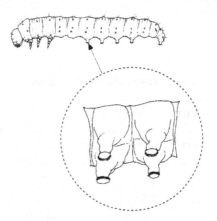

55b.   Abdomen with 6−10 pairs of pseudopods, not tipped with tiny hooks; one pair of pseudopods on second segment 57

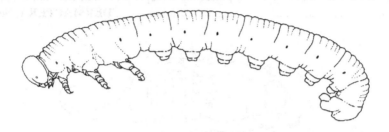

56a.   Head with single ocellus (stemmatum) on each side ....
.... Some larval **HYMENOPTERA**, Section 6.30

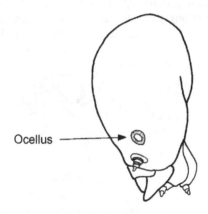

Ocellus

56b.   Head with several ocelli (stemmata) on each side ....
.... Larval **MECOPTERA**, Section 6.24

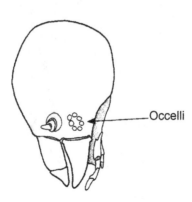

Occelli

57a.   Antennae long and distinct; adult or immature forms 58
57b.   Antennae short; larval forms 71
58a.   Abdomen ending in strong pincerlike forceps; prothorax free ....
.... **DERMAPTERA**, Section 6.13

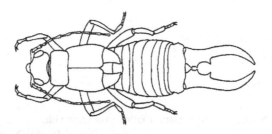

58b.   Abdomen not ending in forceps; prothorax free or fused to mesothorax 59
59a.   Adult abdomen strongly constricted at base, forming a petiole; prothorax fused
with mesothorax .... **HYMENOPTERA**, Section 6.30

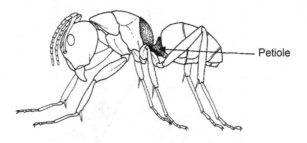

Petiole

59b.   Adult abdomen not strongly constricted at base, broadly joined to thorax 60
60a.   Head produced into beak with mandibles at apex ....**MECOPTERA**, Section 6.24

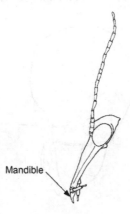

Mandible

60b.   Head not produced into beak 61
61a.   Very small insects with soft body; tarsi two- or three-segmented 62
61b.   Usually very much larger insects; tarsi usually with more than three segments, or body hard and cerci absent 63
62a.   Cerci absent .... **PSOCOPTERA**, Section 6.17

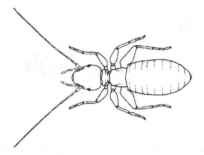

62b.   Cerci of single segment, prominent .... **ZORAPTERA**, Section 6.18

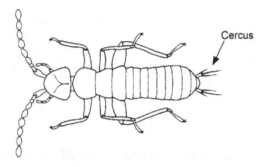

Cercus

63a.   Hind femora enlarged and adapted for jumping; wing pads of immatures inverted, hind wing pads overlapping forewing pads .... **ORTHOPTERA**, Section 6.8

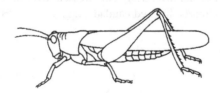

63b.   Hind femora not enlarged or adapted for jumping; wing pads, if present, in normal position 64

64a.   Prothorax much longer than mesothorax; opposable surfaces of front femora and tibiae with long spines adapted for grasping prey (raptorial insects) ....
.... **MANTODEA**, Section 6.10

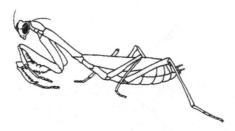

64b.   Prothorax not greatly lengthened; front legs rarely raptorial 67

65a.   Cerci absent; body often strongly sclerotized; antennae usually with two segments .... **COLEOPTERA**, Section 6.22

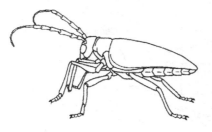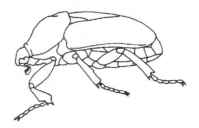

65b.   Cerci present; body usually not strongly sclerotized; antennae usually with more than 15 segments 68

66a.   Cerci with more than three segments 67

66b.   Cerci with one to three segments 69

67a.   Body flattened or dorsoventrally compressed, oval in outline; head deflected downward, with mouthparts directed caudad .... **BLATTODEA**, Section 6.9

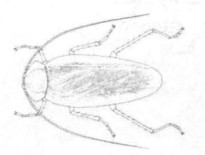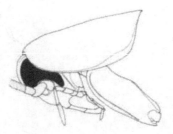

67b.   Body not flattened, elongate in outline; head nearly horizontal, with mouthparts at extreme anterior margin of head 68

68a.   Cerci long, five- to eight-segmented; ovipositor long, swordlike; tarsi five-segmented; found in cold habitats remote from human habitation ....
.... **GRYLLOBLATTODEA**, Section 6.12

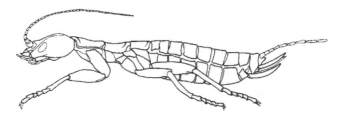

68b.   Cerci short, one- to five-segmented; ovipositor absent; tarsi four-segmented; found in tropical and subtropical habitats or in association with humans..........................
.... **ISOPTERA**, Section 6.14

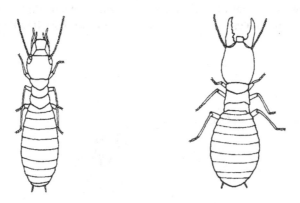

69a.   Tarsi five-segmented (three-segmented in Timema [in Pacific coast states], most antennal segments several times longer than wide); body large and sticklike; not communal or social insects .... **PHASMATODEA**, Section 6.11

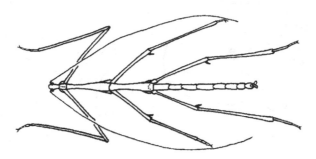

69b.   Tarsi two- or three-segmented; antennal segments beadlike; body usually small, elongate but not sticklike; communal or social insects 70

70a.   Front basitarsi swollen and containing silk-spinning gland for producing web in which insects live communally; cerci conspicuous; sexually dimorphic insects but without castes .... **EMBIIDINA**, Section 6.15

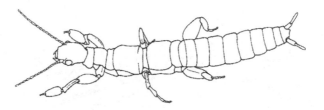

70b.   Front basitarsi not swollen and not producing silk; cerci inconspicuous; polymorphic social insects with castes .... **ISOPTERA**, Section 6.14

71a.   Body cylindrical, caterpillar-like 72

72b.   Body more or less compressed, not caterpillar-like 72

73a.   Head with six ocelli (stemmata) on each side; antennae inserted in membranous area at base of mandibles .... Some larval **LEPIDOPTERA**, Section 6.27

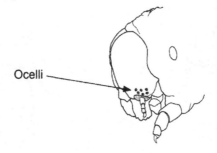

73b.   Head with more than six ocelli on each side; third pair of legs distinctly larger than first pair .... Larval Boreidae **MECOPTERA**, Section 6.24

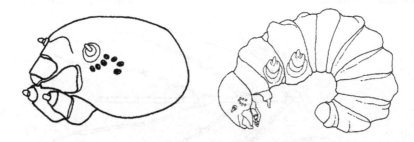

74a.   Mandibles united with maxillae to form sucking jaws...
.... Larval **NEUROPTERA**, Suborder **PLANIPENNIA**, Section 6.25

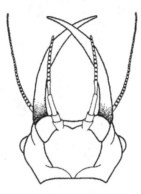

74b.   Mandibles nearly always separate from maxillae .... Larval **COLEOPTERA;**
NEUROPTERA, suborder **RAPHIDIODEA**; suborder **RAPHIDIODEA;**
STREPSIPTERA; DIPTERA

75a.   Body densely covered with scales and hairlike setae; proboscis, if present, coiled
under head .... **LEPIDOPTERA**, Section 6.27

75b.   Body bare, with scattered hairlike setae or waxy coating; mouthparts not coiled
under head 76

76a.   Last tarsal segment bladderlike, without claws; mouth forming a triangular, nonsegmented beak; very small insects .... **THYSANOPTERA**, Section 6.20

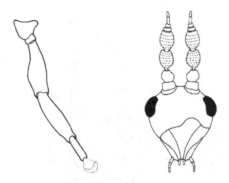

76b.   Last tarsal segment not bladderlike, with distinct claws; other characters different 77

77a.   Prothorax small, hidden when viewed from dorsal aspect ....
.... **DIPTERA**, Section 6.28

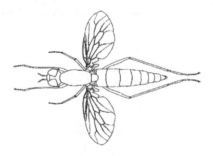

77b.   Prothorax evident when viewed from dorsal aspect 78

78a.   Beak arising from anterior part of head ....
.... **HEMIPTERA**, Suborder **HETEROPTERA** Section 6.21

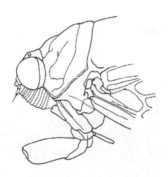

78b.   Beak arising from lower posterior part of head............. **HEMIPTERA**, Suborder
...........................**AUCHENNORHYNCH**A and **STERNORRHYNCH**A Section 6.21

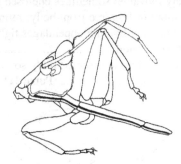

79a.   Legless grubs or maggots; movement by wriggling .... Larval **DIPTERA** (If
aquatic wrigglers, see larvae and pupae of mosquitoes); **HYMENOPTERA;**
LEPIDOPTERA; COLEOPTERA; SIPHONAPTERA; STREPSIPTERA (in body
of wasps or bees with flattened head exposed)

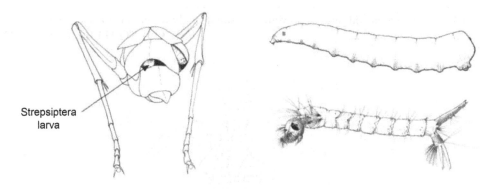

Strepsiptera
larva

79b.   Legless or if legged then each leg with one terminal claw 80
80a.   Small animals with little resemblance to most insects: filamentlike mouthparts
inserted in plant tissue: usually covered with waxy scale, powder, or cottony tufts
.... **HEMIPTERA**, Suborder **STERNORRHYNCHA** Section 6.21

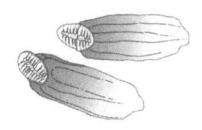

80b.   Body unable to move or able to bend from side to side: mouthparts variable; body enclosed in tight integument, sometimes wholly covering body or sometimes with appendages free, but rarely movable; sometimes enclosed in cocoon pupae 81
81a.   Legs, wings, etc., more or less free from body; biting mouthparts visible 82
81b.   Integument enclosing body holding appendages tightly against body; mouthparts evident as proboscis, without mandibles 84
82a.   Prothorax small, fused with mesothorax; body sometimes enclosed in cocoon .... Pupal **HYMENOPTERA**, Section 6.30

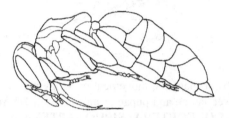

82b.   Prothorax larger and not fused with mesothorax; cocoon development variable 83
83a.   Wing cases with few or no veins; pupation rarely occurs in silken cocoon .... Pupal **COLEOPTERA**, Section 6.22

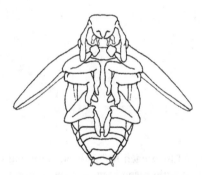

83b.   Wing cases with several branched veins; pupation usually occurs in silken cocoon .... Pupal **NEUROPTERA**, Section 6.25

84a.   Proboscis usually long, rarely absent; four wing cases, one covering each wing; body often in cocoon .... Pupal **LEPIDOPTERA**, Section 6.27

84b.   Proboscis usually short; two wing cases, one covering each forewing; body not in silken cocoon, but often tightly enclosed in hardened last larval integument ....
.... Pupal **DIPTERA**, Section 6.28

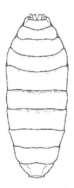 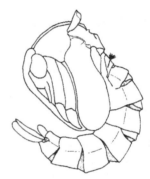

Descriptions of hexapod orders

6

The following discussion of the orders is arranged according to the preceding synopsis. Early classifications of the hexapods put great emphasis on mouthparts and wings. This emphasis is reflected in the various order names.

6.1 Protura (Fig. 6.1)

Protura are a class of entognathous Hexapoda (sometimes considered a primitive order of minute Insecta). The name Protura is derived from Greek (protos = first; oura = tail). Protura are cosmopolitan in distribution and include about 180 named species. Characteristically, proturans are small-bodied (less than 2 mm long) and elongate. The head is prognathous, with entognathous piercing mouthparts and well-developed maxillary and labial palpi. The antenna and compound eye are absent, and the thorax is weakly developed. The legs display five segments, and the pretarsus shows a median claw and empodium. The adult abdomen includes 12 segments; sterna 1—3 possess small, eversible styli, but cerci are absent, and the gonopore is positioned between segments 11 and 12. Development of Protura is anamorphic. That is, the animal emerges from the egg with nine abdominal segments and adds a segment with each of the first three molts. The lifestyle of Protura is cryptic. They inhabit in leaf litter, soil, or moss or may be found beneath rocks or under barks. Protura are most frequently found in Berlese samples.

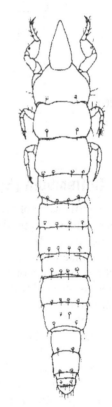

FIGURE 6.1

Proturan.

6.2 Diplura (Fig. 6.2)

Diplura are a numerically small cosmopolitan class of entognathous, epimorphic hexapods whose position in relation to Insecta is questioned. The name is derived from Greek (diploos = two; oura = tail). Diplura consist of about

Insect Collection and Identification. DOI: https://doi.org/10.1016/B978-0-12-816570-6.00006-X

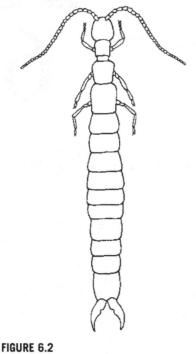

FIGURE 6.2

Dipluran.

700 named species included in four families: Campodeidae, Japygidae, Procampodeidae, and Projapygidae. Diplura are usually small in size, but a few species measure up to 50 mm long. The body is narrow, eyes are absent, and the antennae are moniliform, with each segment containing intrinsic musculature. The legs each include five segments, and the pretarsus displays a pair of lateral claws and occasionally a median claw. The abdomen shows 10 segments; some abdominal segments contain styli and eversible vesicles. Cerci are usually present but variable in development; cerci are annulate in campodeids and forcipate in japygids. The gonostyli and gonopore are positioned between segments 8 and 9. Internal fertilization is not known in Diplura. The male produces a spermatophore that is placed on the substrate. Specimens live beneath logs and rocks; a few records associate Diplura with ants and termites. Campodeids are regarded as phytophagous, and japygids are reported as predaceous and probably use their cerci to capture the prey.

6.3 Collembola [Springtails, Fig. 6.3]

The scientific name of the order is derived from Greek (kolla = glue; embolon = peg) and refers to a tubular appendage (collophore) on the first abdominal sternum that secretes a sticky substance. The common name *springtail* refers to a forked structure called the *furcula* on the fourth abdominal sternum that is held in a place by a clasp-like structure called a *retinaculum* on the third abdominal sternum. When the insect jumps, the forked structure is released with sufficient force against the surface of the ground to propel the animal into the air. A few Collembola lack this device.

Collembola are cosmopolitan, small-to-minute, soft-bodied, wingless hexapods. About 2000 species have been described. The oldest fossil collembolan, *Rhyniella praecursor* (Hirst & Maulik, 1926), was extracted from Devonian beds of Rhynie chert in England. The compound eyes of Collembola are reduced to a few facets, and the antenna contains four to six segments. The mouthparts are reduced and concealed within the head (entognathous). The thoracic segmentation is not always clearly defined, and the legs contain four segments (a coxa, a trochanter, a femur, and a fused tibiotarsus). The abdomen consists of six segments. Metamorphosis is simple. Three commonly collected families are the Poduridae, Entomobryidae, and Sminthuridae. Some species are abundant on the surface of snow and have been given the name *snow fleas*. Most species feed in moist habitats such as decaying vegetation and rotting wood, under bark, in leaf debris, on the surface of ponds and streams, in soil, and on algae, fungi, and pollen. Some species feed on dead or moribund invertebrates in soil. Cannibalism and predation have been documented in some groups. Parthenogenesis has been recorded in a few species. Collembola do not engage in internal fertilization. Males produce and deposit a spermatophore on the substrate. The female straddles the spermatophore, and the sperm moves into the female genital aperture.

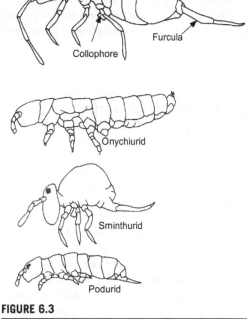

FIGURE 6.3

Collembola.

6.4 Microcoryphia [bristletails, Archaeognatha, Fig. 6.4]

Microcoryphia represents a primitive order of Ectognatha whose members were assigned to the Thysanura. The order name is derived from Greek (micro = small; corypha = head) and refers to the small head. The common name *bristletail* refers to the long cerci and appendix dorsalis. Two modern families are known (Meinertellidac and Machilidae), and about 450 species have been described. The earliest fossil record of Microcoryphia comes from Triassic deposits in Russia (Triassomachilidae). The bristletail body is moderately sized and laterally compressed, with the thorax arched. The compound eyes are well developed, and ocelli are present. The mandible is monocondylic. The coxae of some legs display styli, and the tarsi of all legs are three segmented. All species lack wings, and their ancestors never had wings. The abdomen is two segmented, with styli on some

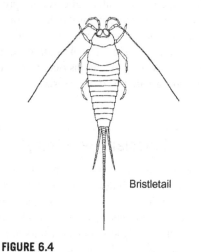

FIGURE 6.4

Microcoryphia.

segments; the appendix dorsalis is longer than the cerci. Bristletails are found under bark and stones and in grass or leaf litter. They feed on algae, moss, and other plant material. Bristletails are typically nocturnal and capable of jumping by rapid flexion of the abdomen.

6.5 **Thysanura [silverfish, firebrats, Fig. 6.5]**

In older classifications, the name Thysanura was applied to include the Microcoryphia (Archaeognatha in part). The name Thysanura is derived from Greek (thysanos = fringe; oura = tail) and refers to an older common name, bristletails, which is more appropriately applied to the Microcoryphia. Thysanura are called *silverfish* because the insects move quickly and the scales of their body shimmer like those of a silverfish. Thysanura represent an order of small- to moderate-

Silverfish

FIGURE 6.5

Thysanura.

sized fusiform and dorsoventrally compressed insects. The body is usually covered with scales that are flattened setae. The mouthparts are adapted for chewing; the mandible displays two points of articulation (dicondylic) with the head, and the compound eye is small or absent; ocelli are sometimes present. The thoracic segments are similar in size and shape; coxal styli are absent. The abdomen is two segmented, and some sterna have styli; the appendix dorsalis and cerci are present and similar in size. Metamorphosis is simple. The most common family is the Lepismatidae. Some species, such as *Lepisma saccharinum* Linnaeus and *Thermobia domestica* (Packard), are found in domestic situations feeding on bookbindings, curtains, wallpaper paste, paper, clothing, and similar articles. Most thysanurans occur outdoors under bark and stones, in grass and leaf litter, or in rotting wood or other debris.

6.6 **Ephemeroptera [mayflies, Fig. 6.6]**

The order name is derived from Greek and refers to the short life (ephemeros) as a winged form (pteron). Mayflies are a small order of Paleoptera consisting of soft-bodied, elongate insects with at least one pair of membranous wings and two or three long, slender tails. Adults do not feed, and their mandibulate mouthparts are not strongly sclerotized. Mouthparts of the nymphal instars are adapted for chewing. The nymphs are aquatic, with gills along the sides of their abdomen. Metamorphosis is simple. Mayflies possess a unique developmental stage called a *subimago*, which is the initial winged form. This is not the

(A)

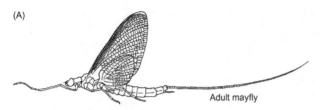

Adult mayfly

FIGURE 6.6

(A and B) Ephemeroptera.

adult stage; the subimago undergoes one molt before the insect is transformed into an adult. Three families are commonly collected: Ephemeridae, Heptageniidae, and Baetidae. Nymphs are found in a variety of aquatic habitats, from fast-flowing streams to the still waters of ponds, where some species burrow into the muck at the bottom. Adults are usually seen near water on vegetation and other objects, but at times, they are attracted in large numbers to lights.

Mayfly nymphs can be collected with an aquatic net or can be handpicked from submerged rocks and vegetation. They should be killed and preserved in 80% alcohol or held and reared to the adult stage in aquariums or quart-sized jars filled halfway with water. Adult mayflies rarely live for more than a few days; nymphs often require an entire year to develop. Unless full-grown nymphs are collected, the collector should be prepared to keep the specimens in captivity for many months. Some species need well-oxygenated water, so a small electric pump to which air hoses can be attached may be necessary. Almost all mayfly nymphs are plant feeders, and an adequate supply of aquatic plants must be provided in the aquarium or rearing container. Some of the vegetation, a stick, or a rock should extend above the surface of the water to provide the subimago with a substrate to which it can cling after emergence.

Subimagos can be preserved in alcohol, as recommended for nymphs, but it is better to hold them in rearing cages until the final molt is completed. Both subimagos and adults can be captured with an aerial collecting net by sweeping

(B)

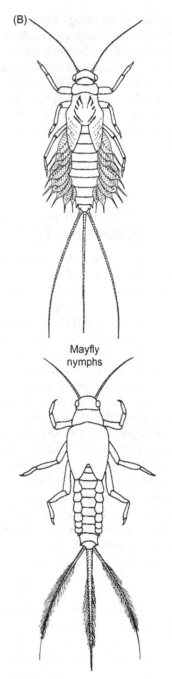

Mayfly nymphs

FIGURE 6.6

(Continued).

or beating the foliage near the water. Mayflies are extremely fragile. Care must be taken when removing them from a net to protect their delicate appendages. Adults may be preserved in alcohol or pinned. However, pinned specimens usually shrivel badly, become brittle, and are difficult to handle.

6.7 Odonata [dragonflies, damselflies, Fig. 6.7]

The order name is derived from Greek and means tooth, referring to the strongly toothed mandibles. The Odonata are ancient, with fossils in North America and Russia dating to the Permian. One fossil species living at that time had a wingspan of more than 70 cm. The Odonata are divided into several suborders, but only three survive today: Anisoptera (dragonflies), Zygoptera (damselflies), and Anisozygoptera. The Anisozygoptera consist of one small family found in the Himalayas and Japan. This group is considered a transitional element between the dragonflies and the damselflies. The adult body and immature characters resemble those of the Anisoptera, and wings are suggestive of the Zygoptera.

All Odonata are predaceous in the immature and adult stages. The dragonflies are generally large-bodied insects that usually keep their wings outstretched when at rest. The damselflies are generally smaller, more delicate insects that usually fold their wings rearward over the abdomen when at rest (Corbet et al., 1960).

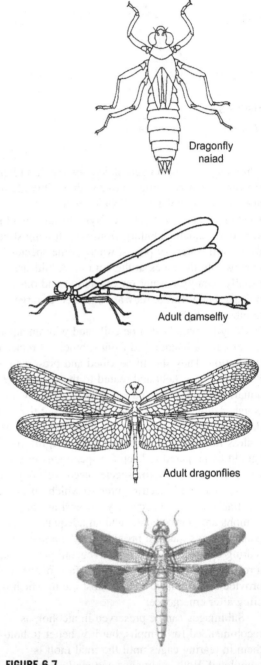

Dragonfly naiad

Adult damselfly

Adult dragonflies

FIGURE 6.7

Odonata.

Dragonflies and damselflies are comparatively large and have two pairs of many-veined wings. The posterior wings are as large as or larger than the front pair. The antennae are short and bristle like, the abdomen is long and slender, the mouthparts are adapted for chewing, and metamorphosis is simple. The aquatic nymphs respire with the aid of gills. Damselflies show their gills as leaf-like structures at the apex of the abdomen. Dragonflies hide their gills in the form of ridges on the rectum. Oxygen is extracted from water drawn into the rectum. Nymphs display an elongate, scoop-like, prehensile labium that can be extended forward rapidly to grasp the prey.

The nymphs may be found in most freshwater habitats and occasionally in brackish water. Some nymphs can tolerate exposure to humid air, and members of at least one genus in Hawaii are terrestrial. Nymphs may be found clinging to aquatic vegetation or in the muck at the bottom of streams or ponds. Specimens are reared easily on a diet of young tadpoles, aquatic insects, or other small aquatic animals. A long stick placed in the rearing tank allows mature nymphs to crawl out of the water for the final molt to the adult stage. Metamorphosis usually occurs at night or early in the morning. Teneral adults avoid water. Nymphs should be preserved in alcohol; large specimens should be killed in boiling water and then transferred to alcohol.

Dragonflies, especially the larger species, are fast fliers and are most easily caught when resting. Adults are typically diurnal, sometimes crepuscular, and occasionally nocturnal. Adults are strong fliers and use sight to locate prey. Prey are captured in flight. Individuals of some species have favorite resting places and fly regular routes. Collectors with patience to wait quietly may be rewarded by insects coming to them. Any strange and sudden movement, as with a collecting net, may cause the dragonfly to dart away. Try to swing the net from behind and a little below the insect. The specimen will be less apt to see the movement until it is too late. Captured specimens may be transferred directly to acetone for half a day, which preserves the colors. Specimens placed in a cyanide jar may lose some colors. Specimens may also be held alive for a day or two without food. The starvation eliminates the contents of the alimentary system before the insect is killed. Decomposing food in the body of a dry specimen may affect some colors. If held alive, adults should be confined to a small space and preferably kept in the dark so that they will not beat and damage their wings. Adults obtained by rearing should be held alive for a time to allow the colors to develop fully. Newly emerged, teneral adults should be held in cages large enough to permit the wings to expand fully and dry.

For permanent collections, pin adults with the wings spread or place them in clear plastic envelopes with the wings folded above the dorsum. In pinned specimens, brace pins may be added on either side of the abdomen for support and to prevent "cartwheeling." This is especially important if the specimens are to be mailed.

6.8 Orthoptera [crickets, grasshoppers, katydids, Fig. 6.8]

The order name is derived from Greek and refers to straight (orthos) wings (pteron). The Orthoptera are a cosmopolitan order of terrestrial insects with more than 20,000 described

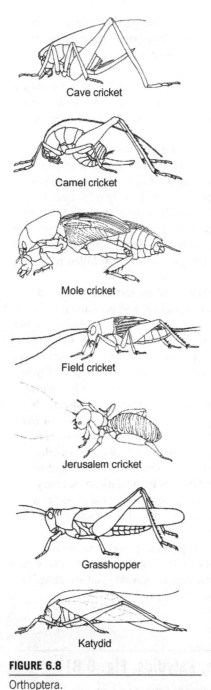

Cave cricket

Camel cricket

Mole cricket

Field cricket

Jerusalem cricket

Grasshopper

Katydid

FIGURE 6.8

Orthoptera.

species. Members of the Orthoptera are anatomically diverse. The mandibles are well developed, and the mouthparts are adapted for chewing. The antennae are multisegmented and sometimes longer than the body. The pronotum is large. The forewings (called *tegmina*) are thickened and characterized by numerous veins; the hind wings also contain many veins but are membranous, fanlike, and folded when in repose. Most Orthoptera display hind legs enlarged and adapted for jumping. The abdomen has 11 segments, and the cerci are well developed. Metamorphosis is simple, and the nymphs resemble the adults. Many Orthoptera communicate acoustically and in this respect differ from the mantids and cockroaches. Older classifications of the Orthoptera included groups such as the cockroaches, mantids, and walking sticks. Currently, these groups are separated, and the Orthoptera consist of about 15 families.

Some crickets are found in domestic situations, but most orthopterans inhabit vegetation such as grass and shrubs upon which they feed. Grasshoppers usually move by walking or jumping, but most adults can fly, often exposing brightly colored hindwings. To capture the adults in a net, watch where they land and bring the net down quickly over the spot. If the tip of the bag is held upright, the grasshopper usually can be coaxed up into the bag. Both adults and nymphs can be captured by sweeping vegetation. Flowerpot cages are suitable for rearing grasshoppers.

Most species of katydids are nocturnal. Many are attracted to light and usually can be caught easily by hand. Crickets are active at night and may be attracted to baits, such as molasses or oatmeal, that have been prepared as a paste and spread in a thin layer. The paste is made by mixing dry oatmeal with a little sugar, dry milk, and water. This same paste dried can be used to feed crickets in captivity, although the diet should be supplemented occasionally with bits of fruits and lettuce.

Adults and nymphs of Orthoptera can be killed and preserved in alcohol, but this is not recommended. Colors are lost in fluids such as alcohol during long-term storage. If specimens are to be stored in alcohol, then a 5% solution of formalin should be added to the alcohol. Specimens should be preserved dry, such as layered in cellucotton. Drying should occur over a period of several hours. This can be accomplished with an oven, a heater, or exposure to the sun. Specimens exposed to the sun should be protected from ants and flesh flies. In tropical regions, it may be necessary to embalm Orthoptera. Larger specimens sometimes decompose. For this reason, the internal organs should be removed. Dissect the internal structures (hindgut, fat body, and internal reproductive system) from the abdomen by making a longitudinal incision along the ventral surface of the abdomen and removing the tissue with forceps. A wad of cotton, moistened with naphthalene, may be placed in the abdominal cavity to retain the shape of the abdomen. For a permanent collection, pinned adults are preferable. For exhibits or demonstrations, winged forms may be mounted with one or both pairs of wings spread. For most collections, the specimens should be mounted with legs, antennae, and wings folded close to the body to conserve space.

6.9 **Blattodea (Blattaria) [cockroaches, Fig. 6.9]**

The Blattodea (Latin, blatt = cockroach) are dorsoventrally compressed insects with the head concealed by the pronotum when viewed from the dorsal aspect. The mouthparts are adapted for chewing, and the antennae are filiform and typically display more than 30 segments. Blattodea are cursorial insects with the hind legs similar in shape and size to the middle legs; many species are extremely rapid runners. Some species lack wings; when wings are present, the forewings are modified into moderately sclerotized tegmina that protect the membranous hind wings. The abdomen contains 10 apparent segments; the cerci usually display many segments. A few species of cockroaches are parthenogenetic. The female produces an ootheca or egg case. In species that deposit the ootheca on the substrate, the egg case is rather hard. Some species are ovoviviparous, with

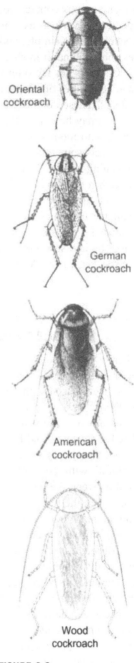

Oriental cockroach

German cockroach

American cockroach

Wood cockroach

FIGURE 6.9

Blattodea.

a membranous ootheca containing eggs retained within the female's brood pouch. A few species are viviparous, with the ootheca incompletely developed. The metamorphosis of cockroaches is simple, and nymphs and adults are found in the same habitats. Nymphs resemble the adult in shape but can differ considerably in coloration.

Cockroaches are omnivorous insects that feed upon plant and animal materials. More than 50 species of cockroaches live in North America of which only 4 are common indoors. These domestic cockroaches are nocturnal and usually avoid light. Some outdoor species may be attracted to light at night, and some species are attracted to pitfall traps baited with molasses. Oatmeal trails may also be effective (Hubbell, 1956). This attraction for sweets sometimes leads cockroaches to invade abandoned beehives to feed on the remaining honey. During the day, most cockroaches hide under the loose bark of trees, beneath rotting logs, or in similar fairly moist habitats. Some species of cockroaches can be found in leaf litter. They can be reared with little difficulty, and many species are available from biological supply houses. Specimens can be layered in cellucotton or pinned. Teneral or recently molted specimens should be allowed to harden before pinning. Large-bodied specimens should have pins placed on both sides of the abdomen to prevent the specimen from "cartwheeling" on the pin.

6.10 Mantodea [mantids, Fig. 6.10]

Mantids are sometimes called *praying mantids* or *soothsayers* (Greek, manti = soothsayer) because their forelegs are held in a supplicatory position resembling prayer. The Mantodea have been placed with the Orthoptera or with the Blattaria in other classifications. Nearly 2000 species have been described. The group is cosmopolitan in distribution but predominantly tropical or subtropical. The mantid body size is typically moderate to large. The head rotates, and live specimens can give a turn to the head that seems quizzical. The pronotum is elongate, narrow, and maneuverable on the mesothorax. The forelegs are raptorial, with the forefemur and tibia extended forward to capture the prey. The prey is held between the femur and the tibia with opposable spines on the venter of the femur and tibia when the forelegs are retracted toward the head. The middle and hind legs are slender and are not modified for running or jumping. The abdomen is 11 segmented, and the cerci consist of many segments. Eggs are protected within an ootheca.

All mantids are predaceous as nymphs and adults. Mantids are usually found on vegetation, where they prey on other insects. During autumn and winter, mantid oothecae can be found attached to shrubs or grass stems. If collected, the oothecae should be kept outdoors or in an unheated garage or porch until spring to prevent the eggs from hatching prematurely. However, if living insects (such as *Drosophila* flies or

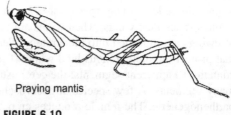

Praying mantis

FIGURE 6.10

Mantodea.

aphids) are available to feed the young nymphs, then eggs can be incubated at any time. Adult mantids usually die during fall soon after the eggs are laid, but sometimes specimens can be kept alive in rearing cages for several weeks indoors by feeding. Mantids are voracious feeders, which suggests that they can be important biological control agents in the garden. Mantid oothecae are sold by garden suppliers who advertise that the mantids will control many species of garden pests.

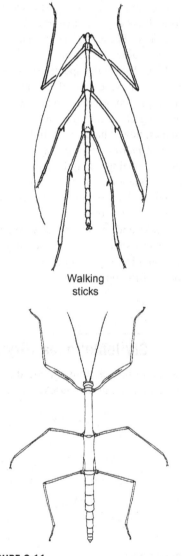

Indeed, the nymphs and adults consume large numbers of insects but only those readily visible on plant foliage. Soil-dwelling or stalk-boring insects are not attacked by mantids. Mantids appear formidable, but with a little experience they can be safely captured by hand. Care should be exercised when handling mantids because their raptorial front legs can cause a painful prick. Specimens may also be collected from vegetation with a sweep net, but some species are difficult to dislodge.

Adults and nymphs of mantids may be preserved in alcohol, but for a permanent collection, adults should be pinned. Soft-bodied specimens preserved on pins should have the body supported with card stock mounted on the pin beneath the specimen. Large-bodied specimens should have pins placed on both sides of the abdomen to prevent the specimen from "cartwheeling" on the pin.

Walking
sticks

6.11 Phasmatodea (Phasmida) [walking sticks, leaf insects, Fig. 6.11]

Phasmids are called *walking sticks* because they are slow-moving insects with elongate bodies that resemble twigs or sticks (Greek, phasm = phantom). Some tropical species are called *leaf insects* because their bodies are dorsoventrally flattened and expanded laterally so that the insect resembles a leaf. Phasmids are cosmopolitan in distribution, and more than 2500 species have been described. Several species are more than 30 cm long. The phasmid's body is often spinose, and the head is typically

FIGURE 6.11

Phasmatodea.

prognathous. Ocelli are usually absent, and the mouthparts are of chewing type (mandibulate). The prothorax of phasmids resembles that of mantids in that it is movable on the mesothorax. However, wings are absent in most species of phasmids. Winged species are most common in tropical areas; when present, the forewings are modified into tegmina, and the hind wings are broad with a uniform branching pattern. The legs are gressorial, with all legs similar in appearance and adapted for walking; the coxae are small and separated. The abdomen is 11 segmented; cerci are not segmented, but sometimes they are long and clasper like in males. Metamorphosis is simple, and the nymphs resemble the adults. Eggs are not enveloped by an ootheca. Instead, they are scattered individually on the ground. Phasmids may remain in the egg stage for more than a year.

The biology and ecology of Phasmatodea have been reviewed (Bedford, 1978; Carlberg, 1986). All species of phasmids are phytophagous and may be located on trees or shrubs upon which they feed. Their slow movements and resemblance to twigs or leaves and cryptic coloration (usually green or brown) make these insects difficult to detect. Sweeping or beating trees and shrubs may yield some individuals. In spring or early summer, newly hatched nymphs may be found, which in the most common species, *Diapheromera femorata* (Say), are only about 0.5 cm long. The nymphs mature in about 6 weeks, attaining a length of 8–10 cm. Their slow-moving, plant-feeding habits enable them to adapt readily to the environment of a rearing cage. Usually, phasmids are leaf-feeding insects. However, in captivity, the newly molted nymphs have been observed eating their cast exuviae.

Adults and immatures may be collected by hand and placed into alcohol or a killing jar. Specimens should be preserved dry. Because phasmids are extremely long bodied, additional brace pins may be necessary to support the abdomen and prevent the specimen from shifting on the pin.

6.12 **Grylloblattodea (Grylloblattaria) [rock crawlers, Fig. 6.12]**

These rare insects, found along the edges of glaciers, have features resembling both crickets (Latin, gryll = cricket) and cockroaches (Latin, blatt = cockroaches). They derive

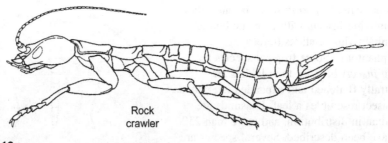

Rock
crawler

FIGURE 6.12

Grylloblattodea.

their common name from their habit of moving about slowly on rocks in cool habitats. The order consists of one family and about 20 cryptozoic species found in cold, wet habitats of North America, Japan, and Russia. Rock crawlers are soft-bodied, elongate, slender, mandibulate insects 15–30 mm long. They are characterized by a prognathous head with compound eyes that are reduced or absent; the ocelli are absent, and the antennae are long with many segments. The prothorax is large, wings are absent, and the legs are cursorial and adapted for running with large coxae. The cerci are long and segmented; the ovipositor is strongly exserted.

Rock crawlers are typically found in rotting logs, beneath stones, and on talus slopes near snow. They are active at low temperatures. Their diet consists of moss and other insects. The life cycle requires several years to complete. Grylloblattoids are of no economic importance, but they are interesting because they are unusual and rarely found in collections. These insects are probably best preserved in 80% alcohol or glycerol.

6.13 **Dermaptera [earwigs, Fig. 6.13]**

The order name is derived from Greek (derma = skin; pteron = wing) and refers to the skin-like appearance of the forewings. The common name is derived from the erroneous superstition that earwigs crawl into the ears of sleeping people. The order Dermaptera is cosmopolitan and currently consists of about 1200 named species. Dermaptera are most common in tropical and warm temperate regions. They were placed among the Orthoptera in nineteenth-century classifications, but Dermaptera represent a distinct group that is probably related to the Grylloblattodea. Dermaptera are elongate insects that measure up to 50 mm long. The head is prognathous, mouthparts are of the chewing type, and the prothorax remains free from mesothorax. The adults usually have two pairs of wings. The forewings are short, leathery, and elytriform or modified into tegmina. The hind wings are large, membranous, fanlike or circular, and folded under the forewings when the insect is at rest. The legs are cursorial, with three segments forming the tarsus. The abdomen is often telescopic, and the cerci are modified into forceps that can pinch if the earwig is handled. Metamorphosis is incomplete, and the nymphs resemble the adults.

Earwigs are typically nocturnal insects that are sometimes collected at lights. During the day, earwigs usually hide in cracks and crevices, under the bark of trees, or in rubbish on the ground. One species, *Anisolabis maritima* (Gene), is commonly collected under stones or driftwood along the Atlantic and Pacific coasts. The eggs are laid in a burrow in the ground, and the

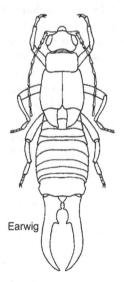
Earwig

FIGURE 6.13

Dermaptera.

female cares for them until they hatch. The families of earwigs found in North America include the Forficulidae, Chelisochidae, Labiduridae, and Labiidae.

Earwigs feed as scavengers on decaying organic matter, but some species may also feed upon the living tissues of flowers, ripening fruits, and vegetables. Earwigs occasionally feed upon aphids and other small insects. Earwigs living in cooler habitats are predominantly herbivorous; those living in warm temperate and tropical regions are predominantly predaceous. Predatory species feed on a variety of insects but seem to prefer soft-bodied larvae. Despite their almost omnivorous feeding habits, or perhaps because of them, earwigs are not strongly attracted to bait traps. It is usually necessary to search by day in likely habitats and to collect the specimens by hand. Adults and nymphs may be killed and preserved in alcohol. However, for a permanent collection, the adults should be pinned or card mounted. Pinned specimens may require a card placed under the body for support.

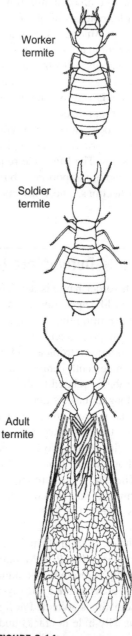

Worker termite

Soldier termite

Adult termite

FIGURE 6.14

Isoptera.

6.14 Isoptera [termites, Fig. 6.14]

The order name is derived from Greek (isos = equal; pteron = wing) and refers to the size and shape of the wings. The common name *termite* is derived from Latin *termes*, which was the medieval name for wood worms. Termites are also called *white ants*, but they may be distinguished from true ants because termites have a broad waist and beaded antennae; winged forms have both pairs of wings in the same size and shape. Termites are a cosmopolitan order including about 2000 species. They are most abundant in tropical and subtropical regions. Termites are small, soft-bodied, polymorphic, and cryptozoic insects. The antennae are moniliform or filiform; the mouthparts are mandibulate. Primary reproductives possess four membranous, similar-sized, net-veined wings that are shed; nymphs, soldiers, and workers lack wings. The abdomen shows 10 apparent segments. The cerci are one- to five segmented, and sclerotized external genitalia are absent in most species. Metamorphosis is incomplete in termites, but each species displays several morphs.

The primary food of termites is cellulose, which is broken down by symbiotic bacteria or flagellate protozoans. Termites feed incidentally on exuviae, dead nestmates, and excrement. The association with symbionts is vital. If these microorganisms are removed from the gut of the termite, the animal will eventually die of starvation. These microorganisms live in the hindgut of the termite. This relationship is precarious because the hindgut and the microorganisms are shed during the molting process. Fortunately, the newly molted individuals receive a new colony of microorganisms by feeding at the hindgut of other termites. This process is called *proctodaeal feeding* or *trophallaxis*.

Termites display a highly developed social system. Castes include primary reproductives (a king and a queen), secondary reproductives (neotenics), soldiers (male and female), and workers (male and female). Reproductives swarm periodically in great numbers. Bonding occurs in the swarm, with a royal pair (king and queen) then excavating a nuptial chamber. Copulation occurs within the chamber, and a new colony is established. After the dispersal flight from the parent colony, the wings of the adults are usually shed, leaving only short stubs that are visible under low magnification. The royal pair is long lived, perhaps as long as 50 years.

Wingless forms are sterile adults of the worker and soldier castes. Workers provide food for the colony, construct new tunnels and chambers, and care for the egg-laying queen. In many groups of termites, the queen's abdomen is so swollen that she is immobile. This condition, called *physogastry*, is typical in some groups. Soldiers defend the colony against attack. Their head is greatly enlarged and usually displays powerful biting mandibles. In some species of termites, the mandibles are supplemented with a beak through which a fluid may be ejected to repel enemies. Soldiers of some species display highly sclerotized heads that are modified to block passage through tunnels of the colony. Such heads are termed *phragmotic*.

Termites are responsible for the economically significant destruction of wood and wood products. Based on the habitat, termites construct several types of homes, including mounds (termitaria), arboreal nests, and subterranean nests. They are a common landscape scene in some tropical and subtropical areas of the world. Termitaria are constructed of earth mixed with salivary gland secretions and excrement and are striking in their architectural diversity and range of size. Some species of termites construct termitaria nearly 7 m tall. Termitaria are exceedingly hard and must be penetrated with iron or steel implements. Arboreal termites include species that live in dead trees, wooden buildings, or furniture not in contact with soil. Subterranean termites live in the soil, sometimes creating mounds protruding a few meters above the surface. Colonies of subterranean termites may be located by digging into these mounds, by prying apart rotting stumps, or by turning over rotting logs and searching in the soil beneath.

Colonics of arboreal termites are more difficult to find, because often there is no external evidence of an infestation. Careful examination may show the entrance holes made by the reproductive adults as they entered the wood, but these holes are usually sealed with cement-like plugs secreted by the termites. If the termites are working close to the outer edge of the wood, surface blisters in the paint or a flaking of the surface of unpainted wood may be a clue to the presence of a colony.

For quantitative studies of termites, Pearce (1990) suggests using a cardboard coaster sandwiched between two glass plates held together with a clip. The cardboard is moistened, and the entire trap is buried in the soil. The termites feed on the moistened cardboard, and the entire trap can be lifted from the soil to assess damage and collect termites without disturbing the trap.

A pick, shovel, or mattock is necessary for opening termite nests and splitting wood to expose specimens. An aspirator can be used to collect termites, but care should be exercised because many termites are fragile and easily damaged. Some termites possess volatile chemicals that can be irritating to the respiratory system of the collector. In such instances, a blow-type aspirator should be used. A moistened camel hair brush is effective in capturing small-bodied termites. Featherweight forceps are excellent for capturing termites. Representatives of each of the castes should be collected because identification often cannot be made from the workers alone. Many individuals of each caste should be collected to appreciate variation in size and coloration characters. Ideally, the king and queen should be captured, but this may be difficult. Do not overlook insects of other orders that often live in the colony and that may mimic termites in color and general appearance. All castes of termites should be killed and preserved in 70%−80% alcohol. Photographs of the nests, casts of the nest architecture, or diagrams of tunnel patterns are also useful kinds of information for specialists.

6.15 Embiidina (Embioptera) [web spinners, footspinners, Fig. 6.15]

These insects have been called Embioptera, but the correct name is Embiidina (Greek, embi = lively) probably due to the fact that they can run backward very rapidly. The common names stem from their ability to spin silk with glands on the feet. Embiidina constitute a widespread group of about 200 named species and an indeterminate number of undescribed species. Embiidina are well represented in tropical and warm temperate regions. Web spinners are elongate, mostly small insects with a prognathous head and chewing mouthparts. The antennae are filiform, and ocelli are absent. Females are wingless, males are winged or wingless, and sometimes both conditions occur in males of one species. When wings are present, the wing venation is reduced. Legs display three tarsomeres. Males, females, and all instars of nymphs spin silk from glands located on the fore basitarsus. The silk emerges from hollow, tubular structures on the ventral surface of the basitarsus and second tarsomere. The abdomen appears 10 segmented, but the 11th segment is reduced; the cerci are two segmented and tactile. All species are gregarious, with nymphs and adults occupying silken galleries. Rapid rearward movement of web spinners is enabled through large depressor muscles in the hind tibiae. Species feed on bark, moss, lichen, dead leaves, and other plant material. Cannibalism may occur, and males are sometimes eaten by females after copulation. Eggs and early instar nymphs are guarded by the female.

Male web spinners are sometimes seen at lights during the night, but females are wingless and are found in the galleries. Galleries may be located among debris and in cracks and fissures or on the surface of bark. Specimens should be preserved in 70%−80% alcohol.

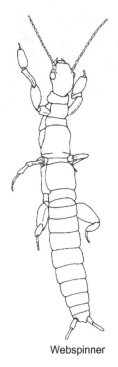

Webspinner

FIGURE 6.15

Embiidina.

6.16 Plecoptera [stoneflies, plaited-winged insects, Fig. 6.16]

The scientific name of the order is derived from Greek (plekein = to fold; pteron = wing) and refers to the habit of fan-like folding of the hindwing beneath the forewing in repose. The common name *stonefly* refers to the habitat in which specimens are often collected. The Plecoptera are a cosmopolitan order containing about 1000 named species. Most genera of stoneflies are found in the Holarctic, where the order radiated or numerically expanded. Adult stoneflies are soft-bodied, moderately sized, and dorsoventrally compressed, with a body seemingly loosely joint. The head shows two to three ocelli, and the antennae are long, many segmented, and slender. The mouthparts are of the chewing type but are sometimes reduced. The prothorax is large and mobile; the mesothorax and metathorax are smaller and subequal in size. Wings are membranous, with numerous veins and crossveins; a few species are brachypterous or apterous (usually males). Legs bear three tarsomeres. The abdomen has 10 segments; the cerci contain one to many segments. Metamorphosis is incomplete.

Most species of stoneflies are diurnal as adults and found near water. Immatures live in water, and the nymphs are called *naiads*. A few species found in the Southern Hemisphere apparently exist in damp terrestrial situations. Plecoptera naiads typically prefer clean, cold, and moving freshwater. They demonstrate intolerance to pollution and narrow tolerance to water- and temperature-related parameters. The life cycle of most species lasts one year, but a few species live longer. Some species apparently modify their life cycle in response to changing environmental conditions. Adults are short-lived. Males die soon after copulation; females die soon after egg laying. Evidence suggests that females must feed as adults on their eggs to develop. Females deposit eggs on the surface of water during flight when the apex of the abdomen touches the water surface. Some species broadcast or deposit eggs beneath the surface of the water or upon submerged objects. Stonefly eggs are small and with an adhesive coating activated by water. Females may lay more than one batch of eggs; total egg production may reach 1000.

Stonefly naiads possess long antennae and cerci; their mouthparts are adapted for chewing. Some North American species display branched gills on the thorax and at the base of the legs. Naiads species found elsewhere may display respiratory gills on the mouthparts, cervix, thorax, or abdomen or within the anus. In North America, stonefly naiads resemble mayfly naiads. However, stonefly naiads differ in the location of the gills and possess two tarsal claws instead of one. Stonefly naiads are found in running water, under stones, or clinging to submerged piles of drifted leaves and other debris. Specimens can be collected with an aquatic net, a dipper, or by hand and dropped into vials of alcohol or held for rearing. To rear Plecoptera, the food preferences of the naiads must be determined. The species most commonly collected are plant feeders, but some species are carnivorous. To reduce the time needed for rearing, some collectors schedule their collecting trips to coincide with the season of adult emergence for a particular species. The adults almost always emerge at night or early in the morning. By visiting streams at night, mature naiads can be collected as they crawl out of the water just before their final molt. These specimens should be held in temporary cages until the teneral adults expand their wings and dry.

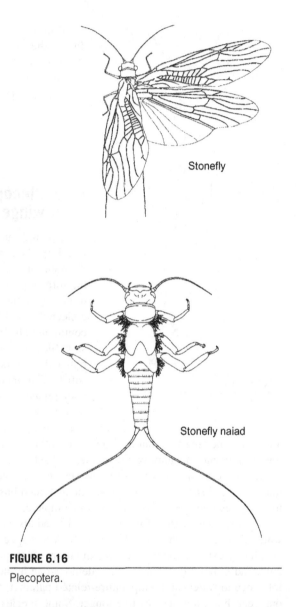

Stonefly

Stonefly naiad

FIGURE 6.16

Plecoptera.

Most insects hibernate or become inactive in cold weather, but adults of many species of Plecoptera emerge during the winter. Their dark forms against the snow make them conspicuous and easy to collect. Species that mature in spring or summer often fly to lights at night. During the day, adults are usually found resting on vegetation or other objects along stream banks and can be collected by sweeping or beating the foliage. Most stoneflies are poor fliers but agile runners. Because they generally are slow to take flight, specimens can often be picked by hand or brushed with a flick of the finger from foliage directly into a vial of alcohol.

Traditionally, adults and naiads have been preserved in alcohol. This curatorial practice arose because specimens shrivel when they are pinned after air drying. Although distortion is an obvious problem, some collectors prefer this method of preservation. Specimens will remain lifelike and not distorted if they are critical point dried. Wings, genitalia, and other structures critical for identification may be dissected and mounted on microscope slides or placed in microvials.

6.17 **Psocoptera [booklice, barklice, Fig. 6.17]**

The name Psocoptera is nonsensical because it is derived from Greek and means "chewing wings." The name stems from the attempts of Shipley (1904) to standardize the endings of names of insect orders. Unfortunately Shipley's name stuck. The alternative order name for Psocoptera is Corrodentia. The common names of Psocoptera are more descriptive. The terms *booklice* and *barklice* refer to the habitats of some frequently encountered species. The Psocoptera form a cosmopolitan order of small- to moderate-sized, soft-bodied insects currently containing about 1800 species. The head is large and movable, the antennae are long and filiform, and the mouthparts are mandibulate; ocelli are present in winged forms but absent in wingless forms. The prothorax is reduced in winged forms, and the meso- and metathorax are fused in wingless forms. The wings are membranous with venation reduced, and the pairs are held together when active; at rest, the wings are held obliquely over the body. Legs display two or three tarsomeres. The abdomen consists of nine segments, and cerci are absent. Members of the Psocoptera resemble psyllids or aphids but differ from them in having chewing mouthparts. Psocoptera undergo incomplete metamorphosis.

Some usually wingless species found on old books or papers have received the common name *booklice*. In damp locations in houses and granaries, booklice may attain colossal numbers.

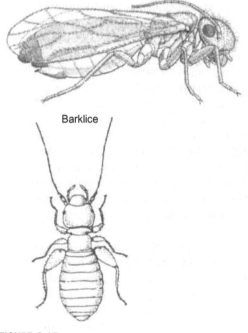

Barklice

FIGURE 6.17

Psocoptera.

Most species of Psocoptera occur outdoors on the trunks and leaves of trees and shrubs, on lichen-covered stones, and on fences. Psocids feed on mold, lichen, pollen, cereals, and starchy materials. Unfortunately, they also feed on dead insects and can cause considerable damage to insect collection unless preventive measures are taken (see Section 3.5.6).

Specimens can be collected by sweeping or beating vegetation, by drying leaf debris in a Berlese funnel, or by picking specimens from tree trunks or rocks with an aspirator or fine brush moistened in alcohol. Adults and nymphs of most species should be preserved in 70%−80% alcohol. Species with scales on their wings should be preserved dry on minuten pins or layered in cellucotton.

6.18 Zoraptera [zorapterans, Fig. 6.18]

Zoraptera constitute one of the smallest orders of Insecta, consisting of one family and about 30 named species. Geographically, zorapterans are widespread. The scientific name is derived from Greek (zoros = pure: a = without; pteron = wing) and is somewhat misleading. When the group name was proposed, entomologists believed that zorapterans were all wingless. Subsequently, it has been learned that some zorapterans are winged and that winged adults cast off their wings. Adults are minute to small-bodied, with a hypognathous head and chewing mouthparts. The adult antenna is moniliform with nine segments: The nymph displays eight antennal segments. The prothorax is well developed in all specimens, but the mesothorax and the metathorax are not differentiated in primitively wingless forms. The wings of zorapterans are membranous with reduced venation, and they are shed at the base. Wingless forms lack compound eyes or ocelli, while winged forms possess compound eyes and ocelli. The legs display tarsi with two segments. The abdomen is composed of 11 segments, the ovipositor is vestigial or absent, and cerci are present but not segmented. Development is gregarious, usually under planks, in piles of old sawdust, in rotting logs, under the bark, or in association with termites. Zorapterans are apparently fungivorous or necrophilous or both.

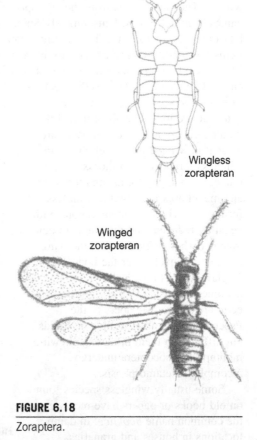

Wingless zorapteran

Winged zorapteran

FIGURE 6.18

Zoraptera.

Only two species are found in North America: They are found in Pennsylvania to Texas and eastward.

Zorapterans have no economic importance. They may be collected with a Berlese funnel, aspirated, or captured with a camel-hair brush while sitting through sawdust or debris. Zorapterans should be preserved in 70%−80% alcohol because they are exceedingly small and not strongly sclerotized.

6.19 Phthiraptera [true lice, Fig. 6.19]

The name Phthiraptera is derived from the Greek (phthir = louse; aptera = without wings). All are secondarily wingless, dorsoventrally flattened, and have reduced eyes and antennae and are ectoparsitic on birds and mammals. Members of this order once consisted of two orders, Mallophaga and Anoplura. As presently constituted, the order Phthiraptera now consists of four suborders with Anoplura (sucking lice) retaining the families from the previous order Anoplura.

The name Anoplura is derived from Greek and is trickier than other Greek combinations (an = without; hoplon = weapon; oura = tail). The name refers to the lack of cerci or caudal appendages on the abdomen. Anatomically, sucking lice resemble

sucking
louse

FIGURE 6.19

(A and B) Phthiraptera.

chewing lice in that both groups are small-bodied and dorsoventrally flattened. The head of sucking lice is somewhat conical, and the compound eye is reduced to a facet or absent. Mouthparts consist of three sclerotized piercing stylets that are concealed when not in use; the maxillary and labial palpi are reduced or absent. All of the thoracic segments are fused, and wings are never present. The tarsi are one segmented, each with a large claw opposable to a thumb-like process on the apex of the tibia. This structure serves to grasp hairs. The abdomen is nine segmented. The female lacks an ovipositor, and cerci are absent in both sexes. Metamorphosis in the suborder Anoplura is simple, and the nymphs resemble adults. All stages are found on the host, including eggs that are glued to the hair.

Anoplura suck blood and are parasites of mammals. Three families in the suborder are found in the United States including Pediculidae (human parasites): the crab louse, *Pthirus pubis* (Linnaeus); the head louse, *Pediculus humanus* capitis De Geer; and the body louse, *P. humanus* Linnaeus, an important vector of typhus and other diseases, Echinophthiridae (marine mammal parasites), and Haematopinidae (parasites of horses, cattle, sheep, hogs, and other animals).

The suborders Rhynchophthirina, Ambylcera, and Ishnocera (chewing lice) are external parasites primarily of birds and rarely of mammals. None are known to attack humans. About 3000 species of chewing lice have been described. Morphologically, they are similar to sucking lice by having small dorsoventrally compressed or flattened, wingless bodies but differ by having mouthparts specifically adapted for chewing. Metamorphosis is paurometabolous, and the miniatures resemble the adults. All stages occur on the host, and chewing lice are rarely found away from their host. Six families are commonly found in North America. Of these, the Philopteridae (parasites on poultry) and Trichodectidae (parasites on cattle, horses, and dogs) are perhaps the most often collected.

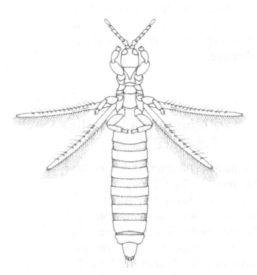

Thrips

6.20 Thysanoptera [*thrips, fringe-winged insects,* Fig. 6.20]

The order name is derived from Greek (thyasnos = fringe; pteron = wing) and alludes to the long marginal fringe found in winged specimens. The common name *thrips* is singular and plural: We do not refer to a "thrip." Thrips are a cosmopolitan order consisting of about 4000 named species. They are related to Hemiptera. Thrips

FIGURE 6.20

Thysanoptera.

are small and slender, but some species may attain a length of 12 mm. The compound eyes usually have large facets, and ocelli are present in winged adults only. Another curious feature of thrips is that the right mandible is absent. The left mandible is adapted for rasping: The labrum and labium form a cone, and the maxillae are adapted for piercing. Wings, when present, are narrow with reduced venation and display a long marginal fringe: Wings are held over the abdomen in repose, but they are not folded. The tarsi are one- or two segmented, and the pretarsus forms an eversible bladder that is used for adhesion. The abdomen of thrips is 10 segmented; an ovipositor is present or absent, but cerci are always absent. Metamorphosis is complex among thrips and shows a transition between simple and complete metamorphosis. Thus, nymphal instars 1−2 are active, feeding, and wingless or wing buds are internal; instar 3 (prepupa) and instar 4 (pupa) are quiescent and nonfeeding, with external wing buds.

Parthenogenesis and viviparity are recognized in some species. Thrips are predominantly phytophagous, a few species are fungivorous, and a few species are predaceous. Eggs are inserted into plant tissue, under bark, or into crevices. Thrips are typically multivoltinc. Most thrips feed on flowers, pollens, leaves, buds, twigs, and other parts of plants. Specimens can be collected by sweeping vegetation or by shaking or beating shrubbery over a white or light-colored pan. Another method involves collecting infested plant parts usually indicated by curled leaves or deformed buds and placing the vegetation in a paper bag. These samples can be examined later, and the thrips can be removed with an aspirator or a fine brush dipped in alcohol. Alternatively, the sample can be run through a Berlese funnel. Both winged and wingless forms are present. Thrips can be collected in 60%−70% alcohol, but alcohol-glycerin-acetic acid (AGA) (see Appendix I) is recommended because it leaves the appendages well extended and makes the specimens easier to mount on microscope slides.

For critical study and permanent collections, thrips should be mounted on microscopic slides. Specimens need not be stained before mounting, but some dark-colored and spore-feeding thrips must be treated with cold 10% sodium hydroxide to lighten the color and remove the body contents. Do not use potassium hydroxide because that compound is too caustic and destroys the wings before the harder parts of the body are sufficiently treated. In some species, the wings are held close to the body by paired setae on either side of the abdomen when the insect is at rest. To spread the wings, insert a needle between the wings and the body at the point where the thorax and abdomen join and gently tease the wings out to a horizontal position. The antennae should also be extended because the number of antennal segments and the position of sensilla on the segments are often critical for identification to species or even to family.

6.21 Hemiptera [true bugs, scales, aphids, cicadas, hoppers, psyllids, and whiteflies, Fig. 6.21]

The order name Hemiptera is derived from Greek (hemi = half; pteron = wing) and most properly alludes to the condition of the forewing of one suborder Heteroptera in which the basal part is thickened, and the distal part is membranous. The etymology of the word bug

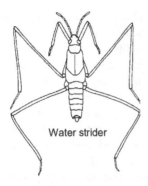

Water strider

Giant water
bug

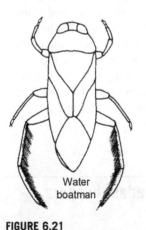

Water
boatman

FIGURE 6.21

(A and B) Hemiptera.

is uncertain. The term bug is used commonly to refer to all insects. However, in a strict sense, only species of the Heteroptera are the "true bugs."

Members of the order Hemiptera are diverse in habits and habitats. Because of this diversity, some classifications once treated Hemiptera and Homoptera as two distinct orders based on wing condition, mouthpart position, and food preference. However, more conventional systematics combine these two orders into Hemiptera with the suborder Heteroptera (Greek, hetero = different; pteron = wing) encompassing the previous Hemiptera and the previous Homoptera being divided into two suborders the Auchenorrhyncha (Greek, auchen = neck; rhynch = beak, snout) and Sternorrhyncha (Greek, stern = breast, chest; rhynch = beak).

Members of the suborder Heteroptera have heads that are usually prognathous, antenna with four or five segments, and beaks typically displaying three or four segments. Forewings are modified into hemelytra, and the tarsi hold two or three segments with two apical claws. Most species are terrestrial, but many Heteroptera live in or on the water. Most Heteroptera are plant feeders, but some species are predaceous on other insects. A few species imbibe the blood of humans and other vertebrate animals and are important vectors of diseases in the tropics.

Collecting methods for Heteroptera depend on the type of habitat. Most species are collected in nets; aquatic nets are used for species associated with water, and sweep nets are used for most plant-feeding and terrestrial predaceous species. Heteroptera living near water may be shot with a stream of alcohol from a laboratory squeeze bottle and then captured with forceps. Terrestrial specimens collected in sweep nets can be taken easily in an aspirator and later transferred to a killing jar. Some species are attracted to lights. Sifting or using a Berlese funnel on leaf litter or on soil around plant roots usually will yield some Heteroptera. Some species are found in association with ants and may mimic the ants in appearance. Species of Cimicidae (bed bugs) are ectoparasites of birds, bats, and other animals, including humans. Cimicids usually feed on their host at night and hide in cracks and crevices during the day. Cimicids are virtually wingless and can be collected easily with small forceps.

With the possible exception of some fragile plant bugs of the family Miridae, most immature and adult Hemiptera can be killed and preserved in 70%–80% alcohol. Adults

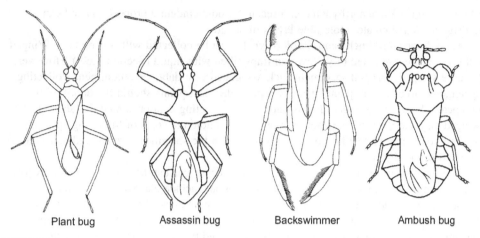

Plant bug   Assassin bug   Backswimmer   Ambush bug

FIGURE 6.21

(Continued).

should not be stored in alcohol for long periods because their colors fade. Soft-bodied specimens may be critical point dried, with excellent results. For permanent preservation, adults should be pinned. Mirids and most Hemiptera can be collected satisfactorily with killing jars, but never put large specimens in the same jar with mirids or other delicate bugs. Small killing tubes, such as those used for parasitic Hymenoptera, are ideal for relatively small Hemiptera. A rumpled piece of tissue paper placed in the jar will help prevent damage to the specimens. Specimens too small or slender for direct pinning should be glued to points on double mounts (Fig. 6.21).

Auchenorrhyncha and Sternorrhyncha forewings are uniform in texture: entirely membranous, as in the aphids, or almost leathery, as in the leafhoppers. The hind wings (absent in male scale insects) are entirely membranous. When at rest, members of the both suborders hold their wings folded roof like over the back. Both groups lack cerci. Immatures resemble adults in most species, and metamorphosis is characterized as simple. However, more complex patterns of development are seen in some Sternorrhyncha. For instance, members of the family Aleyrodidae pass through a pupal stage; males of scale insects pass through a prepupal or pupal stage (or both); the life stages of the Phylloxeridae are even more complex (Stoetzel, 1985).

Anatomically, the Sternorrhyncha are exceedingly diverse. Wingless Sternorrhyncha, such as scale insects, are sometimes difficult to recognize as insects. Most Auchenorrhyncha and some Sternorrhyncha are easily recognized as insects; the head is usually opisthognathous.

Sternorrhyncha and Auchenorrhyncha are terrestrial and feed on plants. Some Homoptera (such as whiteflies, psyllids, and leafhoppers) transmit plant diseases and are regarded as serious economic pests. Other Sternorrhyncha (such as soft scales, armored scales, mealybugs, and aphids) can develop into numerically large, highly dense populations resulting in accumulations of honeydew—a substrate for the development of fungi. Cicadas (Auchenorrhyncha) can reach periodic epidemic proportions and can damage trees and create intolerable levels of sound.

Many Suchenorrhyncha and Sternorrhyncha can be collected with sweep nets. Winged aphids are often attracted to yellow pan traps filled with liquid (Section 1.6.9). However, because swept or trapped specimens lack associated host data, we recommend collecting specimens directly from plants whenever possible. Host plants should be identified by competent botanists and recorded as part of the collecting data presented on labels. Collect winged and wingless forms if both are present and note the color of the living specimens. Natural color may be lost in a preservative or when the specimens are cleared and mounted on microscope slides.

Most specimens of Auchenorrhyncha and Sternorrhyncha can be collected and preserved, at least temporarily, in 70%−80% alcohol. Exceptions include diaspidid scales and whitefly pupae, which should be collected dry on the host plant and placed between pieces of absorbent paper. Do not use plastic bags because they accumulate moisture, which ruins specimens. Aphids, whiteflies, and scale insects should be cleared, stained, or bleached (Section 3.2.8) as needed and then mounted on microscope slides for critical study.

Adult cicadas, leafhoppers, and other relatively hard-bodied Auchenorrhyncha can be collected in killing jars and then pinned. Specimens smaller than 10 mm should be glued carefully to card points. When collecting gall-forming psyllids and aphids, include a sample of the gall. If securely anchored with additional pins, the gall can be kept in the box with the pinned adults. Galls should be treated with a fumigant (naphthalene) to ensure that all of the occupants are dead.

Collectors may be interested in studying the life history of Auchenorrhyncha and Sternorrhyncha or accumulating all the life stages of some species. Aphids and many other plant-feeding forms can be maintained easily in flowerpot cages (Section 1.15). The cages allow all stages of the insect to be observed and collected. Also, the living specimens provide a source of food for mantids and other predaceous insects maintained in captivity.

Collecting Sternorrhyncha can often be rewarding in other ways. For instance, Sternorrhyncha are attacked by many species of parasitic Hymenoptera. Scale insects and mealybugs are particularly susceptible to attack by chalcidoid wasps (Tachikawa, 1981; Viggiani, 1984); leafhoppers are attacked by dryinid wasps (Olmi, 1984a,b); the egg stage of many Auchenorrhyncha is attacked by species of mymarid wasps (Huber, 1986). In some instances, hyperparasites can be reared from some Sternorrhyncha (Viggiani, 1990). Careful dissection of hosts under a microscope can often confirm suspicions of hyperparasitic habits. When parasitic wasps are reared from Auchenorhyncha and Sternorrhyncha, the remains of the host should be preserved with the parasite. Accurate host associations are very important from our understanding of parasitic wasps and their importance in biological control. All collectors can make valuable contributions to the taxonomy of parasitic Hymenoptera through the careful preservation of parasites and their homopterous hosts.

6.22 Coleoptera [beetles, Fig. 6.22]

The order name is derived from Greek (koleos = sheath; pteron = wing) and alludes to the thickened forewings that protect the hind wings. This order contains about 300,000

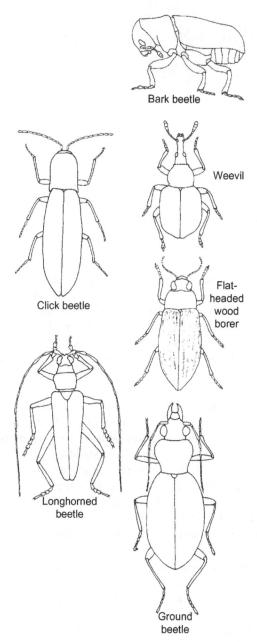

Bark beetle

Weevil

Click beetle

Flat-headed wood borer

Longhorned beetle

Ground beetle

FIGURE 6.22

(A–C) Coleoptera.

described species of insects and is the largest of the insect orders. More than 25,000 described species occur in the United States, with many undescribed species. Approximately 120 families are recognized, which are grouped into three suborders. This order is a favorite among collectors, and many amateurs specialize in collecting select families of beetles.

Beetles have mouthparts adapted for chewing and antennae that are exceedingly variable in segmentation and shape. The prothorax is well developed, the mesothorax is generally reduced, and the abdomen is broadly joined to the thorax. Beetles usually have two pairs of wings although some beetles lack them, and other species have highly modified wings. The forewings are called *elytra* (Greek, elytron = cover, sheath). Elytra are thickened, usually hard or leathery, lack veins, and often are sculptured or display pits and grooves. Elytra usually meet in a straight line dorsally along the middle of the back. The hindwings are membranous and are folded under the forewings when the insect is at rest. Several anatomical types of larvae occur within the order.

Beetles have invaded almost every conceivable aquatic and terrestrial habitat. They feed on plants and fungi. Plant-feeding beetles may be found in association with every part of a plant, from flowers to roots. Some beetles are external feeders; other species mine leaves or bore into the stalk. Many phytophagous species are serious economic pests. In contrast, some

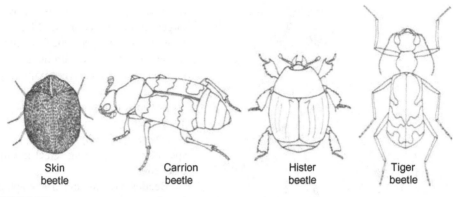

Skin beetle   Carrion beetle   Hister beetle   Tiger beetle

FIGURE 6.22

(Continued).

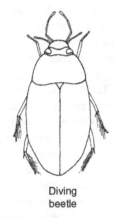

Diving beetle

FIGURE 6.22

(Continued).

beetles are predators of other insects and are viewed as beneficial insects. Predaceous species of Coccinellidae are used in biological control. Some species of beetles develop as parasites, and many species of beetles are scavengers. Some beetles are inquilines (welcome or unwelcome guests) in nests of social insects (termites and ants). Dermestid beetles feed on dead insects and can be a problem in insect collection unless protective measures are taken.

Collecting methods for beetles are almost as diverse as their habits and are limited only by the ingenuity of the collector. Pitfall and other bait traps using fermenting fruit, decaying meat, or excrement attract silphids, staphylinids, nitidulids, scarabs, and other scavengers. Small specimens may be taken with "window" traps (Peck and Davies, 1980). Some scarabs, cerambycids, and many other beetles are attracted to light traps or light sheets. A sifter or Berlese funnel can be used effectively to collect soil-dwelling beetles or beetles found in moss and leaf litter (Bell, 1990; Chandler, 1990; Dybas, 1990; Newton, 1990; Peck, 1990; Tashiro, 1990). An aspirator may be helpful when collecting in grass clumps, bark, and crevices. Species living in ponds and streams can be captured by hand or with an aquatic net. Do not overlook beetles living along the shore in sand and under seaweed or other debris. Some beetles may be captured with an aerial net, but more often beetles can be taken with sweep nets. Sweeping or beating vegetation will gather leaf-feeding and some predaceous beetles. Many species of beetles reside near the base of vegetation. A net of strong construction whose hoop includes a straight surface may be helpful in dislodging specimens from such habitats (Belkin, 1962). A tool for getting under bark or into wood is necessary for collecting species that bore into logs or plant stems.

Immature beetles (grubs or larvae) should be killed by placing them in boiling water for 1–5 minutes, depending on the size of the specimens. They can then be preserved in 70%–80% alcohol. The larvae of some beetles are fatty and should be placed for 20–30 minutes in 30% alcohol with a few drops of acetic acid that has been heated to 70 °C. Larvae can be critical point dried and stored in vials or gelatin capsules or mounted on cards.

Adult beetles may be killed and preserved in alcohol, but we recommend that they be killed in alcohol or in a killing jar and then mounted on pins. Most beetles should be pinned with a no. 1 pin; large beetles should be pinned with a no. 2 or 3 pin. Some pinned specimens may become greasy with time. These specimens can be treated with carbon tetrachloride, xylene, or a similar solvent. The specimens should not be removed from the pin while undergoing treatment, but the labels should be removed from the pin. Specimens of most beetles should have the genitalia removed and prepared (see Section 3.2.8.7) after removal from alcohol before the specimen is pinned. Specimens smaller than 5 mm should be glued to points or cards, using care not to conceal characters on the ventral surface of the body. It is generally inadvisable to use minutens or to spread the wings of beetles. Very thin or very flat cleared beetles may be mounted on microscope slides.

Many Coleoptera are easily reared; this is often desirable because adults may then be associated with immature stages. Mealworms and other larvae that infest stored grain and flour are particular easy to rear and can be used to feed predaceous insects held in captivity.

6.23 **Strepsiptera [twisted-winged parasites, Fig. 6.23]**

The order name is derived from Greek (strepsis = turning; pteron = wing) and refers to the twisted wing condition found in males. The Strepsiptera form a widespread order of about 300 species. Strepsipterans are sometimes grouped with the Coleoptera near the Rhipiphoridae. Adults exhibit strong sexual dimorphism. Females of most Strepsiptera lack eyes, antennae, and legs; the head and thorax are fused, and the general body shape is larviform. Females of some free-living species (some Mengeidae) display compound eyes, a developed head, antennae, and chewing mouthparts. All males resemble a generalized insect, the forewings are reduced (elytraform), and the hind wings are enlarged and membranous with radial venation. Strepsiptera are morphologically unique among insects in lacking a trochanter in the adult leg.

The immature stages of Strepsiptera are parasitic on Thysanura, Blattodea, Mantodea, Orthoptera, Hemiptera, Diptera, and Hymenoptera. Postembryonic development of parasitic species is hypermetamorphic: The first instar larva is called a *triungulinid*. The free-living first instar resembles the triungulin of Coleoptera but differs in lacking a trochanter in all legs, and the antennae and mandibles are not well formed. The triungulinid is very active and searches for a host. After a host is found, the triungulinid enters the host's body and then molts into a legless, worm-like form that feeds and pupates within the integument of the host. The adult male is winged and leaves the host, but the female remains in the host with its body protruding from between the host's abdominal segments. After producing large numbers of the tiny triungulinids, the female dies.

Strepsiptera males may sometimes be found on flowers, apparently searching for females in their hosts. A few species are found under rocks. However, the best way to collect strepsiterans is to capture and rear parasitized hosts (bees, wasps, flies, and bugs). Hosts may be recognized by the small sac-like females protruding from the often-distorted abdomens of their hosts. Males of some species are attracted to light. Males of some Stylopidae are attracted to hosts parasitized by females. These males can be collected by placing a parasitized (stylopized) host in a screened cage (MacSwain, 1949). Specimens should be preserved in 70%–80% alcohol or mounted on microscope slides; males may be mounted on minutens.

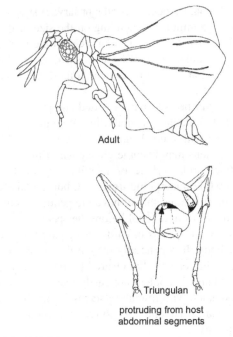

Adult

Triungulan

protruding from host
abdominal segments

FIGURE 6.23

Strepsiptera.

6.24 Mecoptera [scorpionflies, hangingflies, Fig. 6.24]

The order name is derived from Greek (meco = long; ptera = wing). Some Mecoptera are called *scorpionflies* because the upturned genitalia of some males resemble the tail of a scorpion. The name *hangingflies* pertains to one family (the Bittacidae), which resembles craneflies and hang suspended from vegetation by their front and middle legs. Mecoptera are moderate-sized (25 mm long), slender-bodied insects with relatively long wings. The head is prolonged into a rostrum beneath compound eyes, and the mouthparts are mandibulate and positioned at the end of the rostrum. Most Mecoptera are

Scorpionfly

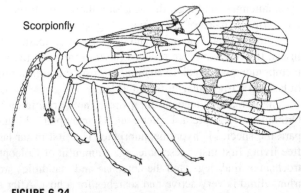

FIGURE 6.24

Mecoptera.

winged; some are brachypterous and a few are apterous. Winged species have four wings similar in size, shape, and venation; the wings are not folded when in repose. The metathorax is fused with the first abdominal tergum, but the first sternum is free. The abdomen is 11 segmented; cerci are two segmented in female and one segmented in male (or rarely absent). Metamorphosis is complete in Mecoptera. The larvae of some species possess compound eyes; the pupa is decticous and exarate.

Scorpionfly adults and larvae feed primarily on living and dead insects, but some adults are attracted to nectar and fermenting fruit. Some species apparently feed on moss. Males are predaceous, but females do not take living prey. Larvae of scorpionflies resemble the larvae of sawflies and some Lepidoptera but differ from the latter by the absence of crochets (tiny hooks) on the prolegs and from the former by having seven or more ocelli.

The larvae of most scorpionflies are found in the soil or in leaf litter and can be collected in a sifter. Most adults are found in heavily wooded areas and can be collected by sweeping or beating the vegetation. Adults are not strong fliers and are relatively easy to net. Adults of the so-called snow scorpionflies of the family Boreidae emerge during winter and often can be picked from the surface of snow with forceps. Members of the Panorpidae (common scorpionflies) should be held until the gut is voided before placing them in a killing jar (Beirne, 1955). Some Mecoptera are collected in light traps, but most species are active only during the day, and flight traps such as the Malaise trap are usually more effective.

Larvae and adults can be killed and preserved in 70%−80% alcohol, but critical point dried specimens are superior. Adults should be mounted on pins or card points. The wings of the adults may be spread if desired, but it is customary to leave them in a natural position.

6.25 Neuroptera [alderflies, antlions, dobsonflies, fishflies, lacewings, owlflies, snakeflies, Fig. 6.25]

The order name is derived from Greek (neuro = nerve; pteron = wing) and refers to the reticulate nature of the wing venation. The common names are generally applied to specific families or groups of Neuroptera based on their appearance or habits. The Neuroptera are a relatively small, cosmopolitan order of endopterygote, neopterous insects that are best represented in tropical regions. Specimens vary in size from small to very large, with a wingspan of more than 100 mm. Neuroptera are soft-bodied; the compound eyes are well developed, but ocelli are usually absent. The antennae are long, multisegmented, and sometimes display a club-shaped enlargement at the apex. Mouthparts are of the mandibulate biting type. The prothorax is movable, and the entire thorax is loosely organized. Neuroptera usually display four large, membranous wings that are subequal in size; venation is abundant and net like. The wing-coupling mechanism is simple; when the insect is at rest, the wings are held roof like over the back. The abdomen is 10 segmented (except in Chrysopidae); cerci are absent. The immatures do not resemble the adults, and pupation usually occurs inside a silken cocoon. The pupa is decticous and exarate; pupal mandibles are well developed.

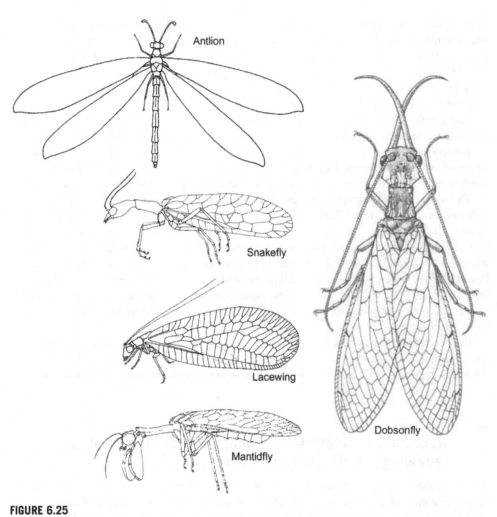

FIGURE 6.25

(A and B) Neuroptera.

Classification of Neuroptera varies among entomologists. Here we recognize three suborders and 14 families. Neuroptera are here considered in the broadest sense to include the dobsonflies (Megaloptera) and snakeflies (Rhaphidioidea). Neuropterans such as dobsonflies and fishflies (Corydalidae), alderflies (Sialidae), brown lacewings (Hemoerobiidae), green lacewings (Chrysopidae), and antlions (Myrmeleontidae) are frequently collected.

Neuroptera are predaceous as larvae and adults. As a consequence, they are highly beneficial to human agricultural efforts. In fact, green lacewings are sold commercially for the control of various pests. Many Neuroptera will bite if handled incautiously when collected. The larvae of several members of the order are aquatic and commonly concealed under stones in streams. These larvae can be collected by hand, in an aquatic net, or with a dipper. Adults of aquatic immatures usually remain on vegetation near water. They are relatively poor fliers and usually can be collected directly into killing jars. Most immatures and adults of terrestrial forms are found on vegetation and can be collected in a sweep net. Many adults of terrestrial and aquatic immatures are attracted to light. Immature myrmeleontids (called antlions) are found partially buried at the bottom of small pits that they dig in sand or dust to trap ants or other insects. These larvae can be collected by blowing strongly on the pit or captured by scooping out the bottom of the pit with a spoon. The rarely collected mantispid larvae are parasitic on spider eggs.

Adult Neuroptera may be killed with cyanide or ethyl acetate. Specimens may be preserved in alcohol, but we recommend mounting the specimens on pins or points. The wings may be spread or left folded over the body in repose. Green specimens lose their color when exposed to light for prolonged periods of time. Immature specimens of medium to small size should be killed and preserved in 70%−80% alcohol; large-bodied immature specimens should be killed in boiling water and then transferred to alcohol. Larvae can be critical point dried and mounted on cards or preserved in vials.

6.26 Trichoptera [caddisflies, Fig. 6.26]

The order name is derived from Greek (thrix = hair; pteron = wing) and refers to the hairy appearance of the wing. Caddisflies are soft-bodied insects with long, slender antennae and two pairs of membranous wings. The wings are clothed with setae and held roof like over the back when the insect is at rest. The larval mouthparts are adapted for chewing; the adult mouthparts are adapted for feeding on liquids. About 5000 species in 20 families are known for the order.

Trichoptera larvae are aquatic and display a pair of hook-like appendages at the apex of the abdomen. Many larvae live in characteristic cases constructed of pebbles, sand grains, twigs, or other materials found in ponds and streams. The abdominal hooks are used to drag the case about as the

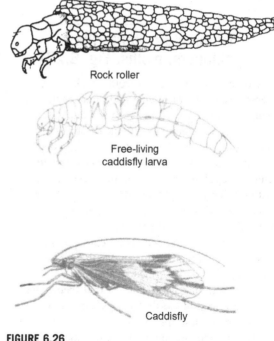

Rock roller

Free-living caddisfly larva

Caddisfly

FIGURE 6.26

(A and B) Trichoptera.

larva feeds. When the larva is ready to pupate, the animal attaches the case to a rock or other fixed object in the water. Case-making larvae are mostly detritus or plant feeders. Some caddisfly larvae do not make cases but instead spin silken webs that are used to capture food drifting in the stream. A few species do not build cases or webs but are free-living and predaceous. When collecting immature Trichoptera, preserve the ease in alcohol along with the larva, pupa, or both.

Adult caddisflies are weak, usually crepuseular dawn or dusk fliers. They are generally found during the day resting on vegetation, bridges, or other objects near ponds and streams. Specimens can be collected with a sweep net or captured directly with a killing jar. Large numbers of adults are often attracted to light, particularly if the light is near water. Some collectors report that blue light is more attractive than other colors (Borror et al., 1989). Larvae and pupae collected from the habitat are valuable because when reared to adult, the immature and adult stages can he associated.

Adults killed and preserved in 70%−80% alcohol are satisfactory, but specimens mounted on pins are preferred. Soft-bodied species tend to shrivel when dried. Critical point drying leaves soft-bodied specimens in life-like condition. The wings of pinned specimens can be spread or left in repose. The abdomen should be bent forward on specimens whose wings are not spread. This permits examination of genitalic characters. Some species can be identified by genitalic characters visible in the late pupal stage. Adults too small for direct pinning should be mounted on minutens.

6.27 **Lepidoptera [butterflies, skippers, moths. Fig. 6.27]**

The order name is derived from Greek (lepis = scale: ptera = wing and refers to the wings usually covered with scales flattened setae). The common name *butterfly* applies to most of the diurnal Lepidoptera with clubbed antennae; the common name *skipper* refers to diurnal Lepidoptera of the family Hesperiidae, with fast, erratic flight and whose antennae are hooked at the apex; and the common name *moth* is applied to nocturnal Lepidoptera whose antennae are not clubbed. The Lepidoptera are a cosmopolitan order of Holometabola consisting of about 150,000 described species placed in about 80 families. More than 11,000 species have been described in North America, but many species await description.

Lepidoptera are among the most popular insects collected by amateur entomologists.

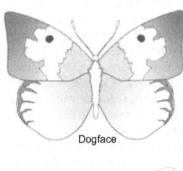

Dogface

Hairstreak

FIGURE 6.27

(A−C) Lepidoptera.

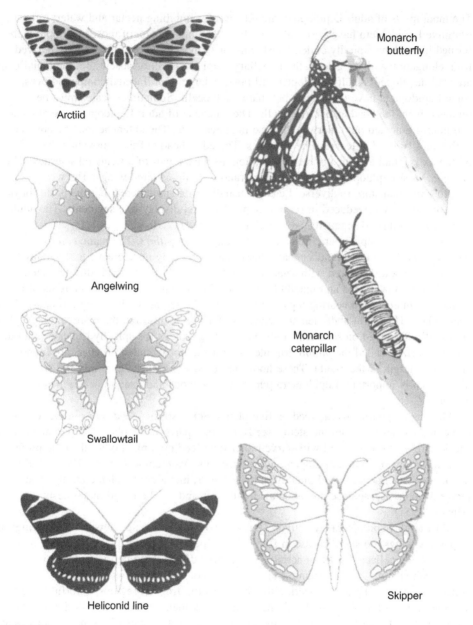

Arctiid

Angelwing

Swallowtail

Heliconid line

Monarch butterfly

Monarch caterpillar

Skipper

FIGURE 6.27

(Continued).

The mouthparts of adult Lepidoptera are adapted for imbibing nectar and water; a few primitive Lepidoptera have vestigial mandibles. The mouthparts of most Lepidoptera are formed into a long, spirally coiled haustellum or proboscis. The coil is actually derived from elongate maxillary galeae; the maxillary palpi are small or absent, but the labial palpi are well developed. Adult Lepidoptera all possess large, multifaceted compound eyes; some Lepidoptera have two ocelli, but many lack ocelli. Sometimes scales must be removed from the head to find the ocelli. The antennae of adult Lepidoptera are elongate and multisegmented but otherwise diverse in appearance. The antennae may be covered with scales or display feather-like branches. The adult head is freely movable; the prothorax is small and membranous and often displays a pair of articulated selerites called *patagia.* Most Lepidoptera have four membranous, scale-covered wings: The wing-coupling mechanisms are diverse. Legs are usually long and tapered, but the posterior or anterior pair may be reduced in some groups: The tarsi display five segments. The adult abdomen contains 10 segments, and cerci are not present.

The larvae of Lepidoptera are commonly called *caterpillars.* The name *caterpillar* is derived from Latin and literally means "hairy cat," probably in reference to the many caterpillars that are covered with setae. Caterpillars usually have a well-differentiated, strongly selerotized head and mandibles adapted for chewing. The thorax consists of three thoracic segments each bearing a pair of legs with five segments: Each leg terminates in a single claw. The larval abdomen also consists of 10 segments, and there are "prolegs" that are usually displayed on segments 3−6 and 10. The prolegs are apically truncate, with the truncate surface called the *planta.* Minute, selerotized hooks called *crochets* are found along the margin of the planta. These hooks are used as holdfast structures and are diagnostically important. Lepidoptera pupae are decticous and exarate or adecticous and obtect.

Most lepidopterous larvae feed on live plants. Some species feed on or in the leaves, and other species bore into the stems, seeds, or other parts of the plant. The larvae of many species are host specific and will starve rather than feed on a plant other than the preferred species. Larvae of some Lepidoptera are predaecous (Montgomery, 1982, 1983). A few larvae are scavengers of dead plant or animal matter, including woolen clothing. The larvae of many Lycaenidae are associated with ants and can be found in ant nests (Hinton, 1951).

Most adult Lepidoptera fly actively, but a few (such as female bagworms) are wingless or possess wings too short for use in flight. Most species of butterflies are diurnal, and they are often collected with an aerial net or captured as they come to rest on flowers, leaves, bait stations, or the ground. Many brightly colored butterflies are found congregating around puddles. During the dry season in tropical regions, colorful butterflies are often found on moisture-laden domestic livestock manure. Collecting during the early morning and late afternoon is not highly productive. Most species of diurnal butterflies are taken from midmorning through midafternoon. Warm, sunny days are best for collecting butterflies, particularly when the wind is not strong.

Adult moths are predominantly nocturnal, and many species may be taken during the first hours after sunset. However, a few moth species are active during the day, and some nocturnal moths may be flushed from their resting places and netted during the day. The most productive and efficient method for collecting moths involves catching or trapping them when they are attracted to light. Bait collecting also is a standard technique for some groups (see Section 1.10).

Lepidoptera are easily reared by amateur entomologists, and the commercial breeding and hybridization of some butterfly and moth species have become widespread. Special equipments or skills are not necessary to rear most foliage-feeding species from the egg stage to adult. Many adult female moths will lay eggs when confined in a small plastic box, a glass jar, or other container with a piece of the food plant or some crumpled paper. For large moths, a paper bag or shoebox can be used as a rearing container. Keep a piece of damp paper towel or sponge in the container to maintain high humidity, especially if the specimens are reared in an air-conditioned or heated building. A water-soaked raisin will provide nourishment for the female and moisture for species that feed as adults.

The rearing of butterflies may sometimes be difficult. Most butterflies require large, brightly illuminated cages containing an ample supply of the natural food to induce them to oviposit. Females caught outdoors will nearly always be fertile, and their eggs will hatch in a week or 10 days unless the species is one that overwinters in the egg stage. If the food plant is unknown, a trial-and-error procedure must be used by offering the newly hatched larvae bits of leaves from a variety of plants that may be possible hosts.

Rearing cages made of screening are appropriate for large moths and most butterflies. Caterpillars require little air, and most species may be reared successfully in closed plastic or metal containers. Excessive moisture is not a problem if the containers are kept at a reasonably constant temperature, and the bottom is lined with a clean paper towel or blotting paper. Only dry foliage should be used as food, and tight containers keep the leaves fresh for several days. Caterpillars need only the water they get by eating fresh leaves. Cleanliness is essential. Uneaten leaves and the accumulation of frass at the bottom of the container must be removed at regular often daily intervals. When mature, the larvae should be provided with suitable conditions for pupation. For example, species that pupate in the ground need several inches of slightly dampened peat, sand, or sawdust. Species that pupate in leaf litter often will do well if allowed to crawl into the folds of a crumpled paper towel. The larvae of some moths will pupate in the corrugations of cardboard.

Preserving the larval instars is often desirable so that they may be identified by positive association with adults reared from the same brood. Larvae should be killed by immersing them for about a minute in boiling water. Larval specimens may be stored in glycerin. Traditionally, they were preserved in 70%−80% alcohol. Large specimens in alcohol should have the alcohol changed after the first 24 hours because the body fluids would have diluted it. As an alternative to alcoholic preservation, larvae may be critical point dried, freeze dried, or inflated and then mounted on pins.

Avoiding the loss of wing and body scales is a difficult problem in collecting adult Lepidoptera. Killing the specimen quickly is essential for preserving scales on the body and wings. Rapidly acting killing agents such as cyanide are recommended but dangerous. Ethyl acetate is not as desirable as a killing agent for Lepidoptera because it is volatile and killing jars must be constantly recharged to work effectively. Some collectors pinch the thorax of the specimen while it is in the net. This is accomplished by holding the thorax of the specimen between the thumb and the index finger and applying just enough pressure to stun the specimen. In this way, the specimen does not flutter its wings or struggle when placed inside the killing jar. The jar should be clean, and absorbent paper should be placed in the bottom. Excessive moisture affects the scales. For large specimens, a widemouthed killing jar is recommended. Do not place large beetles or other large insects in the same jar with butterflies and moths, and do not crowd too many Lepidoptera into one jar. Some collectors prefer to remove the specimens soon after they have been killed and place each specimen temporarily in a separate, labeled envelope; this procedure is especially useful for butterflies. Such papered specimens can later be relaxed and their wings spread. However, it is always advisable to pin the specimens as soon as possible after killing them to minimize damage.

For permanent collections, adult Lepidoptera should be pinned with the wings spread in the standard manner (see Section 3.2.7.2) so that all important features of wing pattern and structure may be seen. Never glue adults to points or place them in alcohol. Pin small microlepidoptera with minutens, spread the wings, and then double mount. Papered specimens or specimens left for long periods in killing jars must be placed in a relaxing chamber before the wings can be spread. Even fresh specimens are easier to handle if placed overnight in a relaxing chamber (see Section 3.2.1).

6.28 Diptera ["true" flies, mosquitoes, Fig. 6.28]

The order name is derived from Greek (dis = twice; pteron = wing) and alludes to the presence of two membranous wings used in flight. Perhaps more than any other order of insects, families of Diptera are called by common names such as mosquitoes, gnats, midges, punkies, no-see-ums, horse flies, bat flies, snipe flies, robber flies, houseflies, bottle flies, and so forth. This is no doubt a consequence of the long-term intimate relationship between flies and humans. Incidentally, by convention we separate the word "fly" from its descriptor when referring to dipterous insects. We combine the word "fly" with its descriptor when referring to nondipterous insects whose names contain that word (e.g., scorpionflies, sawflies). Names notwithstanding the Diptera constitute a cosmopolitan order of holometabolous insects that consists of about 150,000 described species. The order contains about 120 families, and in North America over 18,000 species have been described. Flies are found in virtually all habitats of all zoogeographical realms; they are omnipresent.

Adult Diptera are relatively soft-bodied and recognized by a typically large, manipulable head with large and multifaceted compound eyes. Most flies have three ocelli, some species have two ocelli, and a few species are anocellate. The antennae of flies are highly variable and range from moniliform and multisegmented to three segmented and aristate. The mouthparts of Diptera are adapted for sucking and form a proboscis or rostrum. A piercing-type proboscis is found in predatory and bloodsucking species; a sponging-type proboscis is found in houseflies; a few primitive flies have nonfunctional mandibles. The prothorax and metathorax of Diptera are reduced; the mesothoracic wings are membranous and used in flight; the metathoracic wings have been modified into

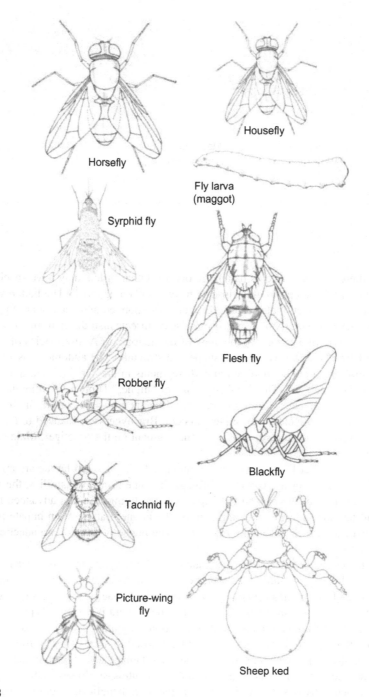

Horsefly

Housefly

Fly larva (maggot)

Syrphid fly

Flesh fly

Robber fly

Blackfly

Tachnid fly

Picture-wing fly

Sheep ked

FIGURE 6.28

(A and B) Diptera.

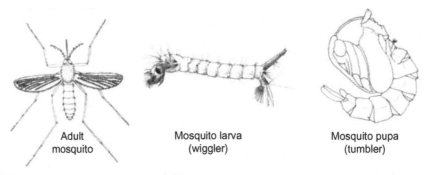

Adult
mosquito

Mosquito larva
(wiggler)

Mosquito pupa
(tumbler)

FIGURE 6.28

(Continued).

club-like halteres that are used as balancing organs. Other orders may have species with
only one pair of wings, but no other insects have knobbed halteres. (The halteres or
hamulohalteres of winged male scale insects are not knobbed and usually are tipped with
one or more hooked setae. Most winged male scale insects also differ from flies in having
a single, long, style-like process at the apex of the abdomen.) A few species of flies lack
wings. The legs of flies are variable in shape and structure. The abdomen has 11 segments
in the ancestral or primitive condition and 10 segments in the derived condition (10 and 11
fuse to form the proctiger). The distal segments of "higher" Diptera are often modified into
a telescopic "postabdomen." Cerci are present at the apex of segment 10; the female cerci
are primitively two segmented (most Nematocera, Brachycera) or reduced to one segment
(throughout Diptera); male cerci consist of one segment on the proctiger. Male genitalia
are complex.

   The larvae of higher Diptera are frequently called *maggots*. All larvae are apodous, but
some species have one or more pairs of prolegs. In some primitive families, the larvae
possess a distinct head capsule and the pupae are free-living. In more advanced flies, as in
the muscoid families, a head is not apparent on the larvae. Larvae often pupate inside the
last larval exuviae, which is called a *puparium*. The pupae of Diptera are adecticous and
exarate or obtect.

   The biological importance of Diptera cannot be overemphasized: Bloodsucking species
transfer many diseases, including malaria; houseflies transfer enteric diseases, and the
nuisance value of biting flies cannot be measured. The larvae of some species cause
myiasis. In terms of agriculture, the impact of some Diptera is notable. Tephritid fruit flies
are serious pests of fruit. Quarantine and eradication programs in some areas cost tens of
millions of dollars yearly in response to the threat posed to agriculture by fruit flies. Leaf-
mining and gall-forming Diptera cause conspicuous damage to foliage and other plant
parts, but their economic impact is questionable. In contrast to these negative impacts on
humans and their activities, many species of Diptera are beneficial as pollinators; other
species are predators or parasites of insect pests. Most species of flies probably are neither

beneficial nor harmful to humans, but often are collected and studied because of the unusual ecological habitats they occupy.

Habitats of Diptera are numerous and varied, but most dipterous larvae live in aquatic or semiaquatic habitats. Perhaps the most unusual habitats involve the seeps of crude petroleum in which larvae of some ephydrid are found (shore flies develop on the hot springs in which some stratiomyid larvae are found). A rain-filled can by the side of the road may harbor mosquito larvae (Culicidae), and stones in a fast-flowing stream may hold black fly larvae (Simuliidae). A dipper or small aquatic net is useful in collecting such larvae. A sweep net or aerial net can be used effectively to collect the adults, which generally remain near the water. Marsh fly adults (Sciomyzidae) are slow flying and rest on emergent vegetation along the margins of freshwater. They often perch with their head directed downward. Marsh fly larvae are obligate feeders of freshwater snails, fingernail clams, land snails, and slugs.

Some Diptera larvae live in plant tissues, where they mine leaves, form galls, or feed on the stems or roots. By placing the host plant, or infested parts of the host plant, in a rearing container, the larvae can be kept until they pupate and the adult flies emerge. Similarly, adult tachinids and other parasitic flies can be obtained by keeping the parasitized host insect in a rearing container. Some species pupate in the soil. To rear these species, place a few centimeters of moist sand or moss in the bottom of the cage into which the mature larvae can burrow for pupation.

Adult flies can be swept from vegetation and collected into a killing jar. The net should remain dry, and the fabric of the net should be made of a lightweight, tight-weave material. Large specimens can be collected into a killing jar with a minimum handling. Small specimens can be collected from the net with an aspirator and then transferred to a killing jar. Alternatively, a small wad of cotton laden with ethyl acetate can be blown down the collecting tube of the aspirator into the bottle. Care should be exercised because the setal and scale patterns may be important in identification. Flies that are scavengers on decaying animal and vegetable matter may be collected in bait traps. Chemical baits that attract some specific groups of flies are available commercially. Light traps or flight traps such as the Malaise trap (see Section 1.6.3), used with or without a bait, usually are extremely successful in capturing adult flies. Some flies are ectoparasites of bats, birds, and other animals. Many of these parasitic flies are wingless and can be collected readily from the host with forceps or an aspirator.

Most flies should be killed dry and mounted. Large specimens should be mounted on pins; small specimens should be mounted on minutens or card points. Mounting should be accomplished within a few hours after the specimens have been killed. If not, the specimens can be held in the freezer compartment of a refrigerator until they can be mounted. Remember that newly emerged adults may be teneral and should be held alive until the wings and colors fully develop. Some species may require more than a day to develop. Small flies, if not mounted on minutens or points very soon after collecting, can be placed in alcohol and run through the Cellosolve-xylene series (see Section 3.2.3) or critical point dried (see Section 3.2.7.6). Flies should not be kept dry layered in boxes or envelopes because their heads, antennae, and legs are easily detached. Dipterous larvae should be killed in boiling water and preserved in alcohol. For permanent collections, mosquito larvae preferably are mounted on microscope slides. Some adult flies can also be preserved satisfactorily in alcohol, but most should never be placed in alcohol. If adults of many families are immersed in liquid, the scales, setae, or bristles critical for identification may detach and the specimens will become useless. The larval exuviae and puparium of reared specimens can be preserved in alcohol or glycerin or mounted with the adult.

6.29 **Siphonaptera [fleas, Fig. 6.29]**

The order name is derived from Greek (siphon = tube; aptera = wingless) and refers to fleas as wingless siphons. This name seems appropriate for reasons explained later. Fleas represent an order of about 1400 species of holometabolous, endopterygote Neoptera divided into about 20 families. Adults may be recognized by the body, which is wingless and typically less than 6 mm long. Fleas are laterally compressed and display bristles and cuticular projections that are directed rearward. These structures are sometimes modified into combs called *ctenidia* that may be found on the gena, prothorax, or metathorax. The compound eyes are absent or replaced by one large lateral ocellus on either side of the head. Mandibles are not present, and the mouthparts are of the piercing−sucking type. The piercing elements are derived from the sclerotized epipharnyx and the laciniae of the maxilla: The maxillary and labial palpi are well developed in adult fleas. The antennae are short, typically with three segments. The antennae are held in depressions called fossae on the side of the head. All fleas are apterous. The legs are spinose, with large coxae, and the hind legs are adapted for jumping. The abdomen contains 10 segments, with tergum I reduced and sternum I absent. A complex sensory apparatus called the *sensillum* is positioned at the apex of the abdomen. Larval fleas display 13 body segments, lack appendages, and show a worm-like appearance. The head capsule is well sclerotized but lacks eyes. The mandibles are brush like, and the antennae are one segmented. The terminal body segment displays anal struts. The larva undergoes three instars before constructing a pupal cocoon. The pupa is adecticous and exarate; pupal wing buds are visible in some species, which suggests that fleas had winged ancestors.

Adults of all flea species are external parasites of mammals and birds. Fleas such as the eat flea. *Ctenocephalides felis* (Bouché), and the dog flea. *C. canis* (Curtis), are well known as household pests. However, fleas are most important to humans because some species can transmit bubonic or sylvatic plague, endemic typhus, and other serious diseases. Some species serve as intermediate hosts of tapeworms.

Adult fleas reflect many features that are adaptations for a parasitic existence on the bodies of their hosts. Adult fleas have lost the need for wings because they live in the fur and feathers of their hosts. Moving around on the body of a host with fur and feathers is difficult, so the flea's body

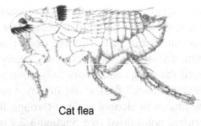

Cat flea

FIGURE 6.29

Siphonaptera.

has become greatly compressed laterally to facilitate movement between hairs and feathers. To protect the flea from damage, its body has become heavily sclerotized. The flea is armed with combs or bristles that engage fur and resist dislodgment from the host during grooming. The antennae are short and fitted into grooves for protection from damage. The piercing–sucking type mouthparts are an adaptation for feeding on warm-blooded vertebrates.

Flea larvae live in the nests of their hosts. The larvae do not suck blood because their mouthparts are not adapted for that kind of feeding. Instead, flea larvae feed on organic debris (including dried blood) within the nest or habitation of the host. Fleas commonly infest animals that live in nests or burrows and seldom are found on cattle, deer, or other hoofed animals. Rodents and other small mammals are frequently heavily infested with fleas, which can be collected by shooting or trapping the host and immediately picking or brushing the fleas from the fur. Because fleas are active, place the host in a jar or bag with a few drops of liquid-killing agent to stun or kill the fleas before examining the fur. Placing the host animal in a refrigerator for an hour will also slow the activity of the fleas. If immediate examination of the host is not possible, the animal should be placed in a bag (each host animal in a separate bag) to confine the fleas, which tend to leave a dead host. We must emphasize that bags and other containers must be scrupulously cleaned between collections. Otherwise, erroneous collection data can be included with collections. In many ways, erroneous collection data are worse than no collection data.

Birds also have fleas, but the fleas that attack birds are usually collected most readily by examining a nest soon after it has been abandoned. In every instance, the species of bird that made the nest should be noted on the data label. The nest can be teased apart or placed in a Berlese funnel or sifter to collect the adult fleas and immature stages. Because the immatures cannot be identified with certainty without associated adults, some of the larvae or pupae should be placed in jars with a quantity of the nest material and reared. For successful rearing, maintain the humidity as high as would normally be found in a nest. The burrows or nests of mammals should also yield many adult and immature fleas.

Siphonaptera adults and larvae may be killed and preserved satisfactorily in 70%–80% alcohol. They should not be preserved in a formalin-based liquid fixative because these fixatives make the process of clearing specimens difficult. Permanent collections display adult fleas mounted on microscope slides. Be sure to include the name of the host on the slide label.

6.30 Hymenoptera [sawflies, ants, wasps, bees, Fig. 6.30]

The order name is derived from Greek (hymen = membrane; pteron = wing) and refers to the four membranous wings. The Hymenoptera are a cosmopolitan order with about 125,000 described species. The Hymenoptera are subdivided into two suborders, the Symphyta (sawflies) and Apocrita (bees, wasps, ants, and parasitic Hymenoptera). The adult head is mobile, usually hypognathous but sometimes prognathous; the head is not fused with the pronotum.

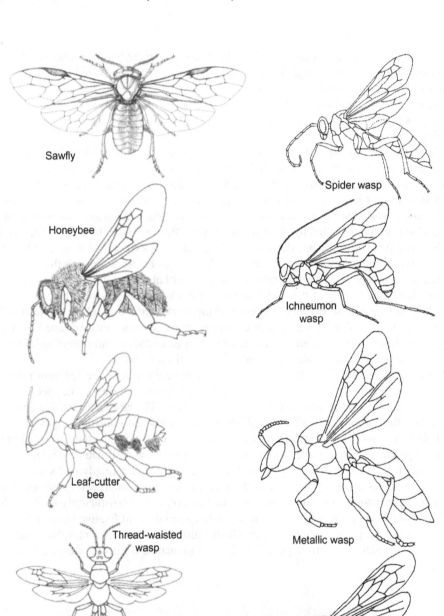

Sawfly

Spider wasp

Honeybee

Ichneumon wasp

Leaf-cutter bee

Metallic wasp

Thread-waisted wasp

Tephiid wasp

FIGURE 6.30

(A–C) Hymenoptera.

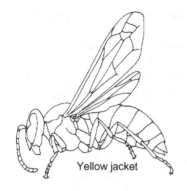

Yellow jacket

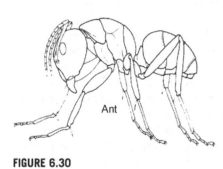

Ant

FIGURE 6.30

(Continued).

The compound eyes typically are large and multifaceted; the eyes are reduced to a few ommatidia in some ants and parasitic species. Three ocelli usually are present, but these are absent in some species. Antennal segmentation and shape are variable, but the antenna typically is *geniculate* (elbowed). The antenna frequently shows sexual dimorphism with the funicle sometimes *ramose* branched in the male and apical segments sometimes differentiated into a club in the female. Mandibles are always present in adult Hymenoptera, and the mouthparts are usually adapted for biting—chewing; some bees show adaptations for chewing—sucking. Most Hymenoptera have four membranous wings, with the forewing usually substantially larger than the hind wing; the forewing and hind wing are connected during flight by hooks called *hamuli*. The forewing venation is complex and extensive in large-bodied species: It is drastically reduced in small—to-minute parasitic species. The hind wing venation forms a few cells at most; otherwise the venation is reduced. Apterous species are distributed throughout the Hymenoptera. Reproductive ants develop wings but shed their wings after nuptial flight: The worker and soldier castes of ants are wingless.

Among the Apocrita, the first abdominal segment is incorporated into the thoracic region and called a *propodeum*. The propodeum and thorax collectively are called the *mesosoma* or alitrunk (in ants). The propodeum is usually separated from the remainder of the abdomen by a constriction of the second abdominal segment called the *petiole* or segments two and three called the *postpetiole* in ants. The abdomen behind the petiole is called the *metasoma* or *gaster* (in ants). The abdomen of Hymenoptera has 10 segments in the ancestral condition; the first segment (propodeum) and second segment petiole typically lack the sternal elements. Female Hymenoptera display a selerotized, elongate, tubular ovipositor (sawflies, parasitic Hymenoptera) or sting (*aculeata*); the cerei are one segmented or modified into two setose patches (*pygostyli*). The male genitalia are complex and sometimes diagnostically important.

Hymenoptera larvae are extremely disparate in terms of their anatomy. The larvae of Symphyta are caterpillar like with thoracic legs and prolegs. The prolegs lack crochets, a character that may be used to separate these larvae from the larvae of Lepidoptera. The larvae of Apocrita are apodous or vermiform. Some species display hypermetamorphic development. That is, the first instar Ian does not resemble subsequent instars. The number of instars among Hymenoptera is variable. Some egg parasites may have only two instars:

Some sawflies may have seven instars. The pupae of Hymenoptera are adecticous and usually exarate, but some species display an obtect condition.

The biology of Hymenoptera is exceedingly diverse. Species are phytophagous, parasitic, predaceous, and gall forming. Among the apocritous larvae, the gut is not connected to the anus until feeding has been completed. Polyembryonic development has been shown in some species, and eusociality has evolved several times among the bees and ants. Parthenogenesis is universal within the order; females possess the diploid ($2\,n$) complement of chromosomes and males possess the haploid (n) complement of chromosomes.

The Hymenoptera, as is the case with the Diptera, have beneficial and noxious species. Hymenoptera are highly beneficial in terms of crop pollination. Many bees are important pollinators of plants, and some are valued for the honey they produce. The parasitic species are of ever-increasing importance as biological control agents of agricultural pests. The plant-feeding forms include some of the most destructive defoliators of forest trees.

Methods of collecting Hymenoptera vary because their habits and habitats are so diverse. Sawfly adults, the so-called because of their saw-like ovipositor, may be captured by interception traps such as the Malaise trap or swept from low vegetation. Most sawfly larvae are external feeders on vegetation, some species mine the leaves or leaf petioles, and a few species cause galls. Outbreaks of sawflies may occur sporadically in an area and offer collectors an excellent opportunity to obtain large samples. From such collections, some specimens can be preserved as immatures and others can be held in rearing cages until adults emerge. Procedures used to rear Lepidoptera (see Section 1.15) can be used to rear sawflies. Sawfly larvae may be preserved in alcohol, inflated, or freeze dried. Many sawflies attach their cocoons to leaves or twigs on the host plant; larvae of some species drop to the ground and pupate in the soil or leaf litter. For this reason, 5−8 cm of moist soil or peat moss should be kept on the bottom of the rearing cages. The adults that emerge may be killed in alcohol but later should be removed and pinned, preferably with the wings extended.

A simple but novel method of collecting paper wasps is to tie a wide-mouth jar containing alcohol to a long pole. The pole is used to position the jar immediately beneath the nest opening. As wasps approach the opening, they become intoxicated and fall into the alcohol (Bentley, 1992). Care must be taken not to touch or bump the nest.

The parasitic Hymenoptera represent a significant portion of the Hymenoptera, yet they are poorly understood. More than 50,000 species have been described, but several times that number remain undescribed. These insects are multitudinous and attack all stages of other insects. Many species of parasitic wasps can be found in Malaise traps. Smaller species associated with soil or duff can be recovered from Berlese samples. Some species are myrmecophilous and are found in ant nests. Some species have only been taken with a sweep net. Collectors can make a valuable contribution to entomology by correctly associating parasites with their hosts. This necessitates rearing adult parasitic Hymenoptera from their hosts. The value of accurate host associations cannot be overemphasized. Policy of the Biosystematics and Beneficial Insects Institute precludes accepting parasitic Hymenoptera for identification unless the host insect has been identified at least to genus. Rearing parasites from their hosts is also a method of associating the sexes with certainty because males and females of many species differ greatly in appearance. The collection labels should include the scientific name of the host insect and the life stage of the host from which the parasite emerged. If the specimens are determined to be polyembryonic (more than one individual produced from one egg), this also should be noted on the label. When possible, the remains of the host should be retained and mounted below the parasite on the pin. Parasitic wasps too small for direct pinning should be glued to points or rectangular cards. Microhymenoptera should not be mounted on minutens.

Mounting parasitic Hymenoptera involves arranging the antennae, legs, and wings so that characters on the body needed for identification are not obscured. Care must be exercised when mounting large-bodied wasps such as ichneumonids. They possess long wings and antennae that, if pointed upward, may be broken when the pin is handled. Tiny specimens (under 2 mm), such as trichogrammatids, may be cleared and mounted on microscope slides, preserved in alcohol, or critical point dried and mounted on cards. Taxonomists in the Systematic Entomology Laboratory prefer that specimens of the Ichneumonoidea be killed and preserved in 95% alcohol; other parasitic Hymenoptera may be preserved satisfactorily in 70%–80% alcohol.

Gall wasps and the galls from which they emerge should be preserved together because identification of many species is based on the gall. If galls are placed in a pinned collection make sure that they are pinned securely, with additional brace pins on each side. Use care when labeling specimens associated with a particular gall. Not all insects that emerge are gall producers; some may be parasites of the gall wasps or inquilines.

Mound-building ants can be collected by pushing an alcohol-filled vial into the excavated center of the ant mound. The vial serves as a pitfall trap for the ants. If ants are not active outside the nest, a small trowel can be used to disturb the ants. Once the ants are active, avoid getting stung or bitten. Ant nests may also yield other kinds of insects besides ants; many of them mimic ants. In collecting ants from a nest, use an aspirator or Berlese funnel and try to obtain specimens of each caste. It may be necessary to return to a nest periodically through the year to find the males or winged females. Ants may be collected and preserved in alcohol; but for a permanent collection, larger specimens should be pinned, and specimens under 5 mm should be glued to points.

Aculeate or stinging wasps and bees may be caught in Malaise traps or by sweeping. Mesh size and other features of Malaise trap design influence the kinds and numbers of Hymenoptera collected (Darling and Packer, 1988). See Section 1.3.5 on how to remove stinging insects from a net. If specimens are taken from a nest, try to collect the nest. If this is not practical, make a note or sketch of the approximate measurements of the nest. Do not overlook the insects that the wasps may have stored in cells in the nest.

Suggested Reading

Insect orders

General taxonomie:

- Essig (1942, 1958)
- Jacques (1947)
- Peterson (1948)
- Chu (1949)
- Matheson (1951)
- Brues et al. (1954)
- Usinger (1956)
- Borror and White (1970)
- Richards and Davies (1977)
- Hollis (1980)
- Parker (1982)
- Ross et al. (1982)
- Arnett (1985)
- Scott (1986)
- Borror et al. (1989)

Arachnids:

- Rohlf (1957)
- Evans et al. (1964)
- Cooke (1969)

Ticks and mites:

- Lipovsky (1951, 1953)
- Strandtmann and Wharton (1958)
- Furumizo (1975)
- Jeppson and Kiefer (1975)
- Krantz (1978)
- McDaniel (1979)
- Evans (1992)
- Triplehorn and Johnson (2005)

Summary

Insects have been on earth for an estimated 280 million years, mites even longer. Considering both numbers of individuals and numbers of species, insects and mites represent a dominant form of life. Clearly, there are more insects and mites alive than any other group of metazoan animals. In terms of biological diversity, insects occupy more habitats and adopt more lifestyles than any other group of organisms; mites rank second. Insects represent humankind's principal competitors for food and fiber; mites often cause considerable damage to plants and food. Insects and mites both directly and significantly impact the health and welfare of humans and other animals. They are at once both beneficial as scavengers, biological control agents, and environmental indicators and detrimental as humans' most damaging pests. Considering their geological age, numbers, biological diversity, and economic importance, the study of insects and mites is compelling. This study can be personally rewarding through understanding the diversity of nature and professionally rewarding by solving the problems these organisms cause. In a philosophical sense, anyone who collects insects or mites as a pastime, studies their life histories, or uses keys to identify them is a student of entomology or acarology. Every student can make a contribution to the understanding of these fascinating and ubiquitous animals. We hope that this text will provide some of the tools necessary to facilitate such a study.

Liquid preservation formulas

I

Distilled water should be used in these formulas if available, but rainwater or bottled drinking water is satisfactory. "Parts" is by volume.

Alcohol—glycerin—acetic acid solution

Commercial ethanol (ethyl alcohol)	8 parts
Water	5 parts
Glycerin	1 part
Glacial acetic acid	1 part

Barber's fluid

Commercial ethanol (ethyl alcohol)	53 parts
Water	49 parts
Ethyl acetate (acetic ether)	19 parts
Benzene (benzol)	7 parts

Embalming fluid

Toluene (or xylene)	60 mL
Tertiary butyl alcohol (TBA)	25 mL
Ethyl alcohol	15 mL
Phenol	5 g
Naphthalene	20 g
Canada balsam in xylene	10 drops

Note: *Embalming fluid contains many toxicants (toluene, phenol, TBA). The original formula calls for paradichlorobenzene (PDB). This carcinogen has been replaced with naphthalene here. PBD kills and repels insects; naphthalene repels insects. Embalming fluid is introduced into the specimen with a hypodermic syringe inserted between sclerites or into the anus. The fluid should be injected into each body region.*

Essig's aphid fluid

Lactic acid	20 parts
Glacial acetic acid	4 parts
Phenol (saturated H_2O solution)	2 parts
Distilled H_2O	1 part

Hoyer's medium

Chloral hydrate	200 g
Distilled water	50 mL
Gum arabic (granules)	30 g
Glycerin	20 mL

There are many variants of Hoyer's Medium. We cite the original. Hoyer's medium is a temporary mountant that breaks down with time. Thus it should not be considered suitable for mounting valuable specimens or important dissections. Chloral hydrate is a controlled substance in the United States and may be difficult to obtain. To prepare Hoyer's medium, dissolve gum arabic in water at room temperature. Add chloral hydrate and allow the mixture to stand for a day or two until all solids have been dissolved. Add glycerin. Filter through glass wool. Store in a glass-stoppered bottle.

Kerosene—acetic acid solution

Commercial ethanol (ethyl alcohol)	10 parts
Glacial acetic acid	2 parts
Kerosene	1 part

Mix in the order given. For very soft-bodied larvae, use half as much kerosene or less. Formerly, this solution was called kerosene, alcohol, acetic acid, and dioxane mixture (KAAD), and one part of dioxane was included.

Guidelines for mounting small and soft-bodied specimens (Systematic Entomology Laboratory)

Aphids, scale insects, and whiteflies cannot be pinned because of their small size and their tendency to shrivel. The following procedures for mounting certain insects and mites for scientific study are preferred by the Systematic Entomology Laboratory (ARS, USDA). A few of the chemicals indicated by an asterisk (*) in these procedures are hazardous. Carefully investigate their properties to ensure their safe use.

1. Place specimen in 10% potassium hydroxide (KOH)* solution and heat gently until body contents are softened, or leave in KOH solution at room temperature for up to 48 hours.
2. Remove specimen from KOH and place in 70% ethanol for 5 minutes. (Note for aleyrodids: if black specimens have not turned brown at this point, bleach them in peroxide−ammonia solution (one drop ammonia to six drops hydrogen peroxide) until brown. Next place in 95% ethanol for 5 minutes, then proceed to step 7.
3. Remove specimen from 70% ethanol and place in Essig's aphid fluid. Make an incision halfway across the body between the second and third pairs of legs. Then squeeze a few times to remove and flush out body contents. If feasible, one or two well-formed embryos should be left in bodies of aphids. Excess wax may be removed by placing specimens in tetrahydrofuran*. *Note*: This is a hazardous chemical and should be used under an exhaust hood and in a very well-ventilated place to avoid inhalation of fumes.
4. Remove from Essig's fluid and place in 95% ethanol for 5 minutes.
5. Remove from the ethanol and place in acid fuchsin stain for about 5 minutes or until properly stained, then place in 70% ethanol for 5 minutes or more to remove excess stain.
6. Remove from 70% ethanol and place in 95% ethanol for 5 minutes.
7. Remove from 95% ethanol and place in clove oil, leave for 5 minutes or until specimen appears nonshiny, dull, and flat.
8. Remove from clove oil and place specimen dorsum upward on a slide in one-to-two drops of Canada balsam. If mounting three or more specimens on one slide, place them in a row right to left with one specimen ventral side up. Keep all specimens neatly horizontal with heads pointed in the same direction.
9. Put coverslip in place and attach labels, preferably so that they may be read with heads of specimens toward you.

Mounting thrips. For detailed study, mount thrips in Canada balsam as described here. Place each specimen by itself centrally on a slide with wings, legs, and antennae spread for easy observation of structures. Most of the specimens should be cleared for optimal appearance of surface detail, but a few specimens should be left in their natural color by omitting steps two and three. Rapid identifications may be made from temporary mounts in glycerin or Hoyer's Medium, but they usually cause distortion. Excess or used fluids may be removed at each step with a pipette.

1. Soak specimen for 24 hours in clean 60% ethanol to remove collecting fluid.
2. Macerate in cold 5% sodium hydroxide* solution for 30 minutes or up to 4 hours for especially dark specimens.
3. Wash briefly in 50% ethanol and then leave in 60% ethanol for 24 hours.
4. Dehydrate through a series of ethanol solutions: 70% for 1 hour, 80% for 2 hours, and 100% for 10 minutes (change alcohol once). Place specimen in clove oil until clear (30 seconds to 10 minutes). Spread appendages carefully at each stage. Dehydration and clearing may be promoted by puncturing thoracic and abdominal membranes in one or two places with a fine needle.
5. Place ventral side uppermost on a 13-mm coverslip in Canada balsam, then lower slide onto coverslip. This method is easier to control than the usual method of lowering coverslip onto slide with forceps.
6. Use two labels on slide, one at each side of specimen, with host, locality, altitude, date, and collector's name on right-hand label and determination and sex data on left-hand label.
7. Cure in oven at 40°C within a few minutes of preparation, or leave for up to 6 weeks for complete curing.

Mounting mites. Mites are most easily mounted if collected in alcohol−glycerin−acetic acid solution. Mount specimens collected in 70%−80% ethanol on slides as soon as possible. The following procedures do not apply to mites of the family Eriophyidae:

1. Place drop of Hoyer's Medium in center of a clean 2.5 by 7.6 cm glass microscope slide.
2. Remove mite directly from host, or pour specimens from collecting vial into small casserole, watch glass, or petri dish. Avoid pouring too much fluid from vial into dish; the less fluid, the easier it is to pick out the mite. The mites may be removed from the host or from the fluid by dipping a needle into the Hoyer's Medium on the slide and then quickly touching the mite with it.
3. Place one specimen in the medium on a slide. Press the specimen to the surface of the slide and spread all legs laterally. Most mites should be mounted dorsoventrally, but males of many species, such as those of the Acaridae and Tetranychidae, should be mounted laterally to allow examination in profile of the specifically characteristic aedeagus. Some mites may require a small body puncture to eliminate the contents. Heavily pigmented mites may be cleared in a solution of lactophenol before mounting. The solution may be heated to hasten clearing.
4. With forceps, carefully place a clean, small (13 mm or smaller) coverslip on the mite in Hoyer's Medium. Gently press the coverslip with forceps to hold the mite in position.

5. Place the slide on a hot plate set at 60°C and remove it rapidly when a single bubble forms in the Hoyer's Medium. Avoid more bubbling or mite will be displaced.
6. Turn the slide so that the anterior part of the mite is directed toward you.
7. Place label on slide to right of coverslip; host, locality, collector, date, and serial number data should be shown on this label.
8. Place slide in oven at 45°C–50°C for 24 hours or longer to cure and solidify the Hoyer's Medium. The slide must be kept horizontal until the medium is firm and there is no danger of the coverslip moving.
9. After removing slide from oven, ringing coverslip with additional Hoyer's Medium helps to prevent the mount from drying out.

Directory state extension service directors and administrators, March 2019

Alabama

Dr. Gary D. Lemme, Director
Alabama Cooperative Extension System
Auburn University, Auburn, AL 36849-5612, United States
334-844-5546; (FAX) 334-844-5544
glemme@aces.edu
Dr. Allen Malone, 1890 Administrator
1890 Extension Programs
Alabama A&M University
P.O. Box 967
Normal, AL 35762, United States
256-372-5943; (FAX) 256-372-5840
aam0057@aces.edu
Dr. Raymon Shange, Interim Assistant Dean
Cooperative Extension
1890 Extension Programs
Tuskegee University
Kellogg Conference Center for Continuing Education
Tuskegee, AL 36088, United States
334-724-4967
rshange@tuskegee.edu

Alaska

Dr. Milan Shipka, Acting Director
Agricultural & Forestry Experiment Station, Cooperative Extension
University of Alaska Fairbanks
P.O. Box 756180
Fairbanks, AK 99775-6180, United States

907-474-7429; (FAX) 907-474-6971
fnatn@uaf.edu

Arizona

Dr. James A. Christenson, Director
Cooperative Extension Service
College of Agriculture and Life Sciences
University of Arizona
Forbes Bldg., Room 301
Tucson, AZ 85721, United States
520-621-7209; (FAX) 907-474-6184
mpshipka@alaska.edu

Arkansas

Dr. Ivory W. Lyles, Director
Cooperative Extension Service
University of Arkansas
2301 South University Avenue
Little Rock, AR 72204, United States
501-671-2001; (FAX) 501-671-2107
rcartwright@uaex.edu
Dr. Muthusamy Manoharan, Dean & Director
School of Agriculture, Fisheries and Human Sciences
1890 Agricultural Programs
University of Arkansas
P.O. Box 4990
1200 N. University Drive
Pine Bluff, AR 71611, United States
870-575-8529; (FAX) 870-575-4629
manoharanm@uapb.edu

California

Dr. Wendy Powers, Associate Vice President
Agriculture and Natural Resources
Cooperative Extension Service
University of California
300 Lakeside Dr., 6th Floor

Oakland, CA 94612-3560, United States
510-987-9033
wendy.powers@ucop.edu

Colorado

Dr. Louis Swanson, Vice President for Engagement and Director of Extension
Cooperative Extension Service
Colorado State University
1 Administration Building
Fort Collins, CO 80523-4040, United States
970-491-6362; (FAX) 970-491-6208
louis.swanson@colostate.edu

Connecticut

Dr. Michael O'Neill, Associate Dean
Cooperative Extension System
College of Agriculture & Natural Resources
University of Connecticut
Young Building, Room 218
1376 Storrs Road, Unit 4066
Mansfield, CT 06269-4066, United States
860-486-6270; (FAX) 860-486-5113
mp.oneill@uconn.edu

Delaware

Dr. Michelle S. Rodgers, Associate Dean and Director
Cooperative Extension
University of Delaware
113 Townsend Hall
Newark, DE 19716, United States
302-831-2504; (FAX) 302-831-6758
mrodgers@udel.edu
Dr. Dyremple B. Marsh, Dean, College of Agriculture & Related Sciences
1890 Extension Programs
Delaware State University
1200 North DuPont Highway
Dover, DE 19901, United States

302-857-6400; (FAX) 302-857-6402
dmarsh@desu.edu

District of Columbia

Mr. William B. Hare, Associate Dean of Land-grant Programs
Cooperative Extension Service
University of the District of Columbia
4200 Connecticut Avenue, NW
Washington, DC 20008, United States
202-274-7133; (FAX) 202-274-7133
whare@udc.edu

Florida

Dr. Nick T. Place, Dean and Director
Florida Cooperative Extension Service
University of Florida
P.O. Box 110210
Gainesville, FL 32611-0210, United States
352-392-1761; (FAX) 352-846-0458
nplace@ufl.edu
Mrs. Vonda Richardson, Associate Director
Cooperative Extension Program
1890 Extension Programs
Florida A&M University
Perry-Paige Building, Room 215
Tallahassee, FL 32307, United States
850-599-3546; (FAX) 850-599-8821
Vonda.Richardson@famu.edu

Georgia

Dr. Laura P. Johnson, Associate Dean and Director
Cooperative Extension Service
College of Agricultural & Environmental Sciences
University of Georgia
Conner Hall, Room 101
Athens, GA 30602-7501, United States
706-542-3824; (FAX) 706-542-8815

lpj4h@uga.edu
Dr. L. Latimore, Jr., Interim Extension Administrator and Assistant Vice President for
Land Grant Affairs
1890 Extension Programs
Fort Valley State University
124 Stallworth Building
Fort Valley, GA 31030, United States
478-825-6827/6320; (FAX) 478-825-6376
latimorm@fvsu.edu

Hawaii

Dr. Jeff Goodwin, Interim Associate Dean and Associate Director for Cooperative
Extension Service
Manoa College of Tropical Agriculture & Human Resources
University of Hawaii
3050 Maile Way, Gilmore 203
Honolulu, HI 96822, United States
808-956-8139; (FAX) 808-956-9712
goodwinj@ctahr.hawaii.edu

Idaho

Dr. Barbara Petty, Director
Cooperative Extension System
University of Idaho
Twin Falls R&E Center
P.O. Box 1827
315 Falls Avenue
Twin Falls, ID 83303-1827, United States
208-736-3603; (FAX) 208-736-0843
bpetty@uidaho.edu

Illinois

Dr. Shelly Nickols-Richardson, Associate Dean and Director
Cooperative Extension Service
University of Illinois
123 Mumford Hall, 1301 West Gregory Drive
Urbana, IL 61801, United States

217-333-2660; (FAX) 217-244-5403
nickrich@illinois.edu

Indiana

Dr. Jason Henderson, Director, Purdue Extension & Associate Dean
Cooperative Extension Service
Purdue University
1140 Agriculture Admin. Bldg.
West Lafayette, IN 47907-114, United States
765-494-8489; (FAX) 765-494-5876
jhenderson@purdue.edu

Iowa

Dr. John D. Lawrence, Vice President
Extension and Outreach
Cooperative Extension Service
Iowa State University
315 Beardshear
Ames, IA 50011-2020, United States
515-294-5390; (FAX) 515-294-4715
jdlaw@iastate.edu

kansas

Dr. Greg Hadley, Director
Cooperative Extension Service
Kansas State University
College of Agriculture
115 Waters Hall
Manhattan, KS 66506, United States
785-532-5820; (FAX) 785-532-6290
ghadley@ksu.edu

Kentucky

Dr. Gary Palmer, Interim Director
Cooperative Extension Service

University of Kentucky
S-107 Ag. Science Bldg. North
Lexington, KY 40546-0091, United States
859-257-4302; (FAX) 859-257-3501
gary.palmer@uky.edu
Dr. Johnnie Westbrook, Associate Extension Administrator
1890 Extension Programs
Kentucky State University
Cooperative Extension Program Facility
400 E. Main Street
Frankfort, KY 40601, United States
502-597-5811; (FAX) 502-227-5933
johnnie.westbrook@kysu.edu

Louisiana

Dr. Gina Eubanks, Associate Vice Chancellor
Agricultural Center
Cooperative Extension Service
Louisiana State University
P.O. Box 25100
Baton Rouge, LA 70894-5100, United States
225-578-4161; (FAX) 225-578-6914
geubanks@agcenter.lsu.edu
Dr. Dawn Mellion-Patin, Vice Chancellor
Extension and Outreach
1890 Extension Programs
Southern University & A&M College
P.O. Box 10010
Baton Rouge, LA 70813, United States
225-771-3532; (FAX) 225-771-2681
dawn_mellion@suagcenter.com

Maine

Dr. Dennis Harrington, Assistant Director and Financial Manager
Cooperative Extension Service
University of Maine
5741 Libby Hall, Room 102
Orono, ME 04469-5741, United States

207-581-3132; (FAX) 207-581-3325
dennis.l.harrington@maine.edu

Maryland

Dr. James Hanson, Assoc. Dean & Assoc. Director
Cooperative Extension Service
University of Maryland
College of Agriculture & Natural Resources
1296 Symons Hall
College Park, MD 20742-5551, United States
301-405-2907; (FAX) 301-405-2963
jhanson1@umd.edu
Dr. Maifan Silitonga, Assoc. Ext. Administrator
1890 Extension Programs
University of Maryland
Eastern Shore
Princess Anne, MD 21853, United States
410-651-6205; (FAX) 410-651-6207
mrsilitonga@umes.edu

Massachusetts

Dr. Patricia Vittum, Associate Director
CAFE and Cooperative Extension Service
University of Massachusetts
212 Stockbridge Hall
Amherst, MA 01003, United States
413-545-6000; (FAX) 413-545-6555
pjvittum@umass.edu

Michigan

Dr. Jeffrey Dwyer, Director
Cooperative Extension Service
Michigan State University
108 Agriculture Hall
East Lansing, MI 48824-1039, United States
517/355-2308
dwyerje@msu.edu

Minnesota

Dr. Beverly Durgan, Dean
Minnesota Extension Service
University of Minnesota
Coffey Hall, Room 240
1420 Eckles Avenue
St. Paul, MN 55108-6070, United States
612-624-2703; (FAX) 612-625-6227
bdurgan@umn.edu

Mississippi

Dr. Gary Jackson, Director
Cooperative Extension Service
Mississippi State University
P.O. Box 9601
Bost Extension Center, Room 201
Starkville, MS 39762, United States
662-325-3036; (FAX) 662-325-8407
gary.jackson@msstate.edu
Dr. Edmund Buckner, Dean and Director
Land Grant Programs
1890 Extension Programs
Alcorn State University
1000 ASU Drive, #690
Lorman, MS 39096, United States
601/877-6136/6137
bucknere@alcorn.edu

Missouri

Dr. Marshall Stewart, Vice Chancellor
Extension and Engagement
Cooperative Extension Service
University of Missouri
108 Whitten Hall
Columbia, MO 65211, United States
573-882-7477; (FAX) 573-882-1955
stewartmars@missouri.edu
Ms. Yvonne Matthews, Interim Dean and Administrator

1890 Extension Programs
Lincoln University
820 Chestnut, P.O. Box 29
Jefferson City, MO 65102, United States
573-681-5109; (FAX) 573-681-5520
matthewy@lincolnu.edu

Montana

Dr. Cody Stone, Extension Executive Director
Cooperative Extension Service
Montana State University
205 Culbertson Hall
P.O. Box 172230
Bozeman, MT 59717-2230, United States
406-994-1750; (FAX) 406-994—1756
cstone@montana.edu

Nebraska

Dr. Charles Hibberd, Dean,
UNL Cooperative Extension
University of Nebraska
211 Agriculture Hall
Lincoln, NE 68583-0703, United States
402-472-2966; (FAX) 402-472-5557
Hibberd@unl.edu

Nevada

Dr. Ivory W. Lyles, Director
Nevada Cooperative Extension
University of Nevada Cooperative Extension
Mail Stop 404
Reno, NV 89557, United States
775-784-7070; (FAX) 775-784-7079
lylesi@unce.unr.edu

New Hampshire

Dr. Kenneth LaValley, Dean and Director
Cooperative Extension Service
University of New Hampshire
59 College Road
103A Taylor Hall
Durham, NH 03824-3587, United States
603-862-4343; (FAX) 603-862-2385
Ken.lavalley@unh.edu

New Jersey

Dr. Brian J. Schilling, Dean and Director
Cooperative Extension Service
Rutgers University
Martin Hall, Room 309
88 Lipman Drive
New Brunswick, NJ 08901-8525, United States
848-932-3591
schilling@njaes.rutgers.edu

New Mexico

Dr. Jon C. Boren, Associate Dean and Director
Cooperative Extension Service
New Mexico State University
P.O. Box 30003
Department 3AE
Las Cruces, NM 88003, United States
505-646-3015; (FAX) 505-646-5975
jboren@nmsu.edu

New York

Dr. Christopher Watkins, Director
Cooperative Extension Service
Cornell University
366 Roberts Hall
Ithaca, NY 14853, United States

607-255-2237; (FAX) 607-255-0788
cbw3@cornell.edu

North Carolina

Dr. Rich Bonanno, Associate Dean & Director
Cooperative Extension Service
North Carolina State University
P.O. Box 7602
Raleigh, NC 27695-7602, United States
919-515-2813; (FAX) 919-
abonann@ncsu.edu 515—3135
Dr. Rosalind Dale, Associate Dean and Extension Administrator
1890 Extension Programs
North Carolina A&T State University
205 Coltrane Hall
Greensboro, NC 27420-1928, United States
336-334-466-; (FAX) 336-256-0810
rdale@ncat.edu

North Dakota

Dr. Greg Lardy, Director
NDSU Extension Service
North Dakota State University
315 Morrill Hall, P.O. Box 5437
Fargo, ND 58105-5437, United States
701-231-7660
gregory.lardy@ndsu.edu

Ohio

Dr. Roger Rennekamp, Associate Dean and Director
Ohio State University Extension
Ohio State University
3 Agricultural Admin. Bldg.
2120 Fyffe Road
Columbus, OH 43210, United States
614-292-1842; (FAX) 614-688-3807
rennekamp.3@osu.edu

Oklahoma

Dr. Damona Doye, Associate Vice President
Cooperative Extension Service
Oklahoma State University
139 Agriculture Hall
Stillwater, OK 74078-6019, United States
405-744-5398; (FAX) 405-44-5339
damona.doye@okstate.edu
Dr. Leslie L. Whittaker, Dean
School of Agriculture & Applied Science
1890 Extension Programs
Langston University
Agricultural Research Building, Room 100-S
P.O. Box 730
Langston, OK 73050, United States
405-466-3836; (FAX) 405-466-3138
wlwhittaker@langston.edu

Oregon

Dr. A. Scott Reid, Vice Provost
Cooperative Extension Service
Oregon State University
Extension Administration
101 Ballard Extension Hall
Corvallis, OR 97331-3606, United States
541-737-2713; (FAX) 541-737-4423
scott.reed@oregonstate.edu

Pennsylvania

Dr. Jeffrey Hyde, Acting Director
Cooperative Extension Service
The Pennsylvania State University
217 Ag. Admin. Bldg.
University Park, PA 16802, United States
814-865-4028; (FAX) 814-865-7815
jeffhyde@psu.edu

Rhode Island

Dr. Deborah Sheely, Associate Dean, Extension and Associate Director
Cooperative Extension Service
University of Rhode Island
College of the Environment and Life Sciences
12 Woodward Hall
9 East Alumni Avenue
South Kingston, RI 02881, United States
401-874-2240; (FAX) 401-874-4017
dsheely@uri.edu

South Carolina

Dr. Thomas Dobbins, Director
Cooperative Extension Service
Clemson University
130 Lehotsky Hall
Clemson, SC 29634, United States
864-656-3382; (FAX) 864-656-5819
tdbbns@clemson.edu
Dr. Delbert T. Foster, V.P. for Land grant Prog, Exec. Dir. 1890 Programs
1890 Extension Programs
South Carolina State University
P.O. Box 7453
Orangeburg, SC 29117, United States
803-536-8189; (FAX) 803-536-3962
dfoster@scsu.edu

South Dakota

Ms. Karla Trautman, Interim Director
Cooperative Extension Service
South Dakota State University
Ag Hall 154, P.O. Box 2207D
Brookings, SD 57007, United States
605-688-4148; (FAX) 605-688-6320
karla.trautman@sdstate.edu

Tennessee

Dr. Robert Burns, Dean
Tennessee Agricultural Extension Service
The University of Tennessee
121 Morgan Hall
Knoxville, TN 37996-4530, United States
865-974-7114; (FAX) 865-974-1068
rburns@utk.edu
Dr. Chandra Reddy, Dean & Director
Research & Administration Extension
1890 Extension Program
Tennessee State University
3500 John A. Merritt Building
Nashville, TN 37209-1561, United States
615-963-7561; (FAX) 615-963-5888
creddy@tnstate.edu

Texas

Dr. Parr Rosson, Interim Director
Texas Cooperative Extension
Texas A&M University
112 Jack K. Williams Administration Bldg.
7101 TAMU
College Station, TX 77843-7101, United States
979-845-7967; (FAX) 979-845-9542
prosson@tamu.edu
Dr. Carolyn Williams, Associate Administrator
1890 Extension Programs
Prairie View A&M University
P.O. Box 3059
Prairie View, TX 77446-3059, United States
936-261-2023; (FAX) 936-261-5143
cjwilliams@pvamu.edu

Utah

Dr. Brian Higginbotham, Associate Vice President
Extension & A.D.

Cooperative Extension Service
Utah State University
4900 Old Main Hill
Logan, UT 84322-4900, United States
435-797-7276; (FAX) 435-797-7220
brian.h@usu.edu

Vermont

Mr. Chuck Ross, Director
Cooperative Extension Service
University of Vermont Extension Systems
601 Main Street
Burlington, VT 05405, United States
802-656-2990; (FAX) 802-656-8642
chuck.ross@uvm.edu

Virginia

Dr. Edwin J. Jones, Director
Cooperative Extension Service
Virginia Polytechnic Institute and State University
Office of the Director
104 Hutcheson Hall, MC 402
Blacksburg, VA 24061, United States
540-231-5229; (FAX) 540-231-4370
ejones1@vt.edu
Dr. M. Ray McKinnie, Interim Dean
College of Agriculture
1890 Extension Programs
Virginia State University
P.O. Box 9081
Petersburg, VA 23806, United States
804-524-5260; (FAX) 804-524-5967
mmckinnie@vsu.edu

Washington

Dr. Michael Gaffney, Acting Director
Cooperative Extension Service

Washington State University
P.O. Box 641046, 411 Hulbert Hall
Pullman, WA 99164-6230, United States
509-335-2933; (FAX) 509-335-2926
mjgaffney@wsu.edu

West Virginia

Mr. Steven Bonanno, Dean and Director
Cooperative Extension Service
West Virginia University
P.O. Box 6031
Morgantown, WV 26506-6031, United States
304-293-5691; (FAX) 304-293-7163
SCBonanno@mail.wvu.edu

Wisconsin

Dr. Karl Martin, Dean & Director
Cooperative Extension Service
University of Wisconsin
432 N. Lake Street
Extension Building, Room 601
Madison, Wl 53706, United States
608-263-2775; (FAX) 608-265-4545
karl.martin@ces.uwex.edu

Wyoming

Dr. Kelly Crane, Associate Director
Cooperative Extension Service
University of Wyoming
College of Agriculture
Room 103, P.O. Box 3354
Laramie, WY 82071-3354, United States
307-766-3563; (FAX) 307-766-3998
kcrane1@uwyo.edu
Territories:

American Samoa

Aufa'i Apulu Ropeti Areta, Extension Program Coord, Coop Extension Srvc
Community & Natural Resources Division
American Samoa Community College
PO Box 5319
Pago Pago, AS 96799, United States
(0111) 684-699-1575/684-733-0760; (FAX) (0111) 684-699-5011
a.areta@amsamoa.edu

Guam

Dr. Lee S. Yudin, Dean & Director
University of Guam
College of Agriculture & Life Sciences
UOG Station
Mangilao, Guam 96923, United States
671-735-2000/1; (FAX) 671-734-6842
lyudin@triton.uog.edu

Micronesia

Dr. Thomas Taro, Vice President and Associate Director
Cooperative Research and Extension
Palau Community College
P.O. Box 9
Koror, PW 96940, Republic of Palau
(0111) 680-488-2726; (FAX) (01111) 680-488-3307
tarothomas@yahoo.com

Northern Marianas

Ms. Patricia Coleman, Interim Dean & Program Leader
Cooperative Research, Extension, and Education Service
Northern Marianas College
P.O. Box 501250
Saipan, MP 96950, United States
670-237-6842; (FAX) 670-234-1270
patricia.coleman@marianas.edu

Puerto Rico

Dr. Eric A. Irizarry-Otaño, Interim Associate Dean
Cooperative Extension Service
University of Puerto Rico
College of Agricultural Sciences
P.O. Box 9030
Mayaguez, PR 00681-9030, United States
787-833-2665
eric.irizarry@upr.edu

Virgin Islands

Mr. Stafford Crossman, Acting State Director
Cooperative Extension Service
University of the Virgin Islands
RR02 Box 10,000
Kingshill, St. Croix, VI 00850, United States
340-692-4093; (FAX) 340-692-4085
scrossm@uvi.edu

Submitting specimens for identification to Systematic Entomology Laboratory Communications & Taxonomic Services Unit

IV

The following text has been taken directly from the USDA-ARS website: https://www.ars.usda.gov/northeast-area/beltsville-md-barc/beltsville-agricultural-research-center/systematic-entomology-laboratory/docs/insect-mite-identification-service/

Systematic Entomology Laboratory (SEL) scientists actively collaborate with members of academia, providing taxonomic support for a broad range of basic and applied biological projects. Every attempt is made to provide timely service; however, our scientists' diversity of commitments and shortage of technical support staff limit the speed at which identification requests can be completed. The information that follows includes suggestions of steps you can take to facilitate the processing of your specimens.

General considerations

Provide tentative IDs. In most cases, you should be able to provide tentative IDs to at least the family level. This information is essential for CTSU staff to assign your specimens to the appropriate specialist. By providing tentative IDs you decrease the amount of time CTSU requires to process your submission.

Sort your specimens. If you are submitting a mixed-lot, please sort your specimens by family. Separating taxa and repackaging them for distribution to multiple specialists can be a time-consuming task, especially for very large, diverse lots.

Contact specialists prior to submission. SEL scientists have a variety of professional commitments that limit the time they have available to perform routine identifications. If time is a factor in your research, you will likely find it advantageous to contact specialists prior to submission and inquire about projected turnaround times.

Submit your specimens to CTSU. If you send specimens directly to scientists, there is a greater probability that your specimens will be misplaced or lost. CTSU maintains a database of all pending SEL identifications, and this system allows us to keep track of submissions within the lab.

Submission requirements

Please adhere to SEL's standard submission requirements.
Include copies of collecting permits when required.

Citations

When identifications are used in publications, please acknowledge the contributing taxonomist(s) and include the taxon or group of taxa identified as well as their professional affiliation. The following is a suggested format for citations:

Name of Taxonomist (taxon), Systematic Entomology Laboratory, Agricultural Research Service, US Department of Agriculture

When available, please send reprints of publications where identifications are reported to CTSU.

Submission requirements

Whether an amateur entomologist, concerned about potential pests in your home or garden, or just curious, SEL's Insect Identification Service is available to all US citizens to assist with specimen identification. The guidelines that follow provide useful tips to help you get the most out of the Insect Identification Service.

General Considerations

SEL provides specimen identifications. SEL does NOT provide extensive biological information along with its identifications, and lab personnel are not authorized to provide specific information regarding pest control strategies. Questions about managing pest species should be directed to your State Cooperative Extension Office.

Not all specimens can be identified. Insects represent the most diverse group of animals on the planet, and our knowledge of this diversity is still very limited. Additionally, many insects are only known from specific life stages (larva, pupa, or adult) or from a single sex. As a result, it is not uncommon for specialists to be unable to provide a complete, species-level, determination.

Submit your specimens to CTSU. If you send specimens directly to scientists, there is a greater probability that your specimens will be misplaced or lost. CTSU maintains a database of all pending SEL identifications, and this system allows us to keep track of submissions within the lab.

Submission requirements

When possible, please adhere to SEL's standard submission requirements. It is especially important that you include a completed copy of SEL's Identification Request Form, ARS-748 (PDF).

We realize that many of the specimen preparation techniques require specialized materials that may not be readily available to the general public. When such materials are unavailable, feel free to improvise. Pill bottles are a good substitute for vials, and rubbing alcohol can be used in place of ethanol. Do NOT use tape or other adhesives to secure specimens to paper.

Please label specimens with the following information and provide a detailed account of the circumstances under which they were collected.

Country: State: County
Town
Host or Specific Location
Date
Collector's Name

The majority of identification requests received by SEL are submitted by APHIS-PPQ, a regulatory division of the USDA charged with safeguarding US agricultural and natural resources from the risks associated with the entry, establishment, or spread of animal and plant pests and noxious weeds. Due to the high volume and variety of submissions that SEL receives from PPQ personnel, additional guidelines have been developed to facilitate the processing of these identification requests.

SEL receives three different types of identification requests from APHIS-PPQ: Digital Images, Standard Port Interceptions, and Domestic Submissions. Special instructions for each type of request are outlined below.

Digital images

Digital images should be submitted ONLY for URGENT interceptions.

Each interception should be submitted separately.

Image dimensions should be roughly 600 × 800 pixels at a resolution of 72 dpi and saved in JPEG format, resulting in files that are 100−200 KB in size.

Files should be named based on the following conventions:

17-Character Interception Number|unique letter.jpg (Example: APHTX061722580001a.jpg)

The Subject line of the e-mail message should include the 17 Character Interception Number and tentative family ID when available. (Example: APHTX061722580001-Pentatomidae)

The body of the e-mail should include the following information:

- Port:
- Interception Number:
- Intercept Date:
- Host:
- Origin:
- Tentative ID:
- Attached Files:

AFTER interception, information has been entered into PestID, submit images to IDService@ars.usda.gov

Standard port interceptions

Urgent interceptions are not subject to SEL's specimen preparation requirements and may be submitted in the most expeditious manner available (typically in ethanol).

All FedEx packages containing Urgent interceptions should be marked with yellow and black "Urgent tape."

Urgent requests should be forwarded to **the appropriate SEL location depending on the taxonomic group of the specimen.**

Prompt and Routine submissions should conform to SEL's submission requirements. Furthermore, all scales and mites must be slide-mounted prior to submission.

Port interceptions do not require form ARS-748; however, they must be accompanied by a copy of the interception record, form PPQ 309.

Domestic submissions

All domestic submissions must be accompanied by SEL's Identification Request Form, ARS-748 (PDF).

While it is acceptable to submit all domestic Urgents and Prompts in ethanol, Routine submissions, such as those resulting from domestic survey projects, need to be prepared in accordance with SEL specimen preparation standards.

If you send domestic Urgents or Prompts directly to SEL scientists stationed at the National Museum of Natural History, please e-mail or FAX a copy of the Identification Request Form to CTSU in Beltsville, MD.

The Systematic Entomology Laboratory (SEL) routinely provides specimen identification assistance as a free service to both governmental and private entities, including federal research and regulatory agencies, state departments of health and agriculture, university researchers, and private citizens. This service is coordinated by SEL's Communications & Taxonomic Services Unit (CTSU) and relies on the expertise of SEL scientists and collaborating specialists. CTSU maintains a relational database allowing for the efficient management of identification assignments. Upon completion, CTSU is responsible for reporting identifications to the submitter and returning specimens when requested.

Submission requirements

Due to the large volume of identification requests that are received by SEL, submission requirements have been instituted to facilitate the processing of ID requests.

The requirements that follow fall into two major categories, documentation and specimen preparation. Timely completion of requested identifications is dependent upon your compliance with our submission standards.

Documentation

Aside from APHIS-PPQ port interceptions, all submissions must be accompanied by a completed copy of SEL's Insect Identification Request Form, ARS-748. Versions of this form are available for download in PDF format. Detailed instructions for completing this form are also available.

Specimen preparation

The ability of SEL scientists and collaborators to provide complete and accurate identifications depends heavily on the quality of the specimens that are submitted for examination. In short, all hard-bodied insects should be pinned or point-mounted while larvae and other soft-bodied specimens should be preserved in 70% ethanol. All specimens should be appropriately labeled with collection locality data. Additionally, all adult Lepidoptera should be submitted with their wings spread to allow for easy examination of wing venation. The information provided in this manual will give you the required specimen preparation techniques. General guidelines for specific taxa can be found in Table A4.1.

Additional Instructions

The following links provide specific instructions for different classes of submitters.

- USDA-APHIS-PPQ
- University Students & Researchers
- Private Citizens

Contact information

Communications & Taxonomic Services Unit
USDA-ARS-Systematic Entomology Laboratory
Building 005, Room 137, BARC-West
10300 Baltimore Avenue Beltsville, MD 20705-2350
e-mail: IDService@ars.usda.gov
phone: 301.504.7041
FAX: 301.504.6482

Table A4.1 Specimen preparation guidelines by taxa.

Taxon	Preferred Preparation Method
Acari (mites & ticks)	Ethanol or Slide
Auchenorrhyncha (planthoppers, treehoppers, froghoppers, etc.)	Pin or Point
Blattaria (roaches)	Pin
Coleoptera (beetles)	Pin, Point, or Ethanol (depending on size)
Collembola (springtails)	Ethanol or Slide (depending on size)
Dermaptera (earwigs)	Pin
Diplura	Ethanol or Slide (depending on size)
Diptera (true flies)	Pin, Point, or Ethanol (depending on size)
Ephemeroptera (mayflies)	Ethanol
Heteroptera (true bugs)	Pin or Point
Hymenoptera (bees, wasps, ants, etc.)	Ethanol, Pin, or Point (depending on size)
Isoptera (termites)	Ethanol
Lepidoptera (butterflies & moths)	Pin or Point
Mantodea (mantids)	Pin
Mecoptera (scorpionflies, earwigflies, etc.)	Pin or Point
Megaloptera (dobsonflies, fishflies, etc.)	Ethanol or Slide (depending on size)
Microcoryphia (bristletails)	Ethanol
Neuroptera (lacewings, antlions, owlflies, etc.)	Pin or Point
Odonata (dragonflies & damselflies)	Pin
Orthoptera (grasshoppers, crickets, & katydids)	Pin
Phasmida (walking-sticks)	Pin
Phthiraptera (lice)	Ethanol or Slide (depending on size)
Plecoptera (stoneflies)	Ethanol
Protura	Slide
Psocoptera (plant lice)	Ethanol or Slide (depending on size)
Siphonaptera (fleas)	Ethanol or Slide (depending on size)
Sternorrhyncha (aphids, scales, whiteflies, psyllids)	Ethanol or Slide (depending on size)
Strepsiptera (twisted-winged parasites)	Ethanol or Slide (depending on size)
Thysanoptera (thrips)	Ethanol or Slide (depending on size)
Trichoptera (caddisflies)	Ethanol
Zoraptera (webspinners)	Ethanol or Slide (depending on size)
Zygentoma (silverfish)	Ethanol

USDA

SYSTEMATIC ENTOMOLOGY LAB
IDENTIFICATION REQUEST

Priority:	Lot Number:
Date Submitted:	Number of Specimens:
Date Needed:	Specimen Disposition: ☐ Return ☐ Keep/Discard

Name:

	Submitter's Reference Number:	Tentative Identification:

Address:

Level of Identification Requested　　　　:

☐ Family　　☐ Genus　　☐ Species

Host:

Reason for Identification:

☐ A – Biological Control

☐ B – Damaging Crop/Plants

☐ C – Suspected Pest of Regulatory Concern

☐ D – Stored Product Pest

Telephone:　　　　FAX:

☐ E – Livestock, Wildlife, or Domestic Animal Pest

☐ F – Danger to Human Health

E-mail:

☐ G – Household Pest

☐ H – Possible Immigrant

Affiliation:

☐ I – Reference Collection

☐ APHIS/PPQ	☐ Private Individual
☐ ARS	☐ Other Federal (US)
☐ Commercial Organization	☐ Other State Agency
☐ US Department of Defense	☐ Private University
☐ Foreign	☐ State Agriculture Agency
☐ US Forest Service	☐ State University

☐ J – Survey

☐ K – Thesis

☐ L – Other (elaborate below)

Collecting Permits:
☐ Required　　☐ Not Required　　If required, please submit copies with specimens.

Submitter is willing to recognize identifier(s) via:
☐ Co-authorship　☐ Citation of relevant publication(s) authored by identifier(s)　☐ Acknowledgement in published work　☐ Other or N/A

Project Description:

Remarks:

USDA

Communications & Taxonomic Services Unit – Systematic Entomology Laboratory
Building 005 – Room 137 – BARC-West
10300 Baltimore Avenue – Beltsville – Maryland – 20705

USDA

OMB 0518-0032-sel-2 (7/2015) -- SEL Identification Request

NOTES:

Glossary

– A –

ABDOMEN Noun. (Latin, *abdomen* = belly). The third or posteriormost division of the insect body. Typically composed of 11 segments in the primitive condition, with a tendency toward reduction to 9–10 visible segments in higher orders. The abdomen of the adult stage lacks legs but contains the organs of reproduction, respiration, digestion, and excretion.

ACEPHALOUS Adj. (Greek, *a* = without; *kephale* = head). Relating to a headless condition. Typically applied to the larval stage of some insects, including Diptera and higher Hymenoptera.

ACULEATE Adj. (Latin, *aculeus* = prickle). 1. A member of the order Hymenoptera, the female members of which bear a sting. 2. A structure invested with spines or minute bristles under the scales of the wing (Microlepidoptera).

ACUMINATE Adj. (Latin, *acuminatus* = point). Tapering to a long point.

ADECTICOUS Adj. (Greek, *a* = without; *dekticos* = biting). Insects in the pupal stage that lack functional mandibles capable of cutting through the pupal cocoon. Compare Decticous.

ADECTICOUS PUPA A form of holometabolous development in which mandibles in the pupal stage are immobile and cannot be used for biting. Compare Decticous Pupa.

ADFRONTAL AREA Lepidoptera larvae: An oblique sclerite laterad of the frons that extends from the base of the antenna to the epicranial suture. A median epicranial suture is formed by the union of oblique sclerites from each side of the head; an epicranial sulcus (notch) is formed when the sclerites do not fuse.

ADFRONTAL SETAE Lepidoptera larvae: A pair of setae found on adfrontal areas.

ADFRONTAL SUTURE Lepidoptera larvae: The suture formed by the medial fusion of adfrontal sclerites.

ADVENTITIOUS VEIN A wing vein that is not part of the regular plan but appears in one species or a single individual; or the result of two or more true veins in line with one another.

AEDEAGUS Noun. (Greek, *aidoia* = genitals; Late Latin. Pl., aedeagi). The penis or intromittent organ of copulation in male insects.

AESTIVATION Noun. (Latin, *aestivus* = of summer). Dormancy in summer. Typically occurs during periods of continued high temperatures in temperate areas or during a dry season in tropical areas. Also compare Diapause.

ALLOMONE Noun. (Greek, *allos* = other; *hormaein* = to excite). A chemical substance (produced or acquired by an organism) that, when contacting an individual of another species in the natural context, evokes in the receiver a behavioral or physiological reaction adaptively favorable to the emitter. Compare Kairomone, Pheromone.

ALULA Noun. (Latin, *ala* = wing. Pl., alulae). In some Diptera and Coleoptera: the expanded axillary membrane of the wing. In Diptera a pair of membranous scales above the halteres. The outermost lobe at the base of the wing.

AMETABOLOUS Adj. Having no metamorphosis, that is, primitively wingless, emerging from the egg in a form essentially resembling the adult, apart from the small size and lack of developed genitalia.

AMPLEXIFORM COUPLING Term applied to a type of wing-locking mechanism in Lepidoptera in which there is no frenulum and the large humeral lobe of the hindwing projects under the forewing.

AMPLIFICATION Verb.(Latin, *amplio* = to make larger). A process using polymerase chain reaction (PCR) to generate thousands to millions of copies of DNA being studied.

AMPLICON Noun. (Latin, *amplio* = to make larger). A segment of DNA resulting from replication or amplification. Arises as a result of artificial or natural means.

ANAL Adj. (Latin, *anus* = anus). In the direction of the anus. Pertaining to the last abdominal segment, which bears the anus. Pertaining to the wing area behind the cubitus and to the veins belonging to it. Compare oral.

ANNULATE Adj. (Latin, *annulus* = ring). 1. Ringed, but not demonstrating primary or secondary segmentation. 2. Bearing rings, or arranged in the form of an annulus or ring. 3. A narrow circle or ring of a color different from the adjacent region. Alternate: Annulated.

ANNULUS Noun. (Latin, *annulus* = ring. Pl., annuli). 1. A ring of membrane, sclerite, or pigment surrounding a joint, segment, spot, or mark. 2. A ringlike marking or a ring of hard cuticle. 3. The sclerotized ring of the head into which the basal segment of the antenna is inserted. 4. An antennal sclerite. Alternate: Antennalis.

ANTENNA Noun. (Latin, *antenna* = sailyard. Pl., antennae). The paired, mobile, segmented sensory appendages located on the head of most arthropods. Missing from Chelicerata; two pairs found in Crustacea; one pair in Insecta. Among hexapods, missing from Protura; present in Diplura and Collembola. In the Insecta, the antenna is the anteriormost appendicular structure of the postembryonic organism, above or before the mouth opening.

ANTERIOR Adj. (Latin, *anterior* = former). Pertaining to structure at or near the head of the insect, or facing in that direction. Ant. Posterior.

ANUS Noun. (Latin, *anus* = anus). 1. The opening at the posterior end of the digestive tract, through which the excrement is passed. 2. Coccidae: A more or less circular opening on the dorsal surface of the pygidium, in various locations, depending on the location of the circumgenital gland orifices.

APHIDOPHAGOUS Adj. (Middle Latin, *aphidis* = *a* plant louse; Greek, *phagein* = to devour). Pertaining to or descriptive of insects that feed on aphids, typically as predators. Example: some Coccinellidae.

APHIS (Acronym: Animal and Plant Health Inspection Service). The branch of the US Department of Agriculture responsible for port inspection and the detection and eradication of unwanted organisms in the United States.

APICAD Adv. (Latin, *apex* = summit; *ad* = toward). Toward the apex or away from the body. Ant. Basad.

APICAL Adj. (Latin, *apex* = summit). At, near, or pertaining to the apex of any structure.

APOCRIA Suborder of Hymenoptera in which segment I of the abdomen is fused to the thorax to form the propodeum in the adult and in which the larvae are apodous.

APODAL Adj. (Greek, *a* = without; *pous* = feet). Pertaining to organisms without feet. Typical Diptera larvae. Alternate: Apodous.

APODEME Noun. (Greek, *apo* = away; *demas* = body). Any rigid, integumental process that projects into the body to form the insect endoskeleton. Apodemes provide surface for the attachment of muscles. See also Phragma.

APODOUS Adj. (Greek, *a* = without; *pous* = feet). Without feet; legless. See also Apodal.

APOLYSIS Noun. In molting, the separation of the old cuticle from the underlying epidermal cells.

APPENDAGE Noun. (Latin, *ad* = to; *pendere* = to hang). Any structure of the body that is attached to the head, thorax, or abdomen by a joint. Example: antennae, legs, wings.

APPENDIX Noun. (Latin, *ad* = to; *pendere* = to hang). Any structural attachment of the body, particularly a short, stumplike vein.

APTEROUS Adj. (Greek, *a* = without; *pteron* = wing). Pertaining to insects without wings; a wingless condition.

APTERYGOTA Old name for primitively wingless and ametabolous insects.

APTERYGOTE   Adj. (Greek, *a* = without; *pteron* = wing). Belonging to the Apterygota, insects whose ancestors were wingless.

AQUATIC   Adj. (Latin, *aquaticus* = water). Organisms living in freshwater or saltwater; saltwater forms are usually described as marine. Compare Terrestrial.

ARBOREAL   Adj. (Latin, *arbor* = tree). Descriptive of organisms living in, on, or among trees.

ARCUATE   Adj. (Latin, *arcuatus* = curved). Pertaining to a structure that is arched or bowlike. Alternate: Arcuatus.

AREOLE   Noun (Latin, *areola* = small space). Lepidoptera: The closed radial cell of the forewing between veins R3 and R4.

ARISTA   Noun. (Latin, *arista* = awn). A specialized bristle or process on the antenna; usually dorsal, rarely apical.

AROLIUM   Noun. (Greek, *arole* = protection. Pl., arolia). A cushionlike medial pad on the pretarsus of many insects. The typical median terminal lobe of the pretarsus, between the bases of the claws. Orthoptera: The terminal pad between the claws.

ARTICULATE   Adj. (Latin, *articulus* = joint). Jointed or segmented.

ASPIRATOR   Noun. (Latin, *aspirare* = to breathe toward). A handheld device for collecting small, terrestrial arthropods. Designs are variable, but typically constructed of a test tube or glass vial that is closed at one end and sealed with a cork or rubber stopper at the other end. Two small glass or metal tubes, of equal diameter, project through the stopper and penetrate the atrium of the larger vial. One small tube is covered with muslin or screen on the end inside the vial; the opposite end of the tube is placed in the mount and lung power is used to generate an air current that draws air and the insect into the tube via the aperture at the end of the other small tube. The device is popular for collecting small insects, particularly microhymenopera, some Hemiptera, and small Diptera. Syn. Pooter.

ATROPHY   Noun. (Greek, *a* = without; *trophe* = nourishment). A term applied usually to tissue, muscles, appendages, parts, or organs that seem diminished in size or lack growth, implicitly from insufficient nourishment or exercise. A structure that is reduced in size or withered.

ATTENUATE   Adj. (Latin, *attenuare* = to thin). Gradually tapering toward the apex or away from the body; appearing drawn out or slender. Alternate: Attenuated.

AUDITORY ORGANS   Organs by which an insect can hear sounds or perceive vibrations.

— B —

BARCODING   Verb. A taxonominc method used to define a restricted portion of a specific gene or genes to identify an unknown organism or part(s) of an organism.

BASAD   Adv. (Latin, *basis* = base). Toward the base of a structure and away from the tip or apex. Ant. Apicad.

BASAL   Adj. (Latin, *basis* = base). Pertaining to the base or point of attachment to or nearest the body. Ant. Distal proximal.

BASITARSUS   Noun. (Greek, *basis* = base; *tarsos* = sole of foot. Pl., basitarsi). The proximal or basal segment of the tarsus.

BIFID   Adj. (Latin, *bis* = twice; *findere* = to split). Cleft or divided into two parts or lobes; forked.

BIFURCATE   Adj. (Latin, *bis* = twice; *furca* = fork). A forked structure or one that is partly divided into two parts.

BIRAMOSE   Adj. (Latin, *bis* = twice; *ramus* = branch). Pertaining to a structure that has two branches or doubled appendages.

BLADE   Noun. (Anglo-Saxon, *blaed* = leaf). 1. Any thin, flat structure like a leaf, sword, or knife. 2. The lacinia of many insects, such as bees. 3. The large, flattened surface of the insect wing.

BP   Abbreviation for base pair. Building blocks of deoxyribonucleic acid consisting of two nucleobases bound by hydrogen bonds. A base pair consists of purines (adenine and guanine) and pyrimidines (cytosine and thymine) with adenine always paired with thymine and guanine always paired with cytosine.

BRACHYPTEROUS   Adj. With short or abbreviated wings.

BUCCAL CAVITY   The mouth opening.

BULLA   Noun. (Latin, *bulla* = bubble. Pl., bullae). 1. A blister or blisterlike structure. 2. Ephemeridae: A stigmalike enlarged part of the costal area of the wing near the apex with many crossveins. The weak-spots on some of the wing veins where they are crossed by furrows. 3. A shieldlike sclerite closing the tracheal aperture in lamellicorn beetles.

– C –

CALLOW   Noun. (Anglo-Saxon, *calu* = bald). 1. The condition immediately following molting when the integument is colored as when hardened. 2. A term sometimes applied to newly emerged (neonate) worker ants. Compare Teneral.

CALLUS   Noun. (Latin, *callum* = hard skin. Pl., calli). 1. A hard lump or moundlike, rounded swelling of the integument, such as a swelling at the base of the wing articulating with the thorax. 2. Heteroptera: The thickened or raised spots on the thorax, especially of Pentatomidae.

CAMERA LUCIDA   An image-deflecting device attached to the eyepiece of a microscope that projects the microscopic image to a mirror and then onto a piece of paper for drawing.

CAMPODEIFORM   Adj. (Greek, *kampe* = caterpillar; *eidos* = form; Latin, *forma* = shape). A term used to describe elongate larval forms that are characterized by well-developed thoracic legs and a progna-thous head. The term calls to mind stages that resemble the genus *Campodea* (Thysanura).

CANNIBALISTIC   Adj. (Spanish, *canibales*, from *caribes* = brave and daring men). Pertaining to insects that feed on members of the same species.

CAPITATE   Adj. (Latin, *caput* = head). Structure with a terminal knoblike enlargement or with a head. Reference to a type of antenna in which the club is abruptly enlarged at the apex and forms a spherical mass.

CARINA   Noun. (Latin, *carina* = keel. Pl., carinae). An elevated cuticular ridge that is not necessarily high or acute. Syn. Keel.

CARNIVOROUS   Adj. (Latin, *caro* = flesh; *vorare* = to devour). Feeding upon the flesh of other ani-mals. Compare Phytophagous. See also Cannibalistic.

CATERPILLAR   Noun. (Late Latin, *cattus* = cat; *pilosus* = hairy). 1. Lepidoptera: The polypod or eruci-form larva. 2. Mecoptera, Symphytous Hymenoptera: The larval stage with abdominal prolegs.

CAUDAD   Adv. (Latin, *cauda* = tail). Toward the tail end, posterior end, or anus of the insect body. Compare Cephalad.

CAUDATE   Adj. (Latin, *cauda* = tail). 1. With tail-like extensions or processes. 2. Hymenoptera (Apocrita): Specialized body form of some endoparasitic ichenumonid larvae, characteristically seg-mented, with long, flexible, caudal appendages. Function of caudal appendages not established, but sometimes progressively reduced in later instars and lost in the last instar.

CELL   Noun. (Latin, *cella* = compartment). 1. Any closed area in an insect wing bounded by veins. In the conventional system, cells derive their names from the vein forming the upper margin (e.g., all cells just below the radius are called radial cells); cells are numbered from the base outward. 2. In a nest or honeycomb: The small chambers in which eggs are placed and the larvae develop.

CEPHALAD   Adv. (Greek, *kephale* = head). Toward the anterior end or head of the body. Compare Caudad.

CEPHALIC   Adj. (Greek, *kephale* = head). Belonging to or attached to the head; directed toward the head.

CEPHALOTHORAX   Noun. (Greek, *kephale* = head; *thorax* = chest). 1. The anterior body tagma of Arachnida and Crustacea; analogous, perhaps homologous, with insect head and thorax.

CERCUS   Noun. (Greek, *kerkos* = tail. Pl., cerci). An appendage (generally paired) of the tenth abdominal segment. Usually regarded as sensory and characteristically slender, filamentous, and segmented. Not segmented in Orthoptera, Phasmatodea, and Dermaptera. Incorrectly applied to appendicular structures on the ninth segment of Coleoptera larvae or eighth and ninth segments of Hymenoptera. See also Urogomphus, Pygostylus.

CHAETA   Noun. (Greek, *chaite* = long hair. Pl., chaetae). A cuticular projection from the integument of insects, typically hairlike or a bristle. The form, number, and arrangement of chaetae are sometimes considered taxonomically useful. Functionally diverse and serving as sensory receptors and components of stridulatory devices.

CHAIN OF CUSTODY   In forensic entomology, a process used to maintain and document the chronological history of the evidence. (Documents should include name or initials of the individual collecting the evidence, each person on entity subsequently having custody of it, date when the item was collected or transferred, agency and case number, victim's or suspect's name, and a brief description of the item.)

CHARACTER   Noun. (Greek, *charassein* = to engrave). A general term used in taxonomic entomology to describe features of insects or insect parts.

CHELICERA   Noun. (Greek, *chele* = claw; *keras* = horn. Pl., chelicerae). 1. The pincerlike first pair of appendages of adult Chelicerata. Structures regarded as homologous with the second antennae of Crustacea. 2. Pinching or grasping claws of Phymatidae in Heteroptera.

CHEMORECEPTOR   Noun. (Greek, *chemeia* = transmutation; Latin, *recipere* = to receive). A sense organ having a group of cells sensitive to chemical properties of matter.

CHEWING MOUTHPARTS   Appendages of the head that are opposable in operation and adapted for the mastication of particulate matter or matrixlike material. Compare Piercing-sucking mouthparts.

CHITIN   Noun. (Greek, *chiton* = tunic). 1. A colorless, nitrogenous polysaccharide, linearly arranged as beta-linked *N*-acetyl-glucosamine units. Widespread in arthropods and plants. First isolated from fungi by Braconot in 1811 and called fungine; first isolated from insects (beetle elytra) in 1823 by Odier. A chemical compound intermediate between proteins and carbohydrates; the chemical formula is $(C_{32}H_{54}N_4O_{21})_x$.

CHLOROFORM   Noun. (CHC_{13}). A clear, colorless, highly flammable, and volatile liquid with an etherlike odor and a burning, sweet taste. Miscible with alcohol, ether, benzene, and oils. Used as a killing agent but stiffens specimens; forms phosgene (carbonyl chloride, COC_{12}), a severe respiratory irritant when stored in dark-colored glass jars.

CHORDA   Noun. (Greek, *chorde* = string). In the lepidopterous wing, the stem of veins R_{4+5}; the part of veins R_{4+5} separating the areole from the basal cell in Lepidoptera.

CHORDOTONAL ORGAN   A scolopophorous sense organ found between the tibia and basitarsus of some Coleoptera, Diptera, and Hemiptera. Used for the detection of substrate vibrations that enable an insect to hear sounds. The cellular elements form an elongate structure attached at both ends to the body wall but not necessarily containing sense rods (scolops). See also Johnston's Organ, Typmanal Organ.

CHORION   Noun. (Greek, *chorion* = skin). The outer shell or covering of the insect egg.

CLAVATE   Adj. (Latin, *clava* = club). Clublike or becoming thicker toward the apex.

CLAVUS   Noun. (Latin, *clavus* = nail). 1. The club of an antenna. 2. The knob or enlargement at the apex of the stigmal or radial veins of the Hymenoptera. 3. The sharply pointed anal area of the hemelytra in Heteroptera, next to the scutellum when folded. 4. A rounded or fingerlike process in the genitalia of male Lepidoptera.

CLAW   Noun. (Anglo-Saxon, *clawu* = claw). A hollow, sharp multicellular organ; generally paired and at the apex of the insect leg.

CLUB   Noun. (Old Norse, *klubbe* = club). Insect antenna: Apical segments of the flagellum that are enlarged or thickened such that they are morphologically differentiated.

CLYPEUS Noun. (Latin, *clypeus* = shield). The anterior sclerite of the insect head below the frons and above the labrum. Highly variable in size and shape. Separated from the frons by the frontoclypeal suture and separated from the labrum by the clypeolabral suture or membrane. Provides area for attachment of cibarial dilator muscles. In some Diptera, divided into an anteclypeus and postclypeus.

COARCTATE Adj. (Latin, *coarctare* = to press together). 1. Contracted or compacted. A term applied to a pupal form in which the future adult is concealed by a thickened, usually cylindrical case or covering (often the hardened skin of the larva). 2. Compressed such that the abdomen and thorax are forced together. 3. Meloid beetles: The third phase of hypermetamorphic development in which the equivalent of the sixth larval instar is heavily sclerotized, develops rudimentary appendages, but is immobile. Compare Obtect.

COARCTATE PUPA Diptera: A pupa that remains enclosed in the old larval integument.

COCOON Noun. (French, *cocon* = cocoon). A covering, composed of silk or other viscid fibers, spun or constructed by the larval stage. In some Lepidoptera: A casing of earth, wood fragments, or other debris. Principally serves as protective device for the pupa.

COI An abbreviation for the cytochrome C oxidase subunit I (cox or COI) mitochondrial gene used as a universal marker with a standardized 658 bp fragment. This region of the gene serves as a readable "barcode" in common use.

COLLAR Noun. (Middle English, *coler* = collar). 1. General: Any structure between the head and thorax. 2. Hymenoptera: The neck. 3. Diptera: May mean the neck, the sclerites attached to the prothorax, the prothorax, or its processes (antefurca). 4. Coleoptera: The narrow, constricted anterior part of the pronotum, generally set off by a groove.

COMB Noun. (Anglo-Saxon, *comb* = comb). 1. Hymenoptera: Brood cells constructed of wax by bees and used to house individual larvae and to store honey. 2. Termite nests: A spongy, dark reddish-brown material made by the workers from excreta from which they construct fungal beds. 3. Lepidoptera (Lycaenidae): Serrate distal margin of rostellum of valva. 4. A row or rows of close-set, short bristles on the distal end of some leg segments.

COMMENSAL Noun. A species that benefits from commensalisms.

COMMENSALISM Noun. Symbiosis in which members of one species are benefited while those of other species are neither benefited nor harmed.

COMPLETE METAMORPHOSIS Metamorphosis in the Holometabola that has four stages: egg, larva, pupa, and adult. Each stage is entirely different from the others.

COMPOUND EYE Paired, lateral aggregation of separate visual elements (ommatidia) located on the head. See also Ocellus.

CONTIGUOUS Adj. (Latin, *contiguus*, from *contingere* = to touch on all sides). Pertaining to adjacent structures; structures with points, margins, or surfaces in contact but not united or fused.

CORIUM Noun. (Latin, *corium* = leather). 1. Heteroptera: The elongate middle section of the hemelytra that extends from the base to the membrane below the embolium. 2. The membrane of the joints in segmented appendages.

CORNEOUS Adj. (Latin, *corneus* = horny). Of a horny or chitinous substance; resembling horn in texture.

CORNICLE Noun. (Latin, *cornu* = a horn). Paired, secretory structures on the abdomen of aphids. Typically located on the posterior margin of the fifth segment; less commonly located on the sixth segment. Variable in shape. Absent or ringlike in Pemphigidae; elongate, and cylinderlike in most families. Surface ranging from smooth to sculptured; apex with membranous lid and sclerotized plate that serves as a valve manipulated by abdominal musculature. Cornicles produce defensive wax secretions, alarm pheromones, but not honeydew.

COSTA Noun. (Latin, *costa* = rib). 1. An elevated ridge that is rounded at its crest. 2. The thickened anterior margin of any wing, but usually of the forewings. 3. The vein extending along the anterior margin of the wing from the base to the junction with the subcostal vein.

COXA   Noun. (Latin, *coxa* = hip. Pl., coxae). The basal segment of the insect leg. The coxa articulates with the pleural wall of the thorax and is attached to the trochanter. Coxae are paired, ventrolateral in position, and found on each of the thoracic body segments.

COXITE   Noun. (Latin, *coxa* = hip. Pl., coxites). 1. The basal segment of any leglike appendage such as found on the abdominal sternites. 2. A rudimentary abdominal limb in Thysanura displayed in the form of paired lateral plates.

CPD   Abbreviation for critical point drying. Normally used to prepare specimens for study under scanning electron microscopy, which preserves surface structure without alteration.

CREMASTER   Noun. (Greek, *kremastos* = hung). 1. The apex of the last segment of the abdomen. 2. The terminal spine or hooked process of the abdomen of subterranean pupae. Used to facilitate emergence from the earth. 3. An anal hook by which some pupae are suspended.

CREPUSCULAR   Adj. (Latin, *crepusculum* = dusk). Pertaining to animals active or flying at dusk or (more rarely) at dawn. Compare Diurnal, Nocturnal.

CROCHET   Noun. (French, *crochet* = small hook). Curved spines or hooks on the prolegs of caterpillars and on the cremaster of pupae.

CROP   Noun. (Middle English, *croppe* = craw). The dilated portion of the alimentary canal behind the gullet. The crop receives and holds food before its passage through the digestive tract.

CROSSVEIN   Noun. Typically short veins between the lengthwise veins and their branches; numerous crossveins exist in net-veined wings. Veins that normally extend more or less crossways in the insect wing.

CRYPTIC   Adj. (Greek, *kryptos* = hidden). Pertaining to a hidden or concealed condition, particularly in relation to protective camouflage.

CTENIDIUM   Noun. (Greek, *ktenos* = comb. Pl., ctenidia). A comblike row of short spines (bristles) most evident on parasitic insects, including Siphonaptera. Found all over the body of Polyctendidae; found only on the prothorax and sometimes the metathorax of Siphonaptera. Aid in retention and movement through host fur.

CUNEUS   Noun. (Latin, *cuneus* = wedge). Heteroptera: The small triangular area at the end of the embolium of the hemelytra. Odonata: The small triangle of the vertex between the compound eyes.

CURSORIAL   Adj. (Latin, *cursor* = a runner). 1. Adapted to running habits. 2. Pertaining to legs that are long and tapered.

— D —

DEBRIS   Noun. (Old French, *debruisier* = to break). The rubbish or remains of anything that has decomposed or physically broken down with time. Compare Detritus, Duff.

DECTICOUS   Adj. (Greek, *dektikos* = biting). Insects in the pupal stage that have functional mandibles capable of being used to free the insect from the pupal cocoon. Compare Adecticous.

DECTICOUS PUPA   Exarate pupa in which the mandibles are modified to help the insect to escape its cell or cocoon. Compare Adeticous Pupa.

DESICCATION   Noun. (Latin, *desiccatus* from *desiccare* = to dry up). Excessive drying by natural loss of moisture or artificial means.

DETRITIVORE   Noun. (Latin, *detritus* = rubbed or worn away; *vorare* = to devour). An organism that feeds upon detritus.

DETRITIVOROUS   Adj. Pertaining to an organism that feeds on detritus. Feeding upon fur and feather detritus.

DETRITUS   Noun. (Latin, *detritus* = rubbed or worn away). Material that remains after disintegration; rubbing away, or the destruction of structure; fragmented material. Any disintegrated or broken matter. Compare Debris, Duff.

DIAPAUSE   Noun. (Greek, *dia* = through; *pausis* = suspending; *diapauein* = to cause to cease). A condition of restrained development and reduced metabolic activity that cannot be directly attributed to unfavorable environmental conditions. Regarded by entomologists to involve a resting period of an insect, especially of larvae in winter. Compare Hibernation, Quiescence.

DIGIT   Noun. (Latin, *digitus* = finger). 1. Insects: Any fingerlike structure. 2. Chelicerates: Distal portion of a chela or chelicera. 3. Immature Diptera: Prothoracic spiracles. Alternate Digitus.

DIMORPHISM   Noun. (Greek, *dis* = twice; *morphe* = shape). A genetically controlled, non-pathological condition in which individuals of a species are characterized by distinctive or discrete patterns of coloration, size, or shape. Dimorphism can be a seasonal, sexual, or geographic manifestation. See also Polymorphism.

DISCAL   Adj. (Latin, *discus* = disk). On the disk or main surface of any part of the body. Ant. Marginal.

DISCAL CELL   Any large cell in the central part (disk) of the wing; usually the first M_2 in Diptera and cell R in Lepidoptera.

DISK   Noun. (Latin, *discus* = disk). 1. The upper, central surface of any anatomical structure or part. 2. The central area of a wing. 3. Orthoptera: The obliquely ridged outer surface of the hind femur in Saltatoria. Alternate Disk.

DISTAL   Adj. (Latin, *distare* = to stand apart). 1. Near or toward the free end of any appendage. 2. The part of a structure farthest from the body. Ant. Basal, Proximal.

DIURNAL   Adj. (Latin, *diurnae* = pertaining to day). Pertaining to activity patterns during daylight only. Compare Crepuscular, Nocturnal.

DNA   Abbreviation for deoxyribonucleic acid, the molecule that carries the genetic instruction that cells need to live, develop, and reproduce.

DORSAD   Adv. (Latin, *dorsum* = back; *ad* = toward). In the direction of the dorsum or back of an insect. Ant. Ventrad.

DORSAL   Adj. (Latin, *dorsum* = back). Of or belonging to the upper surface. Ant. Ventral.

DORSUM   Noun. (Latin, *dorsum* = back). The dorsal surface or plate. Ant. Ventrum.

DRONE   Noun. (Anglo-Saxon, *dran* = drone). Hymenoptera: A male bee or ant.

DUFF   Noun. (Middle English, *dogh* = dough). The partly decayed vegetable matter on the forest floor. Compare Debris, Detritus.

— E —

ECDYSIS   Noun. (Greek, *ek* = out; *dyein* = to enter; *ekdysai* = to strip). The process of shedding the integument during molting.

ECTOPARASITE   Noun. (Greek, *ektos* = outside; *parasitos* = parasite). A parasitic animal that lives on the external parts of its host. Examples: fleas and lice. Syn. Ectoparasitoid. Compare Endoparasite.

EDTA   Abbreviation for ethylenediaminetetraacetic acid. A buffer that can adhere to other molecules.

ELATERIFORM LARVA   1. Any larva that resembles an elaterid (wireworm) larva in form. 2. A slender larva, moderately heavily sclerotized, lacking elaborate ornamentation or sculpture, and bearing three pairs of thoracic legs.

ELYTRON   Noun. (Greek, *elytron* = sheath, from *elyein* = to roll around. Pl., elytra). 1. Coleoptera: Anterior leathery or chitinous wings that cover the hindwings. In repose, elytra typically meet and form a straight line along the middle of dorsum. 2. Orthoptera: The tegmina. 3. Heteroptera: The hemelytra.

EMBOLIUM   Noun. (Greek, *embolos* = wedge). The differentiated costal part of the corium in the forewing or hemelytron of some Heteroptera.

EMPODIUM   Noun. (Greek, *en* = in; *pous* = feet. Pl., empodia). 1. A bristle between the pulvilli of the foot in some Diptera. 2. The single, padlike median structure sometimes present between the insect claws.

ENDOPARASITE   Noun. (Greek, *endon* = within; *parasitos* = parasite). Any organism that develops as a parasite within the body of another organism at the expense and to the detriment of the "host." Examples: many Hymenoptera and some Diptera. Syn. Endoparasitoid. Compare Ectoparasite.

ENDOPTERYGOTA   Noun. (Greek, *endon* = within; *pterygion* = little wing). Insects with complete, complex metamorphosis in which the wings develop internally. Syn. Holometabola; Oligoneoptera; Oligoneuroptera. Compare Exopterygota.

EPISTOMAL SUTURE   (Greek, *epi* = upon; *stoma* = mouth). A suture across the face and separating the frons from the clypeus. The suture forms a strong internal ridge (sulcus). Typically transverse and straight, but arched in some groups of insects and absent in other groups. Syn. Frontoclypeal Suture.

ERUCIFORM LARVA   (Latin, *eruca* = caterpillar; *forma* = shape). Shaped like a caterpillar. Characterized by a well-developed head capsule, thoracic legs, and abdominal prolegs. Seen in Lepidoptera and Symphyta.

ESOPHAGUS   Noun. (Greek, *Oisophagos* = gullet). The part of the foregut between the mouth and the crop. Alternate: Esophagus.

EXARATE PUPA   A form of pupal development in which body appendages (legs and wings) are free from the body. Characteristic of the lower Endopterygota but not restricted to that group. Syn. Free Pupa. Compare Obtect Pupa.

EXOPTERYGOTA   Noun. (Greek, *exo* = outside; *pterygion* = little wing). The Heterometabola (Hemimetabola). A division of the Pterygota in which postembryonic immature insects pass through a simple (sometimes slight) metamorphosis and wing pads develop externally on the body. A pupal stage found in only a few forms (Thysanoptera). Compare Endopterygota. See also Paurometabola.

EXOPTERYGOTE   Adj. (Greek, *exo* = outside; *pterygion* = little wing). Pertaining to insects whose wings develop externally in all stages; insect with incomplete metamorphosis. Compare Endopterygote.

EXOSKELETON   Noun. (Greek, *exo* = outside; *skeletos* = hard). The entire body wall, to the inner side of which muscles are attached; the outside skeleton in insects.

EXUVIAE   Plural noun. (Latin, *exuere* = to strip off). 1. General: The integument of a larva or nymph that has been cast from the body during molting. 2. Diaspinae: The larval integument when cast from the body and incorporated into the scale cover. Incorrectly given as "exuvium" when a singular form of the noun is intended. Alternate Exuvia.

— F —

FACE   Noun. (Latin, *facies* = form, shape, face). 1. Upper or outer surface of any part or appendage. 2. Anterior aspect of the head between compound eyes from mouth to vertex. Usually applied to insects in which the head is vertical. 3. Ephemeroptera: A fusion of front and vertex. 4. Hymenoptera: Generally, area between antennae and clypeus; in bees face extends between the eyes to base of antennae. 5. Diptera: Area between base of antennae, oral margin, eyes, and cheeks.

FACET   Noun. (French, *facette* = small face). Any small face or surface, such as the parts, areas, or lens-like divisions of the compound eye. See also Ommatidium.

FAMILY   Noun. (Latin, *familia* = household). In zoological classification, a level in the taxonomic hierarchy below the order and above the genus. All zoological family names end in "ida"; all zoological subfamily names end in "inae." The familylevel taxon must include a type genus that contains a type species.

FEMUR   Noun. (Latin, *femur* = thigh. Pl., femora). The third and usually the stoutest segment of the insect leg. Articulated with the body via the trochanter and bearing the tibia at its distal margin.

FIXATION   Noun. (Latin, *fixus* = fixed). A histological procedure intended to 1 terminate life process quickly with minimum of distortion to cytological detail; 2 prevent autolysis; 3 prevent microbial action; and 4 increase the refractive index of tissue. See also Histology.

FLAGELLUM Noun. (Latin, *flagellum* = whip. Pl., flagella). 1. Part of the antenna beyond the pedicel (second segment). 2. Any whip or whiplike process. 3. The tail-like process of a spermatozoon.

FORAMEN MAGNUM The opening on the posterior surface of the head that gives passage to structures that extend from the head into the thorax. Syn. Occipital Foramen.

FORCEPS Noun. (Latin, *forceps* = tongs). 1. Hook or pincerlike processes at the apex of the abdomen of many insects. Depending upon the insect species, forceps are used in copulation, defense, predation, or sensory reception. 2. An instrument for grasping or holding delicate objects, similar to tweezers.

FOREGUT Noun. One of the three principal divisions of the alimentary canal. Originates at the mouth (buccal cavity) and terminates at the so-called gizzard. The foregut epithelium is formed from ectodermal invagination and is lined with chitin. Compare Hindgut, Midgut.

FOSSORIAL Adj. (Latin, *fossor* = digger). Structures modified for digging or burrowing.

FRASS Noun. (German, *fressen* = to devour). Solid larval excrement or macerated plant material fashioned by wood-boring insects that is often combined with excrement. Compare Meconium.

FREE PUPA See Exarate Pupa.

FRENULUM Noun. (Latin, *frenulum*, diminutive of *frenum* = bridle. Pl., frenula). The spine (simple in males, compound in females) at the base of the hindwings in many Lepidoptera. The frenulum projects beneath the forewing to unite the wings in flight.

FRONS Noun. (Latin, *frons* = forehead). The sclerite of the head between the arms of the epicranial suture. Typically, the frons bears the median ocellus.

FRONTOCLYPEAL SUTURE See Epistomal Suture.

FUNICLE Noun. (Latin, *funiculus* = small cord). 1. Antennal segments between the ring segments (anelli) or pedicel and the club (clavus). 2. A small cord; a slender stalk.

FURCA Noun. (Latin, *furca* = fork. Pl., furcae). 1. A forklike process. 2. The anal appendage used by Collembola for leaping. 3. The forked processes of the sternum of higher insects; an endosternite. 4. Genitalia of male Lepidoptera: A structure often consisting of paired halves.

— G —

GALEA Noun. (Latin, *galea* = helmet. Pl., galeae). In the generalized maxilla: The lateral sclerite attached to the distal margin of the stipes. Sometimes appearing two-segmented and often hoodlike. Subject to great modifications in Hymenoptera and Diptera; forms the proboscis (coiled tongue) in Lepidoptera. See also Maxilla.

GALL Noun. (Anglo-Saxon, *gaella* = gall). An abnormal growth of plant tissue. Caused by stimuli external to the plant, generally by insects, sometimes by parasitic fungi or other diseases of the plant.

GALL WASP 1. A member of the hymenopterous genus Cynips, which causes gall formation in plants. 2. In general, any hymenopterous insect that induces tumors on a plant.

GASTER Noun. (Greek, *gaster* = stomach; Late Latin, belly). Apocritous Hymenoptera: The posterior seven to eight segments of the abdomen, behind the constricted second segment. In Symphyta the abdomen is broadly attached to the abdomen; in Parasitic Hymenoptera and Aculeata the first abdominal segment has become separated from the remainder of the abdomen by a sclerotized annular constriction. The first (anteriormost) segment is called the *propodeum;* the second segment, which forms the constriction, is called the *petiole;* the remaining segments of the abdomen are collectively called the *gaster.*

GENA Noun. (Latin, *gena* = cheek. Pl., genae). The "cheek" or sclerotized area of the head below the compound eye extending to the gular suture. Odonata: The area between the compound eyes, clypeus, and mouthparts. Diptera: The space between the lower border of the compound eye and the oral margin, merging into the face at the front and limited by the occipital margin behind.

GENERIC NAME   The name of a genus. Always a single word (simple or compound), written with a capital initial letter and employed as a noun in the nominative singular.

GENICULATE   Noun. (Latin, *geniculum* = little knee). Elbowed or sharply bent. A term used to characterize the antennae of ants and some other insects.

GENITALIA   Plural noun. (Latin, *gegnere* = to beget). The internal and external sexual organs of insects.

GILL   Noun. (Middle English, *gile*, of Scandinavian origin).A respiratory organ in the aquatic immature stages of many insects. Usually hollow, thin-walled, lamellar, or filamentous projections from various regions of the body. Gills permit oxygen and other gases in solution to pass from the water into the body. Compare Trachea.

GLOSSA   Noun. (Greek, *glossa* = tongue. Pl., glossae). 1. The median lobe of the maxillae formed by the fusion of the two paraglossae and the two glossae in bees. 2. The paired inner lobes of gnathobases of the labium or fused second maxilla. 3. Loosely used as a synonym for the tongue, especially applied to the coiled structure of the Lepidoptera. See also Ligula.

GREGARIOUS   Adj. (Latin, *grex* = flock). 1. Nonsocial insects that live in societies or communities. 2. Several parasitic insects of one species that develop simultaneously on or in one host. Compare Solitary.

GRESSORIAL   Adj. Having legs fitted for walking.

GRUB   Noun. Scarabaeiform larva; an apodous larva having a tiny head, few sense organs, and a fleshy, rounded body, for example, larve of bees (Hymenoptera: Apoidea) and some Coleoptera.

GYNADROMORPH   Noun. (Greek, *gyne* = female; *aner* = male; *morphe* = form). An insect that has some male features and some female features. A sexual mosaic or intersex.

GYNADROMORPHISM   Noun. (Greek, *gyne* = female; *aner* = male; *morphe* = form). The expression or development of secondary sexual characters or features of both sexes in one individual.

— H —

HABITAT   Noun. (Latin, *habitare* = to inhabit). 1. Ecology: The area within which an organism is found, but not a particular location. The natural region that an organism inhabits or where it was found or taken.

HAIR   Noun. (Anglo-Saxon, *haer* = hair). 1. An epidermal outgrowth unique to mammals. 2. A term applied in insects to any slender, flexible filament largest at the base and often tapering toward the apex. See also Seta.

HALLER'S ORGAN   A sensory organ on tarsus 1 of ticks (acarines).

HALTERES   Plural noun. (Greek, *halter* = weight). Modified hindwings of Diptera that are clublike and consist of a base (scabellum), stem (pedicel), and knob (capitulum). Halteres oscillate during flight and serve as balancing organs to maintain stability during flight. Alternate: Balancing organs.

HAMULUS   Noun. (Latin, *hamulus* = little hook. Pl., hamuli). 1. Any small hook or hooklike process. 2. Orthoptera (treecrickets): Hooklike processes of the genitalia. 3. Siphonaptera: Movable sclerites from the lateral walls of the aedeagal palliolum. 4. Hymenoptera: (a) The small hooklike structures along the anterior margin of the hindwing that link with the posterior margin of the forewing during flight, (b) Penis valves (gonocoxites + gonostyli + volsellae).

HAUSTELLATE   Adj. (Latin, *haurire* = to drain). Pertaining to mouthparts modified for sucking. A term applied frequently to mouthparts of Homoptera-Hemiptera and Lepidoptera. See also Suctorial.

HEAD   Noun. (Anglo-Saxon, *heafod* = head). The first or anteriormost tagma of the insect body. The head bears the mouth structures and antennae and articulates with the thorax. Syn. Cranium. Compare Abdomen, Thorax.

HEMIMETABOLOUS   Adj. (Greek, *hemi* = half; *metabole* = change). Insects with an incomplete metamorphosis. The individual develops with gradual changes in size and shape from a first-instar nymph to an adult. Seen in exopterygote insects. Compare Holometabolous, Paurometabolous.

HEMOLYMPH   Noun. (Greek, *haima* = blood; *lympha* = water). The watery bloodlike or lymphlike nutritive fluid of the lower invertebrates. Insect blood.

HERBIVOROUS   Adj. (Latin, *herba* = green crop; *vorare* = to devour). 1. Pertaining to organisms that feed upon plant tissue. 2. Leaf feeding. 3. Feeding on growing plants.

HIBERNATION   Noun. (Latin, *hibernus* = wintry). A period of suspended development in animals occurring during seasonal low temperatures and short day length. 2. Inactivity during the winter. Compare Diapause.

HINDGUT   Noun. The portion of the intestine extending from the midgut valve to the anus, including the Malpighian tubules and anal glands. Divided into several anatomical regions: ileum, colon, rectum, anus. Syn. Proctodaeum. Compare Foregut, Midgut.

HISTOLOGY   Noun. (Greek, *histos* = tissue; *logos* = discourse). The study of the tissues of organisms at the cellular and subcellular levels.

HOLOMETABOLOUS   Adj. (Greek, *holos* = entire; *metabole* = change). Insects with a complete transformation during metamorphosis. Egg, larval, pupal, and adult stages are distinctly separated by profoundly different morphs at each stage. Compare Hemimetabolous, Paurometabolous.

HOLOTYPE   Noun. (Greek, *holos* = entire; *typos* = pattern). The single specimen selected by the author of a species as its type or the only specimen known at the time of description. The name bearer of a species. The type.

HOMONYM   Noun. (Greek, *homos* = same; *onyma* = name). 1. The same name for two or more different zoological entities. 2. One of two or more names applied to different taxa of animals. Rules for the treatment of scientific-name homonyms are given in the International Code of Zoological Nomenclature. See also Synonym.

HONEYDEW   Noun. 1. A sweetish excretion produced by insects such as aphids, whiteflies, and scale insects. 2. An exudate from the surface of some galls.

HOST   Noun. (Latin, *hospes* = stranger, guest, host). An organism that supplies nutrition or protection essential for the development of another organism (termed a parasite).

HYALINE   Adj. (Greek, *hyalos* = glass). Clear, transparent, or partly so; glasslike, waterlike in color. Alternate: Hyalinus.

HYPANDRIUM   Noun. (Greek, *hypo* = beneath; *aner* = male). 1. The apical abdominal sternal sclerite of many insects (sternum X of mayflies, sternum IX of Psocoptera). 2. The hypoproct sensu Needham; the subgenital plate sensu Snodgrass; the ninth abdominal sternum of male insect's sensu Crampton.

HYPERMETAMORPHOSIS   Noun. (Greek, *hyper* = above; *meta* = after; *morphosis* = shaping). 1. The process in which endopterygote insects as larvae change shape or substance during successive instars as a normal consequence of development. Examples are found in (but not restricted to) Coleoptera (Meloidae), Strepsiptera, Diptera (Acroceridae, Bombyliidae), Lepidoptera (Epipyropidae), and Hymenoptera (Eucharitidae, Perilampidae). 2. The process in which insects pass through a larger-than-expected number of stages, with supernumerary stages interpolated between the mature larva and the adult (archaic).

HYPERPARASITE   Noun. (Greek, *hyper* = above; *para* = beside; *sitos* = food). An organism that develops as a parasite upon another parasite. Syn. Secondary Parasite. Compare Tertiary Hyperparasitism.

HYPOGNATHOUS   Adj. (Greek, *hypo* = under; *gnathos* = jaw). In entomological usage, pertaining to insects with the head vertically oriented and the mouth directed downward. Compare Opisthognathous, Orthognathous, Prognathous.

HYPOPHARYNX   Noun. (Greek, *hypo* = under; *pharyngx* = pharynx). 1. A tonguelike sensory structure projecting from the oral cavity. 2. The upper surface of the labium, which serves as an organ of taste.

HYPOPYGIUM   Noun. (Greek, *hypo* = under; *pyge* = rump. Pl., hypopygia). 1. Adults: The posterior portion of the abdomen; the anus; the ventral sclerite of the anal opening. 2. Diptera: The male sexual organs and terminal segments of the abdomen; propygium. 3. Coleoptera: The last segment behind the elytra.

HYPOSTOMA   Noun. (Greek, *hypo* = under; *stoma* = mouth). 1. Diptera: Part of the head between the antennae, compound eyes, and mouth. 2. Hemiptera: The lower part of the face. 3. Ticks: A dartlike structure arising from the median ventral surface of the basis capituli. 4. The upper lip or labrum of the Crustacea. Alternate: Hypostome.

— I —

IMAGO   Noun. (Latin, *imago* = image. Pl., imagines, imagos). The adult-stage or sexually mature insect.

INQUILINE   Noun. (Latin, *inquilinius* = tenant). 1. An insect that lives as a "guest" of other insects. Typically, an insect that habitually lives within the nests of other species, chiefly bees, ants, wasps, or termites. 2. An insect that lives within the gall developed by another species of insect. Compare Commensal, Parasite.

INSTAR   Noun. (Latin, *instar* = form). A postembryonic immature insect between molts. In the larva, instars are numbered to designate the various periods (e.g., the first-instar larva is the stage between the egg and the first molt). See also Apolysis, Ecdysis, Stadium.

INTEGUMENT   Noun. (Latin, *integumentum* = covering). The outer covering of the insect body. Consists of several layers, including a living epidermal layer and secreted layers that have different functions.

INTERCALARY VEIN   (Latin, *intercalaris* = inserted). Any added or supplementary wing vein.

INTERSEX   Noun. (Latin, *inter* = between; *sexus* = sex). An insect whose sexual characteristics include male and female features. See also Gynadromorph.

INTERTIDAL   Adj. (Latin, *inter* = between; Anglo-Saxon, *tid* = tide). Occurring on the beach between high-water and low-water levels.

INVAGINATION   Noun. (Latin, *in* = into; *vagina* = sheath). A pouch or sac formed by an infolding or indrawing of the outer surface. An infolding or projection toward the inside of the body.

— J —

JOHNSTON'S ORGAN   An auditory organ of the scolopophorous-type located in the second segment (pedicel) of the antenna in most larger orders of insects. Compare Tympanal Organ.

JOINT   Noun. (Old French, *joindre* from Latin, *jungere* = to join). 1. A point or area of articulation between two sclerites. 2. The area of fusion between limb segments. 3. Nonsclerotized cuticula between adjacent sclerotized regions of the integument.

JUGAL LOBE   An area of the forewing projecting posteriad and contacting the hindwing of the insect. Syn. Jugal Region, Fibula.

JUGATAE   Adj. (Latin, *jugum* = yoke). Lepidoptera having a jugum wing-coupling apparatus; any Lepidoptera with a jugum.

JUGUM   Noun. (Latin, *jugum* = yoke. Pl., juga). 1. Heteroptera: Two lateral lobes of the head, one on each side of the tylus. 2. Some Lepidoptera and Trichoptera: A lobe or process at the base of the forewings, overlapping the hindwings and holding the wings together in flight.

— K —

KAIROMONE   Noun. (Greek, *kairos* = opportunistic). A chemical produced or acquired by one organism that mediates behavioral or physiological response in another organism that is favorable to the receiver but not the emitter. See also Allomone, Pheromone.

KEEL   Noun. An elevated ridge or caarina.

KEY   Noun. (Middle English, *key* = key). A taxonomic device by which objects are identified based on decisions for suites of characters or character states. Types: Dichotomous Key, Pictorial Key, Tabular Key.

– L –

LABELLUM   Noun. (Latin, *labellum* = small lip. Pl., labella). 1. The sensitive, ridged, apical mouth structures of some Diptera. 2. A prolongation of the labrum covering the base of the rostrum in Coleoptera and Hemiptera. 3. A small spoon-shaped lobe at the apex of the glossa in the honeybee.

LABIAL GLANDS   The "salivary glands" of insects, opening by a median duct between the base of the hypopharynx and the labium or on the hypopharynx.

LABIAL PALP   (Pl., palpi). The segmented sensory appendages of the insect labium; attached to the palpiger and shorter than the maxillary palp. Compare Maxillary Palp.

LABIUM   Noun. (Latin, *labium* = lip. Pl., labia). The second maxilla, or "lower lip," of the insect mouth. A compound, bilaterally symmetrical sclerite that forms the floor of the mouth in mandibulate insects. Located behind the first maxilla and opposed to the labrum. Sometimes referred to as the "tongue."

LABRUM   Noun. (Latin, *labrum* = lip. Pl., labra). The upper lip, which covers the base of the mandible and forms the roof of the mouth.

LACINIA   Noun. (Latin, *lacinea* = flap. Pl., laciniae). 1. A bladelike sclerite. The mesal endite sclerite attached to the maxillary stipes, or the inner lobe of the first maxilla that is articulated to the stipes and bears brushes of setae or spines. 2. Diptera: A flat lancetlike piercing structure that is never jointed. 3. Psocoptera: The styliform appendage of Ribaga; a hard, elongated rod, slightly bifurcated at its free end and ensheathed by the galea. See also Maxilla.

LAMELLA   Noun. (Latin, *lamella* = small plate. Pl., lamellae). A thin plate or leaflike process; a parademe.

LARVA   Noun. (Latin, *larva* = ghost. Pl., larvae). An immature stage of holometabolous insect; the developmental stage following the egg stage and preceding the pupal stage. Differing fundamentally in form from the adult. In a strict zoological sense, the immature form of animals that undergo metamorphosis. Compare Naiad, Nymph. See also Caterpillar, Grub, Maggot, Slug.

LATERAD   Adv. (Latin, *latus* = side; *ad* = toward). Toward the side and away from the median line.

LATERAL   Adj. (Latin, *latus* = side.) Relating, pertaining, or attached to the side.

LEAFMINER   An insect (usually a larva) that makes a tunnel between the upper and lower surfaces of a leaf.

LEG   Noun. (Middle English, *legge* = leg). Cuticular appendage projecting from the ventrolateral portion of the thoracic wall of the insect body. Six legs are present in the adult and nymphal Paurometabola; the number varies in larvae, depending upon the taxonomic group. Each leg is divided into component parts, including the coxa, trochanter, femur, tibia, and tarsus. Primarily an appendage responsible for locomotion, but other adaptations have evolved, including prey capture, courtship, and excavation. See also Cursorial, Fossorial, Gressorial, Raptorial, Saltatorial.

LIGULA   Noun. (Latin, *ligula* = little tongue. Pl., ligulae). The median sclerite of the labium. The ligula inserts onto the distal margin of the prementum. Usually unpaired (single); sometimes paired. Often regarded as synonymous with "glossa" and "tongue."Corresponds to the united laciniae of the right and left maxillae.

LITTORAL   Adj. (Latin, *litus* = seashore). Dwelling on the seashore or the edge of freshwater.

– M –

MACROPTEROUS   Adj. (Greek, *makros* = large; *pteron* = wing). Long- or large-winged. Wings that are not reduced in size and presumably fully functional in active flight. Compare Brachypterous, Micropterous.

MAGGOT   Noun. (Middle English, *magot* = grub). The larva of Diptera: Without a distinct head, legless, with body usually pointed anteriorly and blunt posteriorly.

MANDIBLE   Noun. (Latin, *mandibulum* = jaw). The anteriormost pair of oral appendages in the insect head. Typically the hardest part of the insect integument; usually stout and highly modified in shape. Toothlike in chewing insects; needle- or sword-shaped in piercing-sucking insects.

MASK   Noun. (French, *masque* = mask). Dragonflies: The modified extensible labium of the nymph, which at rest conceals the mouthparts.

MAXILLA   Noun. (Latin, *maxilla* = jaw. Pl., maxillae. Without qualifying adjective). The second pair of jaws in mandibulate insects. The most persistent mouthparts when the mouth is modified, and represented by some functional part in all insects in which the mouth structures are useful. Second maxillae: The labium or third pair of jaws in a mandibulate insect; composed of five parts. See also Stipes.

MAXILLARY PALP   (Pl., palpi). The palp carried by the stipes on its outer apical surface; consisting of one to seven segments. Presumably sensory in function. Compare Labial Palp.

MECONIUM   Noun. (Latin, from Greek, *mekonion* = poppy juice, from *mekon* = a poppy. Pl., meconia). 1. The liquid substance excreted from the anus by certain holometabolous insects after emergence from the chrysalis or pupa. 2. Parasitic Hymenoptera: The pelletlike excrement of larvae or prepupae. Compare Frass.

MEDIAL   Adj. (Latin, *medius* = middle). Referring to or at the middle of a structure.

MELANIN   Noun. (Greek, *melas* = black). Any one of a group of organic pigments that produces black, amber, and dark brown colors by deposition in the cuticle.

MEMBRANE   Noun. (Latin, *membrana* = membrane). 1. Any thin, transparent, flexible body tissue. 2. The wing tissue between the veins. 3. Heteroptera: The thin, transparent, or translucent apex of the hemelytra, as distinguished from the thickened basal part, the corium.

MEMBRANOUS   Adj. (Latin, *membrana* = membrane.) Tissue that is thin, pliable, and semitransparent; like a membrane.

MESAD   Adv. (Latin, *medius* = middle; *ad* = toward). Toward or in the direction of the median plane of the insect body.

MESAL   Adj. (Latin, *medius* = middle). Pertaining to, situated on, or in the median plane of the body.

MESONOTUM   Noun. (Greek, *mesos* = middle; *noton* = back. Pl., mesonota). The upper surface of the second (middle) thoracic segment (mesothorax) of the insect body.

MESOPLEURON   Noun. (Greek, *mesos* = middle; *pleura* = side. Pl., mesopleura). 1. The lateral surface of the mesothorax. 2. Diptera: The upper portion of the episternum of the mesothorax. 3. Hymenoptera: The sclerite below the insertion of the wings.

MESOTHORAX   Noun. (Greek, *mesos* = middle; *thorax* = chest). The second (middle) thoracic segment, which bears the middle legs and the anterior wings. See also Metathorax, Prothorax.

METAMORPHOSIS   Noun. (Greek, *meta* = change of; *morphe* = form). The transformation in shape or substance during successive stages of development.

METANOTUM   Noun. (Greek, *meta* = after; *noton* = back). 1. The upper surface of the third (posterior) thoracic segment (metathorax). 2. Diptera: The oval, arched sclerite behind the scutellum; best developed in flies with a long, slender abdomen.

METAPLEURON   Noun. (Greek, *meta* = behind; *pleura* = side. Pl., metapleura). 1. In general, the lateral area of the metathorax. 2. Diptera: The pleuron of the metathorax. 3. Hymenoptera: The sclerite behind and below the insertion of the hindwings.

METATHORAX   Noun. (Greek, *meta* = behind; *thorax* = chest). The third thoracic segment, which bears the hind legs and hindwings. The segment is variable: sometimes distinct, sometimes closely united with the mesothorax, and sometimes appearing as a part of the abdomen. See also Mesothorax, Prothorax.

MICROPTEROUS   Adj. (Greek, *mikros* = small; *pteron* = wing). Small-winged. A condition in which the wings of an insect are disproportionately small and presumably less effective as instruments of flight. Compare Brachypterous Macropterous.

MICROTRICHIUM Noun. (Greek, *mikros* = small; *thrix* = hair. Pl., microtrichia). 1. Minute hairlike structures found on the wings of some insects (e.g., Mecoptera and Diptera). Microtrichia resemble small covering setae, but the absence of basal articulation distinguishes them. 2. Plecoptera: Cuticular projections of unknown function.

MIDGUT Noun. The middle portion of the digestive system. Compare Foregut, Hindgut.

MOLT Verb. (Latin, *moutare* = to change). 1. The process by which insects shed elements of the integument during larval, nymphal, or naiadal growth. Noun. 2. The period of transformation when the larva, nymph, or naiad changes from one instar to another. 3. The cast portion of the integument resulting from the processing of molting.

MULTIVOLTINE Adj. (Latin, *multus* = many; Italian, *volta* = turn). Pertaining to organisms with many generations in a year or season. A term often applied to Lepidoptera. Compare Univoltine.

MYIASIS Noun. (Greek, *myia* = fly). Disease or injury of animals, particularly humans, caused by the attack of dipterous larvae.

MYRMECOPHILE Noun. (Greek, *myrmex* = ant; *philos* = loving). A commensal or parasite of ants that inhabits ant nests. Some myrmecophiles are tended by the ants, others prey upon the ants or their brood. Alternate: Myrmecophil.

MYRMECOPHILOUS Adj. Ant-loving, being applied to insects that live in ant (Hymenopera: Formicidae) nests.

– N –

NACREOUS Adj. (French, *nacre* = with a pearly luster). Resembling mother-of-pearl.

NAIAD Noun. (Greek, *naias* = water nymph). The aquatic nymph of the Hemimetabola. A name given by Comstock to distinguish between forms living in water and the terrestrial nymphs of the Paurometabola. See also Nymph.

NAIL Noun. (Anglo-Saxon, *naegel* = nail). A tarsal claw. Specifically the stout, apically pointed claws in predatory Heteroptera. See also Unguis.

NAKED Adj. (Anglo-Saxon, *nacod* = naked). Not clothed; lacking a covering of setae. Applied to the pupa when it is not enclosed in a cocoon or other covering.

NECK Noun. (Middle English, *necke* = neck). The cervix or slender connecting structure between head and thorax of insects that have the head free. Any contraction of the head at its juncture with the thorax.

NECROPHAGOUS Adj. (Greek, *nekros* = dead; *phagein* = to devour). Pertaining to organisms that feed on dead or decaying matter.

NECROPHILOUS Adj. (Greek, *nekros* = dead; *philein* = to love). Pertaining to organisms that are associated with dead and decaying plants or animals.

NECROSIS Noun. (Greek, *nekrosis* = deadness). 1. The condition of decay. 2. The death of cells or tissues. 3. A diseased condition of plant tissues that causes them to turn black and decay.

NECTAR Noun. (Greek, *nektar* = death overcoming; drink of gods). Sweet, frequently scented, substances secreted by flowers, nectarines, or other plant structures. Nectar provides nutrition for many insects, particularly Hymenoptera, Lepidoptera, and some Diptera.

NEOTENY Noun. Retention of juvenile characters in the adult.

NEW SPECIES A species that is purportedly unknown to science and does not have a scientific name. A term appended to the description of formally described and nomenclaturally validated taxa of the species level.

NOCTURNAL Adj. (Latin, *nox* = night). Active during the night. Applied to insects that fly or are active at night. Ant. Diurnal.

NODULE Noun. (Latin, *nodulus* = diminutive of *nodus* = knob). A small knot or swelling.

NOMENCLATURE   Noun. (Latin, *nomen* = name; *calare* = to call). The application of proper scientific names to any taxon based on the prescribed rules.

NOTUM   Noun. (Greek, *noton* = back. Pl., nota). The dorsal or upper surface of a body segment. Particularly applied to the thorax. Syn. Tergum. See also Dorsum. Ant. Sternum.

NYMPH   Noun. (Greek, *nymphe* = chrysalis). An immature insect that emerges from the egg in a relatively advanced stage of morphological development and that undergoes gradual metamorphosis. Nymphs differ from adults in displaying incomplete wings and genitalia. See also Larva, Naiad. Pupa.

— O —

OBTECT   Adj. (Latin, *obtectus* = covered over). Covered; within a hard covering.

OBTECT PUPA   A pupa with the appendages and body fused by a hardening of the exoskeleton. Termed the "theca" in older works. Lepidoptera: A more highly developed specialized pupa, smooth and rounded, with only the fourth, fifth, and sixth segments free. See Coarctate Pupa, Exarate Pupa.

OCCIPITAL   Adj. (Latin, *occiput* = back of head). Pertaining to the occiput, or the back part of the head.

OCCIPUT   Noun. (Latin, *occiput* = back of head). 1. General: The posterior portion of the epicranium between the vertex and the neck; rarely present as a distinct sclerite or clearly demarcated by benchmark sutures. 2. Diptera: The entire posterior surface of the head.

OCELLUS   Noun. (Latin, *ocellus* = little eye. Pl., ocelli). The simple eye of many adult insects, consisting of one biconvex lens. Typically three in number but reduced to one or two in some insects or absent. See also Compound Eye.

OLFACTION   Noun. (Latin, *olfacere* = to smell). Behavior and physiology: The perception of odors, or the sense of smell.

OMMATIDIUM   Noun. (Greek, *ommation* = little eye; *idion* = diminutive. Pl., ommatidia). One of the visual elements that compose the compound eye.

OMNIVOROUS   Adj. (Latin, *omnis* = all; *vorare* = to devour). Feeding generally on animal or vegetable food or on both.

OPERCULATE   Adj. (Latin, *operculum* = lid). Pertaining to a lidlike covering.

OPERCULUM   Noun. (Latin, *operculum* = lid. Pl., opercula). 1. A lid or cover. 2. The lidlike portion of the insect eggshell. Typically over the anterior pole that partly separates and elevates from the body of the eggshell to permit the escape of the nymph, naiad, or larva during eclosion. 3. Aleurodidae: The lidlike structure covering the vasiform orifice. 4. Cicadidae: One of the paired plates covering the timbals. 5. Lecaniine coccids: The anal plates. 6. Diptera: The chitinous lower part of the muscid mouth; the labrum-ephipharynx of Dimmock; the scutes covering the metathoracic stigmata.

OPISTHOGNATHOUS   Adj. (Greek, *opisthe* = behind; *gnathos* = jaw). 1. General zoological usage: Pertaining to animals with retreating jaws. 2. Entomological usage: Pertaining to one of the three principal orientations of the insect head in relation to the body. Characterized by a posteroventral position of the mouthparts resulting from a deflection of the facial region. Seen in most Homoptera. Compare Hypognathous, Prognathous.

ORAD   Adv. (Latin, *os* = mouth; *ad* = toward). Toward the mouth.

ORAL   Adj. (Latin, *os* = mouth.) Pertaining to the mouth. Any structure having to do with the mouth.

ORTHOGNATHOUS   Adj. (Greek, *orthos* = straight; *gnathos* = jaw). Literally, straight jaws. Pertaining to a head whose primary axis is at a right angle to the primary axis of the body. Compare Hypognathous, Opisthognathous, Prognathous.

OVARIOLE   Noun. (Latin, *ovarium* = ovary). An egg tube of the insect ovary. Functionally, three types of ovarioles have been identified in insects: panoistic, polytrophic, telotrophic.

OVARY   Noun. (Latin, *ovarium* = ovary). The enlarged basal portion of the female reproductive system, which typically consists of paired lateral ovaries, each of which is composed of several ovarioles.

OVATE   Adj. (Latin, *ovum* = egg). Egg-shaped. Traditionally viewed as shaped like the egg of a bird because egg shape in insects is highly variable.

OVIDUCT   Noun. (Latin, *ovum* = egg; *ducere* = to lead). The distal tubular portion of the female reproductive system that transmits the egg outside the body.

OVIPARITY   Noun. (Latin, *ovum* = egg; *parere* = to bring forth). Conventional form of development in most insects. The egg is fertilized within the body of the female, with embryonic development resulting in the development of one individual from one egg that has been fertilized by one sperm. Embryogenesis usually proceeds after the egg has been oviposited, but in some instances considerable embryogenesis may occur within the body of the mother.

OVIPAROUS   Adj. Reproducing by eggs laid by the female. See Ovoviviparous, Viviparous.

OVIPOSITOR   Noun. (Latin, *ovum* = egg; *ponere* = to place). The external tubular part of the female reproductive system through which eggs are passed. The ovipositor may be rigid and fixed in length or flexible and telescopic.

OVOVIVIPARITY   Noun. (Latin, *ovum* = egg; *vivus* = living; *parere* = to bring forth). A method of reproduction in which eggs are maintained in the common oviduct (vagina) until eclosion or eclosion occurs soon after oviposition. A common form of development in Diptera (Tachinidae) and some Coccoidea, and represented in most orders of insects. Typically, the wall of the vagina is extensively tracheated to provide oxygen for the developing embryo. There is a trend toward a reduction in the number of eggs laid per female in some groups that utilize ovoviviparity. Compare Parthenogenesis, Polyembryony, Viviparity.

OVOVIVIPAROOUS   Adj. Producing living young by the hatching of the ovum while still within the mother.

OVUM   Noun. (Latin, *ovum* = egg. PI., ova). A female gamete.

– P –

PAEDOGENESIS   Noun. (Greek, *pais* = child; *genesis* = descent). Reproduction by an immature form. A method of reproduction in which ovaries become functional during the larval stage and eggs develop parthenogenetically. Some of the developing eggs result in larvae that themselves become parthenogenetic; other developing eggs result in normal adults. Seen in some Diptera and Coleoptera. Alternate: Pedogenesis. Compare Neoteny.

PALP   Noun. (Latin, *palpare* = to stroke. PI., palpi). A paired, digitiform mouthpart appendage of the maxilla and labium that is tactile or chemosensory in function. See also Labial Palp, Maxillary Palp. Alternate: Palpus.

PARASITE   Noun. (Greek, *parasitos*; Latin, *para* = beside; *sitos* = food). An organism that lives in, on, or at the expense of another organism during at least part of its lifetime.

PARASITOID   Noun. (Greek, *parasitos* = parasite; *eidos* = form). 1. A transitional condition between predation and parasitism in the classical sense. The parasitoid larva is parasitic during the early stages and episitic during later development (Tachinidae). 2. A hymenopterous parasite that eventually kills its host.

PARATYPE   Noun (Greek, *para* = beside; *typos* = pattern). Nomenclature: Any specimen in a series from which a description has been prepared, other than the one specified as the type specimen or holotype of the species. Incorrectly, a specimen that has been compared with the type.

PARTHENOGENESIS   Noun. (Greek, *parthenos* = virgin; *genesis* = descent). Reproduction without fertilization. Development of individuals from egg cells without fertilization by a male gamete. A reproductive phenomenon common in some groups of arthropods, including the Acari and Insecta.

PATAGIUM   Noun. (Latin, *patagium* = boarder. PI., patagia). Lepidoptera: Lobelike structures covering the base of the forewings; often used synonymously with tegula or squamula. Mosquitoes: A sausage-shaped body on each side of the prothorax in front of the first pair of spiracles.

PATHOGEN Noun. (Greek, *pathos* = suffering; *genes* = producing). Any disease-producing microorganism.

PATHOGENIC Adj. (Greek, *pathos* = suffering; *genes* = producing). Disease-causing or disease-producing. A term applied to organisms that cause or carry disease.

PATHOLOGICAL Adj. (Greek, *pathos* = suffering; *logos* = discourse). A diseased or abnormal condition; unhealthy or arising from unhealthy conditions.

PATHOLOGY Noun. (Greek, *pathos* = suffering; *logos* = discourse). The study of diseases.

PAUROMETABOLA Noun. (Greek, *pauros* = little; *metabole* = change). A division of the Heterometabola characterized by a gradual development in which the young resemble the adults in general form and mode of life. Examples include Dermaptera, Orthoptera, Embioptera, Isoptera, Zoraptera, Corrodentia, Mallophaga, Anoplura, Heteroptera, most Homoptera.

PAUROMETABOLOUS Adj. (Greek, *pauros* = little; *metabole* = change). Pertaining to organisms with metamorphosis in which the changes of form are gradual and inconspicuous.

PCR Abbreviation for polymerase chain reaction. A method in molecular biology used to make many copies of a specific DNA sequence that can be used in many different ways.

PECTEN Noun. (Latin, *pecten* = comb). 1. Any comblike structure. 2. Hymenoptera: Curved, rigid setae on the base of the maxilla and labium; compact rows of tibial spines on fossorial wasps; rows of spines on the tarsomeres of pollen-gathering bees; any series of bristles arranged like a comb. 3. Mosquito larvae: Comblike teeth on the breathing tube. 4. Lepidoptera: Stiff, scalelike setae on the antennal scape.

PECTINATE Adj. (Latin, *pecten* = comb). Comblike. Applied to structures, especially antennae, with even processes like the teeth of a comb.

PEDICEL Noun. (Latin, *pediculus* = small foot). 1. Generally, a stalk or stem supporting an organ or other structure. 2. The second segment of the insect antenna, with intrinsic musculature and forming the pivot between the scape and funicle.

PEDIPALP Noun. (Latin, *pes* = foot; *palpare* = to feel, Pl., pedipalpi, pedipalps). Chelicerata: The second pair of appendages on the cephalothorax; used in crushing prey, corresponding to the mandibles in Mandibulata.

PELAGIC Adj. (Greek, *pelagos* = sea). Pertaining to organisms inhabiting the open sea; oceanic.

PENIS Noun. (Latin *penis* = penis. Pl., penes). The flexible, partially membranous, intromittent, copulatory organ of the male insect. See also Aedeagus.

PENULTIMATE Adj. (Latin, *paene* = almost; *ultimus* = last). Adjacent to the terminal segment; next to the last.

PERIPHERAL Adj. (Greek, *peripherein* = to move around). Relating to the outer margin or distant from the center and near the circumference.

PERIPNEUSTIC Adj. (Greek, *peri* = around; *pneustikos* = breathing). Pertaining to the respiratory system of insect larvae with spiracles in a row on each side of the body, with those of the wing-bearing segments closed.

PERITROPHIC MEMBRANE The delicate membrane surrounding the food in the midgut. Present in most insects.

PETIOLE Noun. (Latin, *petiolus* = small foot. Pl., petioles, petioli). 1. A stem or stalklike projection. 2. The slender segment between the thorax and abdomen in certain Diptera and apocritous Hymenoptera. In the latter, a pedicel formed of only one segment, or the first segment of a two-segmented pedicel in ants.

PHARATE Adj. (Greek, *pharos* = loose mantle). A term applied to the condition of an instar within a previous cuticle before ecdysis.

PHARNYX Noun. (Greek, *pharyngx* = gullet). The posterior portion of the mouth and upper part of the throat; a slight enlargement at the beginning of the esophagus. Diptera: Sometimes restricted to the space between the hypopharynx and subclypeal pump, and then equivalent to the subclypeal tube.

PHEROMONE   Noun. (Greek, *phereum* = to carry; *horman* = excite, stimulate). A chemical compound secreted by an animal that mediates the behavior of an animal belonging to the same species. See also Allomone, Kairomone.

PHORESY   Noun. (Greek, *pherein* = to bear, to carry). A form of symbiosis in which one organism is carried on the body of another, larger-bodied organism but the former does not feed on the latter. Commonly manifested as mites on insects and insects on insects.

PHRAGMA   Noun. (Greek, *phragma* = fence. Pl., phragmata). A partition, dividing membrane, or structure. A transverse partition of the endoskeleton; an internal ridge from the endocuticle to which a muscle is attached. Phragmata typically occur at the junction of tergites, pleurites, and sternites, especially in the thorax. See also Apodeme, Suture.

PHYLOGENETIC CLASSIFICATION   A system of classification that purports to be natural, in that all subordinate taxa are members of a higher category (genus, family, order) by genetical relatedness or descent through evolutionary process.

PHYSOGASTRIC   Adj. (Greek, *physan* = to blow up; *gaster* = belly). Pertaining to females with a swollen or abnormally distended abdomen and mainly or entirely membranous with sclerites small and widely separated (e.g., a queen termite).

PHYSOGASTRY   Noun. (Greek, *physan* = to blow up; *gaster* = belly). A phenomenon in which a female displays a pathological enlargement of the abdomen associated with the development of immatures within the body. Females give birth to larvae or adults via "birthing openings" or rupture of the integument. The condition is characteristic of pygmephoroid mites.

PHYTOPHAGOUS   Adj. (Latin).Feeding in or on plants.

PIERCING-SUCKING MOUTHPARTS   Mouthparts with mandibles or maxillae or both (modified into stylets) fitted for piercing plant or animal tissue.

PILE   Noun. (Latin, *pilleus* = felt cap). A setose or furlike covering of thick, fine, short, erect setae that gives a velvetlike appearance.

PILIFEROUS   Adj. (Latin, *pilus* = hair; *ferre* = to carry). Bearing a vestiture of setae or forming a pile.

PILOSE   Adj. (Latin, *pilosus* = hairy). Covered with soft down or short setae; covered with long, flexible setae; covered with long, sparse setae.

PLANIDIUM   Noun. (Greek, *pianos* = wandering; *idion* = diminutive. Pl., planidia). The hypermetamorphic, migratory, first-instar larva of some parasitic insects. Morphologically characterized by a legless condition and somewhat flattened body that often displays strongly sclerotized, imbricated integumental sclerites, and spinelike locomotory processes. A term most appropriately restricted to Hymenoptera (Euchartiidae, Perilampidae, some Ichneumonidae) and Diptera (Tachinidae). Incorrectly used interchangeably with Triungulin.

PLANTA   Noun. (Latin, *planta* = sole of foot. Pl., plantae). 1. The basal segment of the posterior tarsus in pollen-gathering Hymenoptera. 2. The ventral surface of the posterior tarsal segments or anal clasping legs at the end of the caterpillar's abdominal proleg.

PLASTRON   Noun. (French, *plastron* = breast plate). A film of gas held in place by setae or cuticular modifications. Frequently found on the body of aquatic insects.

PLATE   Noun. (Greek, *platys* = flat). 1. Any broad, flat surface, such as a sclerite. 2. Coccids: A thin projection of the pygidium.

PLEOMORPHIC   Adj. (Greek, *pleion* = more; *morphe* = form, shape). Pertaining to pleomorphism.

PLEOMORPHISM   Noun. (Greek, *pleion* = more; *morphe* = form, shape). Structure that is characterized by having many forms. Syn. Polymorphism.

PLEURON   Noun. (Greek, *pleura* = side. Pl., pleura). Lateral sclerites between the dorsal and sternal parts of the thorax. In general, the sides of the body between the dorsum and the sternum.

PLEXUS Noun. (Latin, *plexus* = interwoven). 1. Any complicated network of vessels, nerves, fibers, or tracheae. 2. A knot.

PLICA Noun. (Latin, *plicare* = to fold). 1. General: A fold, convolution, or wrinkle. 2. A longitudinal plait of a wing. 3. The lamellate infolded thickening of the anterior and posterior margin of the abdominal segments.

PLUMOSE Adj. (Latin, *pluma* = feather). Feathered, like a plume. Antennae that have long, ciliated processes on each side of each joint.

POLLEN BRUSH A vernacular term for appendicular structures (typically setae) adapted for collecting pollen from the anthers of flowers. See also Scopa.

POLLINOSE Adj. (Latin, *pollen* = fine flour). Covered with a loose, mealy, often yellow dust, like the pollen of flowers.

POLYEMBRYONY Noun. (Greek, *polys* = many; *embron* = fetus). The form of reproduction in which several embryos develop from one fertilized egg. Seen in Strepsiptera and parasitic Hymenoptera, including Encyrtidae, Platygasteridae, and Dryinidae.

POLYGAMOUS Adj. (Greek, *polys* = many; *gamos* = marriage). Pertaining to male sequestering or copulating with many females. Ant. Monogamous.

POLYGAMY Noun. (Greek, *polys* = many; *gamos* = marriage). The condition in which one male copulates with or inseminates many conspecific females.

POLYMORPHIC Adj. (Greek, *polys* = many; *morphe* = form). Pertaining to a population of individuals that occur in several forms. Differences may be apparent in sex, in season, in locality, or without apparent reason. Alternate: Polymorphous.

POLYMORPHISM Noun. (Greek, *polys* = many; *morphe* = form). 1. The condition of having several forms in one species. 2. The occurrence of different forms of structure or different forms of organs in one individual during different periods of life. See also Dimorphism.

POOTER Noun. A vernacular term coined by an English entomologist for an aspirating device used to collect small, highly mobile insects. Named in honor of F. W. Poos, an American entomologist who employed the device to collect Cicadellidae.

PORE Noun. (Greek, *poros* = channel). 1. A large, isolated puncture. 2. A minute impression that perforates the surface. 3. Any small, round opening on the surface of a structure.

POSTABDOMEN Noun. (Latin, *post* = after; *abdomen* = belly). The posterior segments (including the genital segments) of the female insect abdomen retracted within a genital sinus.

POSTCLYPEUS Noun. (Latin, *post* = after; *clypeus* = shield). 1. General: The posterior or upper part of the clypeus when any line of demarcation exists. 2. Odonata: The upper portion of a transversely divided clypeus. 3. Psocidae: A peculiar inflated structure behind the clypeus. Syn. Eupraclypeus, Nasus, Afternose, Paraclypeus, First Clypeus, Clypeus Posterior.

POSTERIOR Adj. (Latin, *posterior* = latter). A term of position pertaining to a structure situated behind the axis. Toward the rear, caudal, or anal end of the insect. Ant. Anterior

PREDATOR Noun. (Latin, *predaetor* = hunter). Any animal that overpowers, kills, and consumes other organisms (called prey) for food. Compare Parasite.

PREPUPA Noun. (Latin, *prae* = before; *pupa* = puppet. Pl., prepupae). 1. A quiescent instar between the end of the larval period and the pupal period. 2. An active but nonfeeding stage in the larva of the Holometabola. 3. A full-fed larva.

PRETARSUS Noun. (Latin, *prae* = before; Greek, *tarsos* = sole of foot. Pl., pretarsi). The terminal limb segment. In most adult insects and nymphs, the pretarsus usually comprises a pair of lateral claws (ungues) and reduced median parts (arolium, unguitractor plate, median claw). In most larvae, a simple clawlike segment, as a rule withdrawn into the terminal joint or absent.

PRIMORDIAL   Adj. (Latin, *primordium* = beginning). Primitive in development. The first or earliest point in time of a developing structure or lineage.

PROBOSCIS   Noun. (Greek, *proboskis* = trunk. Pl., proboscises, proboscides). The extended mouth structure of an insect. Typically applied to the extensile mouthparts of adult Diptera, the beak of Hemiptera (labium enclosing maxillary and mandibular stylets), the tongue of Lepidoptera (fused galeae and maxillae), and the mouth of long-tongued bees (labium).

PROGNATHOUS   Adj. (Greek, *pro* = before; *gnathous* = jaw). Entomological usage: Orientation of the insect head with a foramen magnum near the vertex and with mandibles directed anterior and positioned at the anterior margin of the head. When viewed in lateral aspect, the primary axis of the head is horizontal. The prognathous condition is displayed by some predaceous forms and other insects living in concealed situations such as between bark and wood or similar confined habitats. Compare Hypognathous, Opisthognathous.

PROLEG   Noun. (Latin, *pro* = for; Middle English, *legge* = leg). Any process or appendage that serves the purpose of a leg. Typically, the pliant, unjointed abdominal legs of caterpillars and some sawfly larvae. Syn. Abdominal Foot, False Leg, Pseudopod.

PRONOTUM   Noun. (Greek, *pro* = before; *noton* = back). The upper (dorsal) surface of the prothorax.

PROPODEUM   Noun. (Greek, *pro* = before; *podeon* = neck. Pl., propodea). Hymenoptera (Apocrita): The first abdominal segment that disassociates from the abdomen and becomes associated with the thorax. In Parasitic Hymenoptera and Aculeata, characterized by anterolateral spiracles and broad attachment to the metanotum anteriad and separated from the remainder of the abdomen by a narrow constriction (the petiole) posteriad.

PROPOLIS   Noun. (Greek, *pro* = for; *polis* = city). A gluelike or resinlike substance elaborated by bees to serve as a cement in situations where wax is not effective in binding a cell.

PROSOMA   Noun. (Greek, *pro* = before; *soma* = body). 1. The cephalothorax of arachnids. 2. The head in apocritous Hymenoptera.

PROTHORAX   Noun. (Greek, *pro* = before; *thorax* = chest). The first thoracic segment bearing the anterior legs but not wings. See also Mesothorax, Metathorax.

PROVENTRICULUS   Noun. (Latin, *pro* = before; *ventriculus* = small stomach). A typically musculated portion of the stomodaeum situated between the crop and circular muscles that provides an anterior constriction for the midgut. Frequently poorly developed or absent from fluid-feeding insects and characterized by well-developed longitudinal folds and acanthae in adult insects that feed on particulate matter. Variably developed in immature insects.

PROXIMAD   Adv. (Latin, *proximus* = next, *ad* = toward). Toward the proximal end of a structure or an appendage.

PROXIMAL   Adj. (Latin, *proximus* = next). The part of an appendage or structure nearest the body or an imaginary midline through the primary axis of the body. Near the base of a structure. Ant. Basal, Distal.

PRUINOSE   Adj. (Latin, *pruina* = hoarfrost). Covered with fine dust, as if frosted; with the brightness of a surface somewhat obscured by the appearance of a plumlike bloom but that cannot be rubbed off.

PSEUDOPOD   Noun. (Greek, *pseudes* = false; *pous* = foot. Pl., pseudopodia). A soft, flexible, footlike appendage characteristic of some dipterous larvae. A proleg.

PTEROSTIGMA   Noun. (Greek, *pteron* = wing; *stigma* = mark. Pl., pterostigmata). An enlarged, pigmented area on the costal margin of the wing near its middle or at the apex of the radius. Found on both wings of Odonata; found on forewing of Psocoptera, Metagloptera, Mecoptera, and Hymenoptera.

PTEROTHORAX   Noun. (Greek, *pteron* = wing; *thorax* = chest). The closely fused meso- and metathorax in most winged insects, such as the wing-bearing thoracic segments in Hymenoptera, Lepidoptera, Thysanoptera.

PTERYGOTA   Noun. (Greek, *pteryx* = wing). A subclass of the Insecta; most members are winged, but some species are secondarily apterous. Compare Apterygota.

PTILINUM   Noun. (Greek, *ptilon* = feather). Schizophorous Diptera: A modification of the frons that forms a saclike, inflatable, pulsatile, cuticular organ on the head of teneral flies. The ptilinum is thrust through an arcuate ptilinal suture just above the base of the antenna. Used to aid emergence from the puparium and burrowing through the soil to reach the surface. After emergence, the ptilinum is retracted. A crescent-shaped sclerite between the ptilinial suture and the antennal socket is called a *lunule*.

PUBESCENT   Adj. (Latin, *pubescere* = to become mature). Covered with soft, short, fine, closely set setae.

PULVILLUS   Noun. (Latin, *pulvillus* = small cushion. Pl., pulvilli). Membranous, padlike structures between the tarsal claws. Cushions of short, stiff setae or other clothing on underside of tarsal joints, rarely fleshy lobes.

PUNCTATE   Adj. (Latin, *punctum* = point). Pertaining to a surface with impressed points, microscopic pits, or punctures.

PUPA   Noun. (Latin, *pupa* = puppet, young girl. Pl., pupae). A phase of complete metamorphosis during which larval anatomical features are destroyed and adult features are constructed. In holometabolous insects, the stage between larva and adult. See also Metamorphosis.

PUPARIUM   Noun. (Latin, *pupa* = puppet, young girl. Pl., puparia). 1. Cyclorrhaphous Diptera: The thickened, hardened, barrel-like third larval instar integument within which the pupa is formed. 2. The covering of certain coccids. 3. Stylopids: The integument of the seventh-instar larva in which the adult female is enclosed.

PYGIDIUM   Noun. (Greek, *pygidion* = narrow buttock, rump. Pl., pygidia). 1. The tergum of the last segment of the abdomen, whatever its numerical designation. 2. Coleoptera: The segment left exposed by the elytra. 3. Diaspididae: A strongly chitinized, unsegmented region terminating the abdomen of the adult female, following after the first four abdominal segments and not to be confused with the true pygidium of other insects.

PYGOSTYLUS   Noun. (Greek, *pyge* = rump; *stylos* = column. Pl., pygostyli). Hymenoptera: Appendicular sensory structures on gastral tergum VIII or IX. Sometimes regarded as homologous with cercus.

— Q —

QUADRANGLE   Noun. (Latin, *quadrangulus* = four-angled). Odonata: A cell in the wing of the Zygoptera bounded by M, Cu, arculus, and a crossvein between M and Cu (similar in position to the triangle of the Anisoptera).

QUADRATE   Adj. (Latin, *quadrans* = fourth part). Four-sided.

QUIESENCE   Noun. A temporary slowing down of metabolism (e.g., estivation or hibernation) brought about by unfavorable conditions of temperature, humidity, or other climatic factors, by lack of some essential vitamins or other food substances, by failure to secrete particular hormones, or by a disturbance of the balance between opposing hormones. Compare Diapause.

— R —

RACE   Noun. (French, *race* = family). An anthropological term used to indicate a population of a species with constant characters that are not quite specific, usually occurring in a different faunal region from the type and thus geographical; sometimes incorrectly considered synonymous with subspecies.

RADIAL   Adj. (Latin, *radius* = ray). 1. Arranged like rays starting from a common center. 2. Of or pertaining to the radius or radial wing vein.

RAPD   Abbreviation for rapid amplification of polymorphic DNA. A form of PCR in which segments of DNA are randomly amplified. Does not require specific knowledge of the DNA sequence of the organism under study.

RAPTORIAL   Adj. (Latin, *raptor* = robber). Adapted for seizing prey; predacious. Modified for seizing prey, as of forelegs.

RECTUM   Noun. (Latin, *rectus* = straight). The posterior part of the terminal section of the proctodaeum, commonly applied to the posterior intestine as a whole.

RETICULATE   Adj. (Latin, *reticulatus* = latticed). Covered with a network of lines, veins, or ridges; meshed, netted.

RETINACULUM   Noun. (Latin, *retinaculum* = tether. Pl., retinacula). 1. Lepidoptera: The loop into which the frenulum of the male is fitted. 2. Hymenoptera: Horny scales that move the sting or prevent its hyperextension from the body. 3. Coleoptera: The middle, toothlike process of the larval mandible. 4. Collembola: See Tenaculum.

RETRACTILE   Adj. (Latin, *retractus* = withdrawn). Pertaining to a structure that is capable of being produced and drawn back or retracted.

RETRACTOR   Noun. (Latin, *retrahere* = to draw back). Any structure used to withdraw; specifically, a muscle.

REVISION   Noun. (Latin, *revisio* = a seeing again). A comprehensive taxonomic treatment of a higher taxonomic category such as a genus, tribe, or family. Revisions include, but are not restricted to, descriptions of new taxa, redescriptions of poorly characterized taxa, clarification of nomenclatural problems, proposal of new synonymy, review of existing synonymy, assessment of previous taxonomic work, statements of geographical distribution, phenology, and summaries of biological information. Compare Synopsis.

RING JOINT   Insect antenna: The much shorter proximal segment(s) distad of the pedicel; ringlike in form and smaller than following segments. See also Annulus.

RIPARIAN   Adj. (Latin, *ripa* = riverbank). Descriptive of organisms frequenting rivers or their shores.

RIPICOLOUS   Adj. (Latin, *ripa* = riverbank). Pertaining to organisms inhabiting river banks.

ROSTRUM   Noun. (Latin, *rostrum* = beak). 1. General: A snoutlike prolongation of the head. 2. Hemiptera: The beak; a joined sheath formed by the labium to enclose the stylets or trophi. 3. Coleoptera: The rigid extension of the head in the Rhynchophora.

RNAlater   An aqueous, nontoxic tissue storage chemical that negates the need to immediately process tissues or to hold tissues in liquid nitrogen for later processing. RNAlater stored tissues do negatively impact the quantity or quality of RNA extraction.

RUDIMENT   Noun. (Latin, *rudimentum* = the beginning). 1. The beginning of any structure or part before it has developed. 2. Structure not developed beyond an elementary or incomplete stage. 3. Vestige.

RUGOSE   Adj. (Latin, *ruga* = wrinkle). Wrinkled.

– S –

SALIVARY GLANDS   Glands that open into the mouth or at the beginning of the alimentary canal and secrete a digestive, irritant, or viscid material called *saliva*. See also Labial Glands.

SALTATORIAL   Adapted for jumping usually describing the hind legs.

SANGUINVOROUS   Adj. (Latin, *sanguis* = blood; *vorare* = to devour). Pertaining to insects that are blood-eating; blood-feeding.

SAPROPHAGOUS   Adj. (Greek, *sapros* = rotten; *phagein* = to eat). Pertaining to insects that feed on dead or decaying animal or vegetable matter.

SARCOPHAGOUS   Adj. (Greek, *sarx* = flesh; *phagein* = to devour). Flesh-eating; feeding on flesh.

SCALE   Noun. (Anglo-Saxon, *sceala* = shell). 1. General: Any small, flat, cuticular projection from the integument. Typically, a unicellular outgrowth of the body wall, wing, or appendage of various shapes;

often a modified seta. 2. Isoptera: The stump of the shed wing. 3. Coccoids: The cover of a scale-insect, consisting of exuviae and glandular secretions; the waxy covering of a male lecaniid. 4. Lepidoptera: The flattened, highly modified, and ridged setae or macrotrichia on the wing; scales overlap and form the wing covering. 5. Diptera: The alula.

SCANSORIAL   Adj. (Latin, *scandere* = to climb). Formed for climbing. Applied to insects (such as lice) whose legs are adapted for climbing on mammalian hair.

SCAPE   Noun. (Greek, *skapos* = stalk). The basal segment of the antenna.

SCATOPHAGOUS   Adj. (Greek, *skat* = dung; *phagein* = eating). Feeding upon dung or excrement, merdivorous. Syn. Coprophagous.

SCAVENGER   Noun. (Middle English, *scavenger* = an officer with various duties). An animal that feeds on dead and decaying animal or vegetable matter. Compare Predator.

SCLERITE   Noun. (Greek, *skleros* = hard). Any hard portion of the insect integument separated from similar areas by suture, sulcus, or apodeme.

SCLEROTIZATION   Noun. (Greek, *skleros* = hard). The process of hardening and darkening (tanning) of the exoskeleton after ecdysis. Sclerotizing.

SCOPA   Noun. (Latin, *scopa* = broom. Pl., scopae). 1. Any brush, tuft, mat, or pile of setae found on the body or appendages that is used to collect pollen. 2. Hymenoptera: Bees: The pollen brush located on the posteriomedial surface of the hind tibia and basitarsus. Megachilidae: The mat of setae on the metasomal venter. Ichneumonidae: A tuft of setae found on the hind coxa of females. Symphyta: An enlarged, sometimes setose, apicoventral flange on the gonostylus. Lepidoptera: A fringe of long scales along the posterior margin of abdominal segment VIII.

SCOPULA   Noun. (Latin, *scopula* = little broom, small brush. Pl., scopulae). 1. A brush. 2. Any small, dense tuft of setae, frequently bristlelike, stiff, of similar length and shape.

SCROBE   Noun. (Latin, *scrobis* = ditch). 1. A cuticular impression or groove formed for the reception, protection, and concealment of an appendage. 2. Orthoptera: The pits in which the antennae are set. 3. Rhynchophora: Grooves at the sides of the rostrum to receive the antennal scapes; also applied to grooves on the sides of mandibles. 4. Hymenoptera: The usually circular impressions on the frons, in which the scapes revolve. Syn. Scrobal Impressions.

SCUTELLUM   Noun. (Latin, *scutellum* = a little shield. Pl., scutella). 1. Any small, shieldlike sclerite. 2. Heteroptera: The triangular part of the mesothorax, generally placed between the bases of the hemelytra, but in some groups overlapping them. 3. Coleoptera: The triangular sclerite between the elytra. 4. Hymenoptera: Sclerite posterior of the transcutal suture. 5. Diptera: A subhemispherical part cut off by an impressed line from the mesonotum. Compare Scutum.

SCUTUM   Noun. (Latin, *scutum* = shield. Pl., scuta). 1. The second dorsal sclerite of the meso- and metathorax. 2. The middle division of the notum in ticks.

SEGMENT   Noun. (Latin, *segmentum*, from *secare* = to cut). 1. Any natural or apparent subdivision of the body. 2. Any subdivision of an arthropod appendage separated from similar structural elements by areas of flexibility, or associated with muscle attachments. Syn. Arthromere, Embryonic, Metamere, Podite, Somite.

SEJUGAL FURROW   A line that separates the propodosoma and hysterosoma in acarines.

SENSE ORGAN   1. Any specialized, innervated structure of the body wall receptive to external stimuli; most insect sense organs are innervated setae. 2. Any structure by which an insect can see, smell, feel, or hear.

SENSILLUM   Noun. (Latin, *sensus* = sense. Pl., sensilla). 1. A simple sense organ or sensory receptor found on various appendages and tagmata of the insect body. 2. A complex, bilaterally symmetrical, sexually dimorphic structure found on the tenth tergum of Siphonaptera. Covered with short, tapering spines (termed *microtrichia*), circular pits that give rise to longer setae (termed *trichobothria*), and a domelike cupola. Functions as a compound sense organ.

SESSILE   Adj. (Latin, *sedere* = to sit). 1. Pertaining to any body structure or appendage broadly attached to or resting on a base without a constriction, stalk, or pedicel. 2. Pertaining to the abdomen broadly attached to the thorax.

SETA   Noun. (Latin, *seta* = bristle. Pl., setae). Hollow, cuticular structures developed as extensions of the epidermal layer of the integument. Often slender and hairlike. Syn. Bristle, Hair, Microtrichium. See also Chaeta.

SILK GLANDS   1. Lepidopterous caterpillars: The greatly elongated, tubular labial or salivary glands. 2. Head glands or rectal glands in other silk-spinning insects. 3. Glands that secrete liquids that harden into silk on exposure to the air.

SOLITARY   Adj. (Latin). Occurring singly or in pairs, i.e., not in colonies.

SPECIES   Plural noun. (Latin, *species* = particular kind). 1. Species or kind. The primary biological unit, debatably an actual thing or a purely subjective concept. 2. A static moment in the continuum of life; an aggregation of individuals similar in appearance and structure, mating freely, and producing young that themselves mate freely and bear fertile offspring resembling each other and their parents, including all varieties and races. Abbreviated "sp." for one species, "spp." for two or more species.

SPECIMEN   Noun. (Latin, *specere* = to look, to behold). A single example of anything (e.g., the individual unit of a collection, a single insect of any species, or a part of an insect properly preserved).

SPERMATHECA   Noun. (Greek, *sperma* = seed, *theke* = case. Pl., spermathecae). A sac, duct, or reservoir within the female that receives spermatozoa during copulation. Variable in number and morphology and capable of storing spermatozoa for periods up to several years.

SPERMATOPHORE   Noun. (Greek, *sperma* = seed; *pherein* = to bear, carry). A packet or capsule composed of sperm mixed with secretions from accessory glands. Typically manufactured by elements of the male reproductive system and transmitted to the female during copulation. Provides nutrients to the female and/or serves as a physical or chemical barrier to subsequent copulation, insemination, or fertilization.

SPIRACLE   Noun. (Latin, *spiraculum* = air hole. Pl., spiraculae, spiracles). A breathing pore or aperture. Typically paired lateral holes in the pleural wall of insect body segments through which air enters the tracheae.

SPUR   Noun. (Anglo-Saxon, *spora* = spur). Any spinelike appendage of the integument that articulates with the body wall. Frequently found on the leg segments, such as the tibia.

STADIUM   Noun. (Latin, from Greek *stadion, stadios* = fixed, stable. Pl., stadia). The interval between the molts during larval development. See also Instar.

STAGE   Noun. (Old French, *estage* = dwelling, habitation). Any defined period in the development of an insect: the egg stage, larval stage, pupal stage, adult stage. Compare Instar.

STERNITE   Noun. (Greek, *sternon* = chest). 1. The ventral portion of a sclerotized ring or body segment. 2. A subdivision of a sternal sclerite or any one of the sclerotic components of a definitive sternum.

STERNUM   Noun. (Latin, *sternum* = breast bone. Pl., sterna). 1. The entire ventral division of any segment. 2. The ventral surface of the insect thorax between the coxal cavities.

STIGMA   Noun. (Greek, *stigma* = mark. Pl., stigmata). 1. A spiracle or pore of the respiratory system. 2. Hymenoptera: A pigmented spot along the costal margin of a wing, usually at the end of the radius. 3. Diptera: A colored wing spot near the tip of the auxiliary vein. 4. Lepidoptera: The specialized patch of black scales on the forewings of Hesperidae.

STING   Noun. (Anglo-Saxon, *stingan* = to sting). Aculeate Hymenoptera: A sclerotized, tapered, tubular shaft developed at the tip of the abdomen to inject venom into prey and for defense.

STIPES   Noun. (Latin, *stipes* = stalk. Pl., stipites). 1. The second segment of the maxilla in the insect head. Broadly attached to the cardo basad, bearing the movable palpus laterad, and attached to the galea and lacinia distad. Collectively, the cardo and stipes are called the *coxopodite* in the generalized mouthpart. Modified into a piercing structure in some Diptera and into a lever for flexing the proboscis in others. 2. Either of the pair of forceps in the male genitalia of aculeate Hymenoptera (sagittae). 3. The stalk of an elevated eye. 4. Distal portion of an embolus in spiders.

STRIDULATE   Verb. (Latin, *stridor* = grating). Insects: to make a creaking, grating, or hissing sound or noise by rubbing two ridged or roughened surfaces against each other. Example: Crickets.

STRIDULATING ORGANS   Parts of the insect structure that are used for making sounds. In general, one part is a filelike area and the opposing part is a scraper or rasp.

STRIGIL   Noun. (Latin, *strigilis* = comb, from *stringere* = to scrape, graze). 1. A tibial comb. 2. A scraper. 3. A structure adapted for cleaning the antenna. 4. Corixidae: The peculiar structure on the abdomen dorsally (sometimes shaped like a comb) opposed by a hair-fringed concavity at the base of the metatarsus. Hymenoptera, Lepidoptera: A curved, comblike movable spur on the distal end of the foretibia.

STRUCTURE   Noun. (Latin, *structura*, from *structum* = to build, to arrange). 1. The arrangement of constituent elements (parts, cells, tissues, organs) of a body. 2. Any organ, appendage, or part of an insect.

STYLE   Noun. (Greek, *stylos* = pillar; Latin, *stylus* = pricker. Pl., styli, styles). 1. A short, cylindrical appendage. 2. Aphididae: The slender tubular process at the end of the abdomen. 3. Coccidae: A long spinelike appendage at the end of the abdomen of the male; the genital spike. 4. Diptera: The ovipositor; the single immovable organ immediately below the forceps. Male Tipulidae: a thickened, jointed arista at or near the apex of the third antennal segment. 5. In the plural form (styli) applied to small, usually pointed, exarticulate appendages; most frequently found on the terminal segments of the abdomen. See also Stylus.

STYLET   Noun. (Latin, *stylus* = pricker). 1. A small style or stiff process. 2. Phthiraptera, some Hemiptera, Diptera: One of the mouthparts modified for piercing. 3. Hemiptera: A median dorsal element in the shaft of the ovipositor, formed of the united second valvulae. 4. Hymenoptera: First valvulae. Aculeata: Fused portions of second valvulae.

STYLIFORM   Adj. (Latin, *stylus* = pricker; *forma* = shape). 1. Shaped like a stylus; terminating in a long, slender point. Example: The antenna of some Diptera.

STYLUS   Noun. (Latin, *stylus* = pricker. Pl., styli). 1. A small, pointed, nonarticulated process. 2. Collembola: The dorsal, tapering process of the ventral abdominal appendages, freely movable at its base; a small appendage attached to the coxa of the middle and hind legs. 3. Coccidae: The outer sheath of the male genitalia. 4. Female Odonata: A small rod-shaped projection at the tip of the lateral gonapophyses of the ovipositor; in male nymphs, a short acute process on the ventral surface of the ninth abdominal segment. See also Style.

SUBCOXA   Noun. (Latin, *sub* = under; *coxa* = hip. Pl., subcoxae). The basal part of the limb base when divided from the coxa. The subcoxa usually is incorporated into the pleural wall of the body segment.

SUBFAMILY   Noun. (Latin, *sub* = under; *familia* = household, from *famulus* = servant). A division in zoological classification containing a group of closely related genera that differ from allied groups. Subfamily names end with "-inae."

SUBGENITAL PLATE   An apical sternal sclerite or process that covers and protects the gonopore. The hypandrium sensu Snodgrass. Syn. Subgenital Shield.

SUBGENUAL ORGAN   A chordotonal organ in the proximal part of the tibia that is sensitive to vibrations.

SUBIMAGO   Noun. (Latin, *sub* = under; *imago* = likeness. Pl., subimagoes). 1. A stage between the pupa and the adult. 2. The stage in mayflies just after emergence from the pupa and before the final molt.

SUCTORIAL   Adj. (Latin, *sugere* = to suck). Mouthparts adapted for sucking. See also Haustellate.

SULCUS   Noun. (Latin, *sulcus* = furrow. Pl., sulci). A furrow or groove; typically a groove in the insect integument.

SUPERPARASITISM   Noun. (Latin, *super* = over; *para* = beside; *sitos* = food). 1. A biological phenomenon in which a female parasite oviposits more eggs on or in a host than can hatch and successfully develop to maturity. 2. Oviposition on or in a host previously parasitized by a conspecific female. 3. Parasitism of one host by more larvae than can survive to maturity.

SUTURE   Noun. (Latin, *sutura* = seam). 1. A seam or line of contact between two sclerites or hardened body parts that makes those body parts immovably connected. 2. The division of the distinct parts of the body wall. 3. The line of juncture of the elytra in Coleoptera or of the tegmina or hemelytra in other orders. Compare Sulcus. See also Apodeme, Phragma.

SYMBIONT   Noun. (Greek, *symbionai* = to live together). One of the partners in a symbiotic relationship. Example: An insect or other arthropod living in symbiosis with termites or ants. Alternate Symbiote.

SYMBIOSIS   Noun. (Greek, *symbionai* = to live together). A living together in more or less intimate association by organisms of different species. The association is not necessarily for mutual benefit. See also Commensalism, Parasitism.

SYMPATRIC   Adj. (Greek, *syn* = with; *patra* = homeland). Pertaining to organisms (usually species) that have overlapping geographical distributions.

SYMPHYTA   Suborder within the Hymenoptera, including sawflies and horntails. Adults possess an abdomen broadly sessile at its base and without a marked constriction.

SYNONYM   Noun. (Greek, *syn* = with; *onymia* = name). Two or more different names for the same thing. In zoological nomenclature, a different scientific name given to a species or genus previously named and described. See also Homonym.

SYNOPSIS   Noun. (Greek, *syn* = together; *opsis* = a view). A taxonomic publication that briefly summarizes current knowledge of a group. New information is not necessarily included in a synopsis.

SYNOPTIC COLLECTION   A group of taxa that has been authoritatively identified and that serves as the basis of comparison for the identification of other specimens.

SYSTEMATICS   (Greek, *systema* = a whole made of several parts). The practice of classifying organisms. See also Taxonomy.

SYSTEMATIST   Noun. (Greek, *systema* = a whole made of several parts). In zoology, a person who studies taxa and develops classifications of forms or groups according to biological relationships, phenetic affinities, or both. Compare Taxonomy.

– T –

TAGMA   Noun. (Greek *tagma* = corps. PI., tagmata). A group of successive segments (somites) of a metamerically organized animal that forms a distinct region of the body (head, thorax, abdomen).

TARSUS   Noun. (Greek, *tarsos* = sole of the foot. PI., tarsi). 1. The foot or jointed appendage attached at the apex of the tibia, bearing the claws and pulvilli. 2. The distal part of the insect leg, consisting of one to five segments.

TAXONOMIST   Noun. One who studies taxonomy. See Taxonomy.

TAXONOMY   Noun. (Greek, *taxis* = arrangement; *nomos* = law). 1. The system of arranging organisms into groups based on anatomical detail or biological relationship to each other. 2. The practice of identifying, describing, and naming organisms based on the rules of zoological nomenclature. Compare Systematics.

TEGULA   Noun. (Latin, *tegula* = tile. PI., tegulae). Hymenoptera: Lepidoptera, Diptera: An articular sclerite on the mesothorax positioned at the base of the wing. Typically a small, scalelike sclerite and at the extreme base of the costa of the forewing. Misused for alula in Diptera. See also Patagium.

TENACULUM   In Collembola, a minute organ with two divergent prongs, situated medially on the ventral surface of the third abdominal segment, serving to hold the furcula in place.

TENERAL   Adj. (Latin, *tener* = delicate). Describing the imago or adult shortly after emergence from the nymphal or pupal stage when the integument is not hardened or its color has not matured. Compare Callow.

TENTORIAL PITS   External depressions in the head at the base of the tentorial arms. The anterior tentorial pits are located in the subgenal suture or in the epistomal suture; the posterior tentorial pits are located in the lower ends of the postoccipital suture.

TENTORIUM Noun. (Latin, *tentorium* = tent. Pl., tentoria). The endoskeleton of the insect head, consisting of two or three pairs of apodemes coalescing internally. The tentorium provides rigidity and strength to the head, supports the brain and foregut, and provides attachment surface for some muscles in the head.

TERGITE Noun. (Latin, *tergum* = back). 1. A dorsal sclerite that is part of a body segment or sclerotized ring of a generalized body segment. 2. A sclerotized subdivision of a body tergum. Compare Tergum. Ant. Sternite.

TERGUM Noun. (Latin *tergum* = back. Pl., terga). The upper or dorsal surface of any body segment of an insect, whether it consists of one or several sclerites. Compare Tergite. Ant. Sternum.

TERMINAL Adj. (Latin, *terminus* = end). Pertaining to the apex or extreme tip of a structure or appendages. Ant. Basal. Syn. Apical.

TERMITOPHILOUS Adj. (Latin, *termes* = woodworm; Greek, *philein* = to love). Termite-loving. A term applied to insects and other guests habitually living in a termite colony with and among the termites. Compare Myrmecophilous.

TERRESTRIAL Adj. (Latin, *terra* = earth). Organisms living on or in the land. Compare Aquatic.

TERTIARY HYPERPARASITISM A form of hyperparasitism in which hyperparasitic individuals attack one another. Conceptually divided into interspecific tertiary hyperparasitism (allohyperparasitism) and intraspecific tertiary hyperparasitism (autohyperparasitism). See also Hyperparasite.

TESTES Plural noun. (Latin, *testis* = testicle. Sing., testis).The male reproductive gland that produces spermatozoa. Usually composed of many tubular structures called *follicles.*

THIGMOTACTIC Adj. (Greek, *thigema* = touch; *taxis* = arrangement). Contact-loving or pertaining to species that tend to live in close proximity or in contact with a surface or in a crevice.

THORAX Noun. (Greek, *thorax* = chest. Pl., thoraces). The second or middle region of the insect body. The thorax bears true legs and wings and is composed of three segments (pro-, meso-, and metathorax). When the prothorax is free (e.g., Coleoptera, Orthoptera, and Hemiptera), the term *thorax* is sometimes used for that segment only. In Odonata (where the prothorax is small and not fused with the larger mesothorax and metathorax), the term *thorax* is commonly used for these segments, excluding the prothorax.

TIBIA Noun. (Latin, *tibia* = shin. Pl., tibiae). The fourth division of the leg, articulated proximally with the femur and distally with the basitarsus. The tibia is typically long and slender. See also Leg.

TIMBAL Noun. (French, *timbale* = kettledrum). The shell-like drum in cicadas, at the base of the abdomen, used to produce sound. Syn. Tympanum. Alternate: Tymbal.

TONGUE Noun. (Anglo-Saxon, *tunge* = tongue). A term applied to the coiled mouth of Lepidoptera, the lapping organ of flies, the ligula of bees and wasps, and sometimes to the hypopharynx of other insects.

TOXOGNATH Noun. (Greek, *toxikon* = poison; *gnathos* = jaw). A poison claw in centipedes. The modified first pair of legs in centipedes and not a fang as in Chelicerata.

TRACHEA Noun. (Latin, *trachia* = windpipe. Pl., tracheae). An element of the respiratory system. Typically, tracheae are internal, elastic, spirally ringed air tubes in the terrestrial insect body. Compare Gill.

TRACHEATE Adj. (Latin, *trachia* = windpipe). Supplied with tracheae. A general term applied to all arthropods that breathe by means of spiracular openings into a system of tubes that extend to all parts of the body.

TRANSECT Noun. (Latin, *trans* = across; *secare* = to cut). A narrow area, of various widths, within which quantitative samples of physical and biological data are collected.

TRANSPIRATION Noun. (Latin, *trans* = across; *spirare* = to breathe). The act or process of exhaling or passing off liquids as vapor. In plants, the passing of water vapor through the stomata.

TRANSVERSE Adj. (Latin, *transversus* = across). 1. Pertaining to structures that are wider than long. 2. Structure extending across or cutting the longitudinal axis at right angles. Ant. Longitudinal.

TRANSVERSE SUTURE   1.A division across the body; more particularly across the dorsum of the tho-rax. 2. Diptera: The suture between the prescutum and the scutum of the mesothorax.

TRAP   Noun. (Middle English, *trappe*; Anglo-Saxon, *treppe* = trap). A device or machine designed or adapted for the collection of insects. Types: flight intercept trap, light trap, pit trap, Malaise trap.

TROCHANTER   Noun. (Greek, *trochanter* = runner). A segment of the insect leg. The trochanter is sometimes divided between the coxa and the femur, sometimes fused with the femur. Typically, the smallest segment of the insect leg.

TRUNCATION   Noun. (Latin, *truncatus* = cut off). A transverse, abrupt termination of a structure or appendage.

TUBERCLE   Noun. (Latin, *tuberculum* = a small hump. Pl., tuberculi). 1. A small, rounded protuber-ance. 2. Hymenoptera: A projection in adult Sphecoidea (wasps), consisting of rounded lobes on the dorsolateral margin of the pronotum. 3. Lepidoptera caterpillars: Body structures (of this character) sometimes bearing setae. Syn. Verruca.

TUBERCULATE   Adj. (Latin, *tuberculum* = a small hump). Covered or furnished with tubercles; formed like a tubercle. Alternate: Tuberculose, Tuberculous.

TUFT   Noun. (Middle English, from Old French, *tufe*; Greek, *zopf* = clump of hair, pigtail). Several closely arranged, elongate, flexible structures (setae, cuticular evaginations) whose bases are juxtaposed but whose apices are free and widely spaced.

TUMID   Adj. (Latin, *tumidus* = swollen). Pertaining to bodies or structures that are swollen or enlarged.

TURBINATE   Adj. (Latin, *turbo* = whirl). Top-shaped, nearly conical. Turbinate differs from pyriform in being shorter and more suddenly attenuated at the base.

TYMPANAL ORGAN   A complex auditory organ found in the wing, abdomen, and thorax of Hemiptera, Lepidoptera, and Neuroptera. Elements include sensory cells adjacent or attached to a thin, flexible, chitinous membrane (the tympanum). A chitinous frame surrounds the tympanum, and a tra-chea or air sac is adjacent to the inner surface of the tympanum. Compare Chordotonal Organ, Johnston's Organ, Subgenual Organ.

TYMPANUM   Noun. (Greek, *tympanon* = drum; Pl., tympana, tympanums). Any stretched membrane like the covering of a drum. Specifically applied to the membrane covering the auditory organs in Orthoptera and some Lepidoptera.

TYPE   Noun. (Latin, *typus* = pattern). Taxonomy: 1. The single specimen or member of a series of speci-mens from which a species is described. 2. The species upon which a genus is founded or that is selected as the type of a genus. 3. A holotype. See also Paratype.

TYPEGENUS   Families: The specific genus on which the family is founded. The typegenus is not neces-sarily the oldest genus. When the family takes its name from the typegenus, the family name follows nomenclatural changes.

TYPESPECIES   A species that is the type of a genus.

– U –

UNGUIS   Noun.(Latin, Pl. ungues).Clawlike structure of the maxilla.

UNISEXUAL   Noun. (Latin, *unus* = one; *sexus* = sex). 1. Parthenogenetic species in which only females exist. 2. Female-only generations in species that cyclically produce males. A biological feature widely distributed within the Insecta but best studied in aphids and cynipid wasps.

UNIVOLTINE   Adj. (Latin, *unus* = one; Italian, *volta* = time). Pertaining to populations or species with one generation per year or season. Compare Multivoltine.

UNSCLEROTIZED   Adj. (Middle English, *un* = not; Greek, *skleros* = hard). Pertaining to tissue not having sclerotin. In general, describes parts of the insect integument that are not hardened by the depo-sition of sclerotin.

UROGOMPHI   Plural noun.(sing., urogmphus). In larval Coleoptera, usually paired processes from the posterior end of the tergum of the ninth abdominal segment.

URTICATING HAIRS   (Latin, *urtica* = nettle). Some caterpillars and adult insects: Setae or chaetae connected with cutaneous poison glands, through which venom issues. Barbed setae that cause discomfort presumably induced by mechanical irritation.

– V –

VANNAL REGION   The wing area containing the vannal veins, or veins directly associated with the third axillary. When large, the vannal region is usually separated from the remigium by the plica vannalis, often forming an expanded fanlike area of the wing.

VANNAL VEINS   The veins associated at their bases with the third axillary sclerite and occupying the vannal region of the wing. The "anal" veins except the postcubitus.

VANNUS   Noun. (Latin, *vannus* = fan). A large fanlike expansion of the posterior part of the insect wing.

VARIETY   Noun. (Latin, *varietas* = variety). A taxonomically ambiguous term "used to designate variants in size, structure, and color and varying ranks of all these," or to indicate a form for which the taxonomic status is uncertain. Varieties have no standing in zoological nomenclature.

VEIN   Noun. (Latin, *vena* = vein). Any chitinous, rodlike, or hollow tubelike structure supporting and stiffening the wings in insects, especially those extending longitudinally from the base of the wing to the outer margin.

VENATION   Noun. (Latin, *vena* = vein). The complete system of veins in an insect wing.

VENOM   Noun. (Latin, *venenum* = poison). Hymenoptera: Toxic fluid consisting of biologically active peptides, secreted from accessory glands and injected into prey or hosts by the sting or ovipositor. Venoms induce death or paralysis and cause pain, edematous reactions, and swelling of skin.

VENTER   Noun. (Latin, *venter* = belly). 1. The abdomen. 2. The ventral surface of the abdomen.

VENTRAD   Adv. (Latin, *venter* = belly; *ad* = to). Toward the venter; in the direction of the venter. Ant. Dorsad.

VENTRAL   Adj. (Latin, *venter* = belly). Pertaining to the undersurface of a structure. Ant. Dorsum.

VENTROLATERAL   Toward the venter on the side. Pertaining to the venter and the side.

VERRUCA   Noun. (Latin, *verruca* = wart. Pl., verrucae). Lepidopterous larvae: Somewhat elevated portions of the cuticle bearing tufts of setae (Imms). A wart or wartlike prominence.

VERTEX   Noun. (Latin, *vertex* = top). The top of the head between the eyes, front, and occiput.

VERTICAL BRISTLES   Diptera: Two pairs of bristles (an inner and an outer pair) more or less behind the upper and inner margins of the eyes; vertical cephalic bristles.

VESTIGIAL   Adj. (Latin, *vestigum* = trace). Small or degenerate. Descriptive of the remains of a previously functional part or organ, like remnants or vestiges. Descriptive of a structure in the process of disappearing.

VESTITURE   Noun (Latin, *vestitus* = garment). Clothing; the general surface covering of insects. Typically composed of scales, setae, or spines.

VISCERA   Plural noun. (Latin, *viscera* = bowels). The internal organs of the body.

VISCID   Adj. (Latin, *viscum* = mistletoe). Sticky; covered with a shiny, resinous, or greasy substance.

VIVIPARITY   Noun. (Latin, *vivus* = living; *parere* = to beget). A method of reproduction in which living young (not eggs) are extruded from the female body. Found in many groups of insects and varies in complexity. Offspring obtain nutrients directly from "milk glands" within the mother. Egg contains little or no yolk, lacks a chorion, and develops within an ovariole, and follicle cells supply some nutrients. Similar to ovoviviparity, in that there is a reduction in the number of offspring produced per female.

VIVIPAROUS   Adj. (Latin, *vivus* = living; *parere* = to beget). Pertaining to organisms that bear living young; opposed to organisms that lay eggs. Compare Oviparous, Ovoviviparous.

VULVA   Noun. (Latin, *vulva* = vulva). Insects: The orifice of the vagina in the female. Ticks: The female genitalia.

— W —

WAX GLAND   1. Any gland that secretes a wax product in a scale, string, or powder. 2. Coccidae: The circumgenital and parastigmatic glands. 3. Hymenoptera: Apoidea: Abdominal glands that produce beeswax. 4. Neuroptera (Coniopterygidae): Glands on head, thorax, and abdomen that produce a meal-like wax. Wax glands have been identified on all tagmata and appendages.

WHORL   Noun. (Anglo-Saxon, *hweorfan* = to turn). 1. A ring or circle of setae surrounding a body or appendage segment. 2. A circular arrangement of structures that radiate outward from an imaginary center like the spokes of a wheel.

WING   Noun. (Middle English, *winge*, from Old Norse, *vaengr*). Paired, often membranous and reticulated cuticular expansion of dorsolateral portion of the mesothorax and metathorax. Forewings (primaries) attached to mesothorax; hindwings (secondaries) attached to metathorax. Specifically adapted as organs of flight in insects or modified to protect the pair of wings involved in flapping flight. First detected in Carboniferous fossils, and present in most species.

WORKER   Noun. (Middle English, *werk, weore* = work). 1. The neuter or sterile individuals within the colony of a social species. 2. Isoptera: The sexually undeveloped forms, except the so-called soldiers. 3. Hymenoptera: Individuals anatomically female that lack the capacity to reproduce or lay unfertilized eggs that produce only males. Workers are responsible for nest building, brood care, and colony defense.

— X —

XENOBIOSIS   Noun. (Greek, *xenos* = guest, stranger; *biosis* = living). A commensalistic relationship in which one ant species lives as a guest in the nest of another species in mutual toleration.

XERIC   Adj. (Greek, *xeros* = dry). Pertaining to dry conditions or an absence of moisture.

XYLOPHAGOUS   Adj. (Greek, *xylon* = wood; *phagein* = to eat). Feeding in or upon woody tissue.

— Z —

ZOOGEOGRAPHY   Noun. (Greek, *zoon* = animal; *ge* = earth; *graphein* = to write). The study of distribution, distributional patterns, and association or groupings of plants or animals in space, realm, or region.

ZOOPHAGOUS   Adj. (Greek, *zoon* = animal; *phagein* = to eat). Animal-eating; feeding on animals or animal products. Compare Herbivorous.

Bibliography

A'Brook, J., 1973. Observations on different methods of aphid trapping. Ann. Appl. Biol. 74, 253–267.

Acree Jr., F., Turner, R.B., Gouck, H.K., Beroza, M., Smith, N., 1968. L-Lactic acid: a mosquito attractant isolated from humans. Science (Washington, D.C.) 161, 1346–1347.

Acuff, V.R., 1976. Trap biases influencing mosquito collecting. Mosq. News 36, 173–196.

Agosti, D., 2001. Ants: Standard Methods for Measuring and Monitoring Biodiversity. Smithsonian Institution Press, Washington, DC [Excellent example of specific procedures for surveying the diversity of one family of insects.].

Akar, H., Osgood, E.A., 1987. Emergence trap and collecting apparatus for capture of insects emerging from soil. Entomol. News 98, 35–39.

Alderz, W.C., 1971. A reservoir-equipped Moericke trap for collecting aphids. J. Econ. Entomol. 64, 966–967.

Almand, L.K., Stering, W.L., Green, C.L., 1974. A collapsible truck-mounted aerial net for insect sampling. Tex. Agric. Exp. Stn. Misc. Publ. 1189, 1–4.

Anderson, G.S., 1999. Forensic entomology in death investigations. In: Fairgreave, S. (Ed.), Forensic Anthropology Case Studies From Canada. Charles C Thomas, Springfield, IL.

Andreyev, S.V., Martin, B.K., Molchanova, V.A., 1970. Electric light traps in research on the protection of plants against insect pests. Entomol. Rev. 49, 290–297.

Apperson, C.S., Yows, D.G., 1976. A light trap for collecting aquatic organisms. Mosq. News 36, 205–206.

Arensburger, P., Buckley, T.R., Simon, C., Moulds, M., Holsinger, K.E., 2004. Biogeography and phylogeny of the New Zealand cicada genera (Hemiptera: Cicadidae) based on nuclear and mitochondrial DNA data. J. Biogeography 31, 557–569.

Arnett, R.H., 1985. American Insects. Van Nostrand Reinhold, New York.

Arnold, A.J., 1994. Insect suction sampling without nets, bags or filters. Crop Prot. 13, 73–77.

Asghar, U., Malik, M.F., Anwar, F., Javed, A., Raza, A., 2015. DNA extraction from insects using different techniques: a review. Adv. Entomol. 3, 132–138. Available from: https://doi.org/http://dx.orgi/10.42236/ae.2015.34016.

Atkins, M.D., 1957. An interesting attractant for *Priacma serrata* (LeC.) (Cupesidae: Coleoptera). Can. Entomol. 89, 214–219.

Azrang, M., 1976. A simple device for collecting insects. Entomol. Tidskr. 97, 92–94.

Bailey, W., 1991. Acoustic Behavior of Insects. Routledge, Chapman & Hall, London.

Baker, J.R., 1958. Principles of Biological Microtechnique. John Wiley & Sons, Methuen, MA, London, New York.

Balogh, J., 1958. Lebensgemeinschaften der Landtiere; ihre Erforschung unter besonderer Berücksichtigung der zoozonologischen Arbeitsmethoden. Verlag Ungar. Akad. Wiss. Budapest; Akad. Verlag, Berlin. (In German.) [An extensive compilation, with 27 pp. of references.]

Banks, C.J., 1959. Experiments with suction traps to assess the abundance of Syrphidae (Diptera), with special reference to the aphidophagous species. Entomol. Exp. Appl. 2, 110–124.

Banks, N., 1909. Directions for Preserving and Collecting Insects. U.S. Natl. Mus. Bull. 67. [Mostly of historical interest, but describes well the old methods and contains much general information about insects.]

Banks, W.A., Lofgren, C.S., Jouvenaz, D.P., Stringer, C.E., Bishop, P.M., Williams, D.F., et al., 1981. Techniques for Collecting, Rearing, and Handling Imported Fire Ants. U.S. Dept. Agric., Agric. Res. Serv. AAT-S-21.

Barber, H.S., 1931. Traps for cave-inhabiting insects. J. Elisha Mitchell Sci. Soc. 46, 259−266. pl. 23.

Barber, M.C., Matthews, R.W., 1979. Utilization of trap nests by the pipe-organ mud dauber, *Trypargilum politum* (Hymenoptera: Sphecidae). Ann. Entomol. Soc. Am. 72, 258−262.

Barnard, D.R., 1979. A vehicle-mounted insect trap. Can. Entomol. 111, 851−854.

Barnard, D.R., Mulla, M.S., 1977. A non-attractive sampling device for collection of adult mosquitoes. Mosq. News 37, 142−144.

Barnes, H.F., 1941. Sampling for leather jackets with orthodichlorobenzene emulsion. Ann. Appl. Biol. 28, 23−28.

Barr, A.R., Smith, T.A., Boreham, M.M., White, K.E., 1963. Evaluation of some factors affecting the efficiency of light traps in collecting mosquitoes. J. Econ. Entomol. 56, 123−127.

Barrett Jr., J.R., Deay, H.O., Hartsock, J.G., 1971. Striped and spotted cucumber beetle response to electric light traps. J. Econ. Entomol. 54, 413−416.

Bartelt, R.J., Vetter, R.S., Carlson, D.G., Baker, T.C., 1994. Influence of pheromones on *Carpophilus mutilatus* and *C. hemipterus* (Coleoptera: Nitidulidae) in a California date garden. J. Econ. Entomol. 87, 667−675.

Bartnett, R.E., Stephenson, R.G., 1968. Effect of mechanical barrier mesh size on light trap collection in Harris County, Texas. Mosq. News 28, 108.

Batiste, W.C., Joos, W., 1972. Codling moth: a new pheromone trap. Econ. Entomol. 65, 1741−1742.

Beatty, G.H., Beatty, A.F., 1963. Efficiency in caring for large Odonata collections. Proc. North Cent. Branch Entomol. Soc. Am. 18, 149−153.

Beaudry, J.R., 1954. A simplification of Hubbell's method for trapping and preserving specimens of *Ceuthophilus* (Orthoptera, Gryllacrididae). Can. Entomol. 86, 121−122.

Beavers, J.B., McGovern, T.P., Beroza, M., Sutton, R.A., 1972. Synthetic attractants for some dipteran species. J. Econ. Entomol. 65, 1740−1741.

Becker, P., Mouve, J.S., Peralta, F.J.A., 1991. More about Euglossine bees in Amazonian forest fragments. Biotropica 23 (4 pt. B), 586−591.

Bedford, G.O., 1978. Biology and ecology of the Phasmatodea. Annu. Rev. Entomol. 23, 125−150.

Belding, M.J., Isard, S.A., Hewings, A.D., Irwin, M.E., 1991. Photovoltaic-powered suction trap for weakly flying insects. J. Econ. Entomol. 84, 306−310.

Belkin, J.N., 1962. The Mosquitoes of the South Pacific (Diptera, Culicidae), 2 vols. University of California Press, Berkeley, CA [Vol. 1 includes introduction to methods.].

Bell, R.T., 1990. Insecta: Coleoptera Carabidae adults and larvae. In: Dindal, D.L. (Ed.), Soil Biology Guide. John Wiley & Sons, New York, pp. 1053−1092.

Belton, P., 1962. Effects of sound on insect behavior. Proc. Entomol. Soc. Manit. 18, 22−30.

Belton, P., Kempster, R.H., 1963. Some factors affecting the catches of Lepidoptera in light traps. Can. Entomol. 95, 832−837.

Belton, P., Pucat, A., 1967. A comparison of different lights in traps for *Culicoides* (Diptera: Ceratopogonidae). Can. Entomol. 99, 267−272.

Bentley, J.W., 1992. An alcohol trap for capturing vespids and other Hymenoptera. Entomol. News 103, 86−88.

Beroza, M., 1970. Current usage and some recent developments with insect attractants and repellents in the USDA. In: Beroza, M. (Ed.), Chemicals Controlling Insect Behavior. Academic Press, New York, pp. 145−163.

Beroza, M., 1972. Attractants and repellents for insect control. Pest Control: Strategies for the Future. Wash. Acad. Sci, Washington, DC, pp. 226−253.

Beroza, M., Green, N., 1963. Materials Tested as Insect Attractants. U.S. Dept. Agric., Agric. Handb. 239.

Beroza, M., Paszek, E.C., Mitchell, E.R., Bierl, B.A., McLaughlin, J.R., Chambers, D.L., 1974. Tests of a 3-layer laminated plastic bait dispenser for controlled emission of attractants from insect traps, Environ. Entomol. 3. pp. 926–928.

Berte, S.B., 1979. An improved method for preserving color patterns in pinned insects. Entomol. News 90, 147–148.

Bidlingmayer, W.L., 1967. A comparison of trapping methods for adult mosquitoes: species response and environmental influence. J. Med. Entomol. 4, 200–220.

Birch, M. (Ed.), 1974. Pheromones. North-Holland, Amsterdam, 495 pp.

Bisanti, M., Ganassi, S., Mandrioli, M., 2009. Comparative analysis of various fixative solutions on insect preservation for molecular studies. Entomol. Exp. Appl. 130, 290–296.

Blaimer, B.B., Lloyd, M.W., Guillory, W.X., Brady, S.G., 2016. Sequence capture and phylogenetic utility of genomic ultraconserved elements obtained from pinned insect specimens. PLoS One. Available from: https://doi.org/10.137/journal.pone.0161531.

Blakeslee, T.E., et al., 1959. *Aedes vexans* and *Culex salinarius* light trap collection at five elevations. Mosq. News 19, 283.

Bland, R.C., Jacques, H.E., 1978. How to Know the Insects, third ed Wm. C. Brown, Dubuque, IA [Successor to Jacques, H. E., 1947, 2nd ed.].

Blum, M.S., Woodring, J.P., 1963. Preservation of insect larvae by vacuum dehydration. J. Kans. Entomol. Soc. 36, 96–101.

Blume, R.R., Miller, J.A., Eschle, J.L., Matter, J.J., Pickens, M.O., 1972. Trapping tabanids with modified Malaise traps baited with CO_2. Mosq. News 32, 90–95.

Borgmeier, T., 1964. How to prepare minuten-pin double mounts of small Diptera. Stud. Entomol. 7, 489.

Borror, D.J., White, R.W., 1970. Afield Guide to the Insects of America North of Mexico. Houghton Mifflin, Boston, MA.

Borror, D.J., Triplehorn, C.A., Johnson, N.F., 1989. An Introduction to the Study of Insects, sixth ed Saunders College Publishing, Philadelphia, PA.

Bosic, J.J., 1997. Common Names of Insects & Related Organisms. Entomological Society of America, College Park, MD.

Boudreaux, H.B., 1979. Arthropod Phylogeny With Special Reference to Insects. John Wiley & Sons, New York.

Bowles, D.E., Stephan, K., Mathis, M.L., 1990. A new method for collecting adult phryganeid caddisflies. Entomol. News 101, 222–224.

Bradbury, W.C., Morrison, P.E., 1975. A portable electric aspirator for collecting large insects. Can. Entomol. 107, 107–108.

Bram, R.A., 1978. Surveillance and Collection of Arthropods of Veterinary Importance. U.S. Dept. Agric., Agric. Handb. 518.

Brandenburg, R.L., Villani, M.G. (Eds.), 1995. Handbook of Turfgrass Insect Pests. Entomological Society of America, Landham, MD.

Breyev, K.A., 1963. The effect of various light sources on the numbers and species of blood-sucking mosquitoes (Diptera, Culicidae) collected in light traps. Entomol. Obozr. 42, 280–303 (In Russian; translated into English in Entomol. Rev. 42: 155–168.).

Briggs, J.B., 1971. A comparison of pitfall trapping and soil sampling in assessing populations of two species of ground beetles (Col.: Carabidae). Rep. E: Mailing Res. Stn. 1970, 108–112.

Brindle, A., 1963. Terrestrial Diptera larvae. Entomol. Rec. J. Var. 75, 47–62 [Includes collecting methods.].

Broadbent, L., 1949. Aphid migration and the efficiency of the trapping method. Ann. Appl. Biol. 35, 379–394.

Broadbent, K., Doucaster, J.P., Hull, R., Watson, M.A., 1948. Equipment used in trapping and identifying alate aphids. Proc. R. Entomol. Soc. Lond., Ser. A, Gen. Entomol. 23, 57.

Broce, A.B., Goodenough, J.L., Coppedge, J.R., 1977. A wind-oriented trap for screwworm flies. J. Econ. Entomol. 70, 413–416.

Brockway, P.B., et al., 1962. A wind directional trap for mosquitoes. Mosq. News 22, 404–405.

Brown, A.W.A., 1954. Studies on the responses of the female *Aedes* mosquito. Pt. IV. The attractiveness of coloured clothes to Canadian species. Bull. Entomol. Res. 45, 67–68.

Brown, R.D., 1973. Funnel for extraction of leaf litter organisms. Ann. Entomol. Soc. Am. 66, 485–486.

Brues, C.T., Melander, A.L., Carpenter, F.M., 1954. Classification of insects. Bull. Mus. Comp. Zool. 73, 1–917.

Burbutis, P.P., Stewart, J.A., 1979. Blacklight trap collection of parasitic Hymenoptera. Entomol. News 90, 17–22.

Burditt, A.K., 1982. *Anastrepha suspensa* (Loew) (Diptera: Tephritidae) McPhail traps for survey and detection. Fla. Entomol. 65, 367–373.

Burditt, A.K., 1988. Western cherry fruit fly (Diptera: Tephritidae): efficacy of homemade and commercial traps. J. Entomol. Soc. B: C 85, 53–57.

Buriff, C.R., 1973. Recapture of released apple maggot flies in sticky board traps. Environ. Entomol. 2, 757–758.

Burrells, W., 1978. Microscope Technique: A Comprehensive Handbook for General and Applied Microscopy. Fountain Press, London, Halsted Press, New York; Intl. Publ. Serv. New York.

Byers, G.W., 1959. A rapid method for making temporary insect labels in the field. J. Lepid. Soc. 13, 96–98.

Cade, W., 1975. Acoustically orienting parasitoids: fly phonotaxis to cricket song. Science (Washington, DC) 190, 1312–1313.

Callahan, P.S., Sparks, A.N., Snow, J.W., Copeland, W.W., 1972. Corn earworm moth: vertical distribution in nocturnal flight. Environ. Entomol. 1, 497–504.

Campion, D.G., 1972. Some observations on the use of pheromone traps as a survey tool for *Spodoptera littoralis*. Cent. Overseas Pest Res. Rep. 4.

Cantrall, I.J., 1939–1940. Notes on collecting and studying Orthoptera. Ward's Combined Entomol. Nat. Sci. Bull. 13 (3), 1–5 (4): 1–6; (5): 4,5; (6): 5–7; (7): 4,5 (all pp. unnumbered). [Much of the same material is in Cantrall, I. J., 1941.].

Cantrall, I.J., 1941. Notes on collecting and preserving Orthoptera. Compendium of entomological methods, pt. II. Ward's Combined Entomol. Nat. Sci. Bull. 13 (4), 1–9.

Carayon, J., 1969. Emploi du noir chlorazol en anatomie microscopique des insectes. Ann. Soc. Entomol. Fr. (n.s.) 5, 179–193 [On the use of the dye chlorazol black to color cuticle in microscopical preparations.].

Carlberg, U., 1986. *Phasmida:* a biological review (Insecta). Zool. Anz. 216, 1–18.

Carlson, D., 1971. A method of sampling larval and emerging insects using an aquatic black light trap. Can. Entomol. 103, 1365–1369.

Carroll, J.F., 1988. Short-duration dry ice sampling for American dog ticks (Acari: Ixodidae) in Maryland: a comparison with dragging. J. Entomol. Sci. 23, 131–135.

Castner, E., Byrd, J. (Eds.), 2001. Forensic Entomology: The Utility of Arthropods in Legal Investigations. CRC Press, Boca Raton, FL.

Carvalho, A.O.R., Viera, L.G.E., 2000. Comparison of preservation methods of *Atta* spp. (Hymenoptera: Formicidae) for RAPD analysis. Ann. Soc. Entomol. Bras. 29 (3). Available from: https://doi.org/10.1590S0301-80592000000300011.

Castalanelli, M.A., Severtson, D.L., Brumley, C.J., Szito, A., Foottit, R.G., Grimm, M., et al., 2010. A rapid non-destructive DNA extraction method for insects and other arthropods. J. Asia-Pac. Entomol. 13, 243−248.

Catts, E.P., 1970. A canopy trap for collecting Tabanidae. Mosq. News 30, 472−474.

Chamberlin, J.C., 1940. A Mechanical Trap for the Sampling of Aerial Insect Populations. U.S. Dept. Agric., Bur. Entomol. Plant Quar. ET-163.

Chandler, D.S., 1990. Insecta: Coleoptera Pselaphidae. In: Dindal, D.L. (Ed.), Soil Biology Guide. John Wiley & Sons, New York, pp. 1175−1190.

Chapman, J.A., Kinghorn, J.M., 1955. Window flight traps of insects. Can. Entomol. 87, 46−47.

Chapman, T.W., Willsie, A.P., Kevan, P.G., Willis, D.S., 1990. Fiberglass resin for determining nest architecture of ground-nesting bees. J. Kans. Entomol. Soc. 63, 641−643.

Cheng, L., 1975. A simple emergence trap for small insects. Pan-Pac. Entomol. 50, 305−307.

Chiang, H.C., 1973. A simple trap for certain minute flying insects. Ann. Entomol. Soc. Am. 66, 704.

Chick, A., 2016. Insect Microscopy. The Crowood Press, Ramsbury, England.

Christiansen, K.A., 1990. Insecta: Collembola. In: Dindal, D.L. (Ed.), Soil Biology Guide. John Wiley & Sons, New York, pp. 965−995.

Chu, H.F., 1949. How to know the immature insects. An Illustrated Key for Identifying the Orders and Families of Many Immature Insects With Suggestions for Collecting, Rearing, and Studying Them. Wm. C. Brown, Dubuque, IA.

Clark, J.D., Curtis, C.E., 1973. A battery-powered light trap giving two years' continuous operation. J. Econ. Entomol. 66, 393−396.

Clark, W.H., Blom, P.E., 1979. Use of a hand sprayer as a collecting technique. Entomol. News 99, 247−248.

Clark, W.H., Gregg, R.E., 1986. Housing arthropods and other invertebrates stored in alcohol. Entomol. News 97, 237−240.

Clarke, J.F.G., 1941. The preparation of slides of the genitalia of Lepidoptera. Bull. Brooklyn Entomol. Soc. 36, 149−161.

Clifford, C.W., Roe, R.M., Woodring, J.P., 1977. Rearing methods for obtaining house crickets, *Acheta domesticus*, of known age, sex, and instar. Ann. Entomol. Soc. Am. 70, 69−74.

Clinch, P.G., 1971. A battery-operated vacuum device for collecting insects unharmed. N. Z. Entomol. 5, 2830.

Coffey, M.D., 1966. Studies on the association of flies (Diptera) with dung in southeastern Washington. Ann. Entomol. Soc. Am. 59, 207−218.

Cogan, B.H., Smith, K.V.G., 1974. Insects: Instructions for Collectors No. 4a, fifth ed., rev Brit. Mus. (Nat. Hist.) Publ, p. 705.

Colless, D.H., 1959. Notes on the culicine mosquitoes of Singapore. VI. Observations on catches made with baited and unbaited trap-nets. Ann. Trop. Med. Parasitol. 53, 251−258.

Coluzzi, M., Petrarca, V., 1973. Aspirator with paper cup for collecting mosquitoes and other insects. Mosq. News 33, 249−250.

Comstock, J.H., 1940. An Introduction to Entomology, ninth ed Comstock, Ithaca, NY.

Cook, E.F., 1954. A modification of Hopkin's technique for collecting ectoparasites from mammal skins. Entomol. News 15, 35−37.

Cooke, J.A.L., 1969. Notes on some useful arachnological techniques. Bull. Br. Arachnol. Soc. 1, 42–44.

Coon, B.F., Pepper, J.O., 1968. Alate aphids captured in air traps at different heights. J. Econ. Entomol. 61, 1473–1474.

Copeland, T.P., Imadate, G., 1990. Insecta: Protura. In: Dindal, D.L. (Ed.), Soil Biology Guide. John Wiley & Sons, New York, pp. 911–933.

Corbet, P.S., 1965. An insect emergence trap for quantitative studies in shallow ponds. Can. Entomol. 97, 845–848.

Corbet, P.S., Longfield, C., Moore, N.W., 1960. Dragonflies. Collins, London.

Coulson, R.N., Franklin, R.T., Crosley Jr., D.A., 1970. A self-maintaining window trap for collecting flying insects. Entomol. News 81, 164.

Cushing, C.E., 1964. An apparatus for sampling drifting organisms in streams. J. Wildl. Manage. 28, 592–594.

Dales, R.P., 1953. A simple trap for tipulids (Dipt.). Entomol. Mon. Mag. 89, 304.

Darling, D.C., Packer, L., 1988. Effectiveness of Malaise traps in collecting Hymenoptera: the influence of trap design, mesh size, and location. Can. Entomol. 120, 787–796.

Darling, D.C., Plowright, R.C., 1990. HPLABEL: a program and microfont for the generation of date/locality labels using a laser printer. Entomol. News 101, 143–146.

Davidson, J., Swan, D.C., 1933. A method for obtaining samples of the population of Collembola (Symphyleona) in pastures. Bull. Entomol. Res. 24 (pt. 3), 351–352.

Davies, J.B., 1971. A small mosquito trap for use with animal or carbon dioxide baits. Mosq. News 31, 441–443.

Davis, E.W., Landis, B.J., 1949. An Improved Trap for Collecting Aphids. U.S. Dept. Agric. Bur. Entomol. Plant Quar., ET-278.

Dean, M.D., Ballard, W.O., 2001. Factors affecting mitochondrial DNA quality from museum preserved *Drosophilia simulans*. Entomol. Exp. Appl. 98, 279–283.

DeBarro, P.J., 1991. A cheap lightweight efficient vacuum sampler. J. Aust. Entomol. Soc. 30, 207–208.

Debolt, J.W., Jay, D.L., Ost, R.W., 1975. Light traps: effect of modifications on catches of several species of Noctuidae and Arctiidae. J. Econ. Entomol. 68, 186–188.

DeFoliart, G.R., 1972. A modified dry-ice-baited trap for collecting haematophagous Diptera. J. Med. Entomol. 9, 107–108.

DeJong, D.J., 1967. Some problems connected with the use of light traps. Entomophaga 3, 25–32.

Dessauer, H.C., Cole, C.J., Hafner, M.S., 1990. Collection and storage of tissues. In: Hillis, D.M., Mortiz, C., Mable, B.K. (Eds.), Molecular Systematics. Sinauer Associates, Inc, Sunderland, MA.

Dethier, V.G., 1955. Mode of action of sugar-baited fly traps. J. Econ. Entomol. 48, 235–239.

Dethier, V.G., 1992. Crickets and Katydids, Concerts and Solos. Harvard University Press, Cambridge, MA.

Dietrick, E.J., 1961. An improved backpack motor fan for suction sampling of insect populations. J. Econ. Entomol. 54, 394–395.

Dietrick, E.J., Schlinger, E.I., Van den Bosch, R., 1959. A new method for sampling arthropods using a suction collecting machine and modified Berlese funnel separator. J. Econ. Entomol. 52, 1085–1091.

Dillon, N., Austin, A.D., Bartowsky, E., 1996. Comparison of preservation techniques for DNA extraction from hymenopterous insects. Insect Mol. Biol. 5 (1), 21–24.

Dindal, D.L. (Ed.), 1990. Soil Biology Guide. John Wiley & Sons, New York.

Dioni, W., 2014. Safe microscopic techniques for amateurs: slide mounting. Onview Books. Oxfordshire, England.

Doane, J.F., 1961. Movement on the soil surface of adult *Ctenicera aeripennis destructor* (Brown) and *Hypolithus bicolor* Esch. (Coleoptera: Elateridae) as indicated by pitfall traps, with notes on captures of other arthropods. Can. Entomol. 93, 636–644.

Dodge, H.R., 1960. An effective, economical flytrap. J. Econ. Entomol. 53, 1131–1132.

Dodge, H.R., Seago, J.M., 1954. Sarcophagidae and other Diptera taken by trap and net on Georgia mountain summits in 1952. Ecology 35, 50–59.

Dominick, R.B., 1972. Practical freeze-drying and vacuum dehydration of caterpillars. J. Lepid. Soc. 26, 69–79.

Dresner, E., 1970. A sticky trap for Mediterranean fruit fly survey. J. Econ. Entomol. 63, 1813–1816.

Dryden, M.W., Broce, A.B., 1993. Development of a trap for collecting newly emerged. Ctenocephalides felis (Siphonaptera: Pulicidae) in homes. J. Med. Entomol. 30, 901–906.

Dunn, G.A., Reeves, R.M., 1980. A modified collection net for catching insects under cloth bands on trees. Entomol. News 91, 7–9.

Dybas, H.S., 1990. Insecta: Coleoptera Ptiliidae. In: Dindal, D.L. (Ed.), Soil Biology Guide. John Wiley & Sons, New York, pp. 1093–1112.

Eads, R.B., Smith, G.C., Maupin, G.O., 1982. A CO_2 platform trap for taking adult *Dermacentor andersoni* (Acari: Ixodidae). Proc. Entomol. Soc. Wash. 84, 342–348.

Eastop, V., 1955. Selection of aphid species by different kinds of insect traps. Nature (Lond.) 176, 936.

Edmondson, W.T., Winberg, G.G. (Eds.), 1971. Manual for Estimating Secondary Production in Fresh Waters. Blackwell Sci. Publ. (IBP Handb. 17), Oxford.

Edmunds Jr., G.F., McCafferty, W.P., 1978. A new J. G. Needham device for collecting adult mayflies (and other out-of-reach insects). Entomol. News 80, 193–194.

Edmunds Jr., G.F., Jensen, S.L., Berner, L., 1976. The Mayflies of North and Central America. University of Minnesota Press, Minneapolis, MN, pp. 8–26 [Methods of collecting and preservation.].

Edwards, J.G., 1963. Spreading blocks for butterfly wings. Turtox News 41, 16–19.

Edwards, D., Leppla, N., 1987. Arthropod Species in Culture in the United States and Other Countries. Entomological Society of America, Lanham, MD.

Edwards, S.R., Bell, B.M., King, M.E., 1981. Pest Control in Museums: A Status Report (1980). Assoc. Syst. Collect. vii, 34, a-iii, A-30, b-iii, B-20, c-i, C-5, d-v, D-35, e-iii, E-14, f-iii, F-29, g-iii, G-10 pp. [Includes insecticides and their properties, regulations for their use, identification of pests, list of agencies, etc.]

Ellington, J.J., Kiser, K., Cardenas, M., Duttle, J., Lopez, Y., 1984a. The Insectavac: a high-clearance, high-volume arthropod vacuuming platform for agricultural ecosystems. Environ. Entomol. 13, 259–265.

Ellington, J.J., Kiser, K., Ferguson, G., Cardenas, M., 1984b. A comparison of sweepnet, absolute and insectavac sampling methods in cotton ecosystems. J. Econ. Entomol. 77, 599–605.

Elliott, J.M., 1970. Methods of sampling invertebrate drift in running water. Ann. Limnol. 6, 133–159.

Ellis, K.A., Surgeoner, G.A., Ellis, C.R., 1985. Versatile program for generating insect labels on an IBM-PC microcomputer. Can. Entomol. 117, 1447–1448.

Emden, H.F. van (Ed.), 1972. Aphid Technology. Academic Press, London [Collection and preservation on pp. 1–10, also much data on sampling, extraction, trapping, etc., in other chapters.].

Erzinclioglu, Y.Z., 1983. The application of entomology to forensic medicine. Med. Sci. Law 23, 57–63.

Essig, E.O., 1942. College Entomology. Macmillan, New York.

Essig, E.O., 1958. Insects and Mites of Western North America. Macmillan, New York.

Ettinger, W.S., 1979. A collapsible insect emergence trap for use in shallow standing water. Entomol. News 90, 114−117.

Evans, G.O., 1992. Principles of Acarology. CAB International, Wallingford, CT.

Evans, L.J., 1975. An improved aspirator (pooter) for collecting small insects. Proc. Br. Entomol. Nat. Hist. Soc. Trans. 8, 8−11.

Evans, G.O., Sheals, J.G., MacFarlane, D., 1964. The terrestrial Acari of the British Isles: an introduction to their morphology, biology and classification, Introduction and Biology, vol. 1. Brit. Mus. (Nat. Hist.), London, pp. 61−88 [Techniques treated on.

Everett, R., Lancaster Jr., J.L., 1968. A comparison of animal- and dry-ice-baited traps for the collection of tabanids. J. Econ. Entomol. 61, 863−864.

Fahy, E., 1972. An automatic separator for the removal of aquatic insects from detritus. J. Appl. Ecol. 6, 655−658.

Fallis, A.M., Smith, S.M., 1964. Ether extracts from birds and CO_2 as attractants for some ornithophilic simuliids. Can. J. Zool. 42, 723−730.

Ferguson, L.M., 1990. Insecta: Microcoryphia and Thysanura. In: Dindal, D.L. (Ed.), Soil Biology Guide. John Wiley & Sons, New York, pp. 935−949.

Ferro, M.L., Park, J.-S., 2013. Effect of propylene glycol concentration on mid-term DNA preservation of Coleoptera. Coleopts Bull. 67 (4), 581−586.

Fessenden, G.R., 1949. Preservation of Agricultural Specimens in Plastics. U.S. Dept. Agric. Misc. Publ. 679.

Finch, S., Skinner, G., 1974. Some factors affecting the efficiency of water-traps for capturing cabbage root flies. Ann. Appl. Biol. 77, 213−226.

Fincher, G.T., Stewart, T.B., 1968. An automatic trap for dung beetles. J. Ga. Entomol. Soc. 3, 11−12.

Fischer, R.W., Jursic, F., 1958. Rearing houseflies and roaches for physiological research. Can. Entomol. 90, 1−7.

Flaschka, H., Floyd, J., 1969. A simplified method of freeze-drying caterpillars. J. Lepid. Soc. 23, 43−48.

Fleming, W.E., et al., 1940. The Use of Traps Against the Japanese Beetle. U.S. Dept. Agric. Circ. 594.

Foote, R.H., 1948. A Synthetic Resin Imbedding Technique. U.S. Public Health Serv. CDC Bull. 1948 (July−September): 58−59.

Ford, R.L.E., 1973. Studying Insects, A Practical Guide. Warne, London & New York.

Frampton, M., Droege, S., Conrad, T., Prager, S., Richards, M.H., 2008. Evaluation of specimen preservatives for DNA analyses of bees. J. Hymenopt. Res. 17 (2), 195−200.

Fredeen, F.J.H., 1961. A trap for studying the attacking behavior of black flies *Simulium arcticum* Mall. Can. Entomol. 93, 73−78.

Freeman, T.J., 1972. Laboratory Manual—Slide Mounting Techniques. Calif. Dept. Agric. Lab. Serv. Entomol, Sacramento, CA.

Frost, S.W., 1952. Light Traps for Insect Collections, Survey, and Control. Pa. Agric. Exp. Stn. 550.

Frost, S.W., 1957. The Pennsylvania insect light trap. J. Econ. Entomol. 50, 287−292.

Frost, S.W., 1958. Insects attracted to light traps placed at different heights. J. Econ. Entomol. 51, 550−551.

Frost, S.W., 1964. Killing agents and containers for use with insect light traps. Entomol. News 75, 163−166.

Frost, S.W., 1970. A trap to test the response of insects to various light intensities. J. Econ. Entomol. 63, 1344–1346.

Fukatsu, T., 2002. Acetone preservation: a practical technique for molecular analysis. Mol. Ecol. 8 (11). Available from: https://doi.org/10.1046/j.1365-294x.1999.00795.x.

Furumizo, R.T., 1975. Collection and isolation of mites from housedust samples. Calif. Vector Views 22, 19–27.

Gary, N.E., Marston, J.M., 1976. A vacuum apparatus for collecting honeybees and other insects in trees. Ann. Entomol. Soc. Am. 69, 287–288.

Gerberich, J.B., 1945. Rearing tree-hole organisms in the laboratory. Am. Biol. Teach. 7, 83–85.

Gerking, S.D., 1957. A method of sampling the littoral macrofauna and its application. Ecology 38, 219–226.

Gier, H.T., 1949. Differential stains for insect exoskeleton. J. Kans. Entomol. Soc. 22, 79–80.

Gilbert, M.T.P., Moore, W., Melchlor, L., Worobey, M., 2007. DNA extraction from dry museum beetles without conferring external morphological damage. PLoS One 2 (3), E272. Available from: https://doi.org/10.1371/journal.pone.0000272.

Gillies, M.T., 1969. The ramp-trap, an unbaited device for flight studies of mosquitoes. Mosq. News 29, 189–193.

Gillies, M.T., Snow, W.F., 1967. A CO_2-baited sticky trap for mosquitoes. Mosq. News 29, 189–193.

Gist, C.S., Crossley, D.A., 1973. A method for quantifying pitfall trapping. Environ. Entomol. 2, 951–952.

Glasgow, J.P., Duffy, B.J., 1961. Traps in field studies of *Glossina pallidipes* Austen. Bull. Entomol. Res. 52, 795–814.

Glen, D.M., 1976. An emergence trap for bark-dwelling insects; its efficiency and effects on temperature. Ecol. Entomol. 1, 91–94.

Glick, P.A., 1939. The Distribution of Insects, Spiders, and Mites in the Air. U.S. Dept. Agric. Tech. Bull. 673.

Glick, P.A., 1957. Collecting Insects by Airplane in Southern Texas. U.S. Dept. Agric. Tech. Bull. 1158.

Gojmerac, W.L., Davenport, E.C., 1971. Tabanidae (Diptera) of Kegonsa State Park, Madison, Wisconsin: distribution and seasonal occurrence as determined by trapping and netting. Mosq. News 31, 572–575.

Golding, F.D., 1941. Two new methods of trapping the cacao moths (*Ephestia cautella*). Bull. Entomol. Res. 32, 123–132.

Golding, F.D., 1946. A new method of trapping flies. Bull. Entomol. Res. 37, 143–154.

Goldstein, P.Z., Desalle, R., 2003. Calibrating phylogenetic species formation in a threatened insect using DNA from historical specimens. Mol. Ecol. 12 (7), 1993–1998.

Goma, L.K.H., 1965. The flight activity of some East African mosquitoes (Diptera, Culicidae). I. Studies on a high steel tower in Zika forest, Uganda. Bull. Entomol. Res. 56, 17–35.

Goodenough, J.L., Snow, J.W., 1973. Increased collection of tobacco budworm by electric grid traps as compared with blacklight and sticky traps. J. Econ. Entomol. 66, 450–453.

Goonewardene, H.F., Townshend, B.G., Bingham, R.G., Borton, R., 1973. Improved technique for field use of female Japanese beetles as lures. J. Econ. Entomol. 66, 396–397.

Gordh, G., Hall, J.C., 1979. A critical point drier used as a method of mounting insects from alcohol. Entomol. News 90, 57–59.

Gordon, W.M., Gerberg, E.J., 1945. A directional mosquito barrier trap. J. Natl. Malar. Soc. 4, 123–125.

Governatori, M., Bulgarini, C., Rivasi, F., Pampiglione, S., 1993. A new portable aspirator for Culicidae and other winged insects. J. Am. Mosq. Control Assoc. 9, 460–462.

Graham, H.M., Glick, P.A., Hollingsworth, J.P., 1961. Effective range of argon glow lamp survey traps for pink bollworm adults. J. Econ. Entomol. 54, 788–789.

Grandjean, F., 1949. Observation et conservation des tres petits arthropodes. Bull. Mus. Natl. Hist. Nat., Ser. 2 21, 363–370.

Granger, C.A., 1970. Trap design and color as factors in trapping the salt marsh greenhead fly. J. Econ. Entomol. 63, 1670–1672.

Gray, J.S., 1985. A carbon dioxide trap for prolonged sampling of *Ixodes ricinus* L. populations. Exp. Appl. Acarol. 1335–1344.

Gray, T.G., Ibaraki, A.I., 1994. An economical cage for confining and collecting emerging insects and their parasites. Can. Entomol. 126, 447–448.

Gray, T.G., Shepherd, R.F., Struble, D.L., Byers, J.B., Maher, T.F., 1991. Selection of pheromone trap and attractant dispenser load to monitor black army cutworm *Actebia fennica*. J. Chem. Ecol. 17, 309–316.

Green, C.H., 1994. Bait methods for tsetse fly control. Adv. Parasitol. 34, 229–291.

Greenberg, B., Kunich, J.C., 2002. Entomology and the Law: Flies as Forensic Indicators. Cambridge University Press, London.

Greenslade, P.J.M., 1964. Pitfall trapping as a method for studying populations of Carabidae. J. Anim. Ecol. 33, 301–310.

Greenslade, P., 1973. Sampling ants with pitfall traps: digging-in effects. Insectes Soc. 20, 343–353.

Greenslade, P., Greenslade, P.J.M., 1971. The use of baits and preservatives in pitfall traps. J. Aust. Entomol. Soc. 10, 253–260.

Gressitt, J.L., Sedlacek, J., Wise, K.A.J., Yoshimoto, C.M., 1961. A high-speed airplane trap for airborne organisms. Pac. Insects 3, 549–555.

Grigarick, A.A., 1959. A floating pan trap for insects associated with the water surface. J. Econ. Entomol. 52, 348–349.

Grimaldi, D., Engel, M.S., 2005. Evolution of the Insects. Cambridge University Press, Cambridge.

Grimstone, A., 1972. A Guidebook to Microscopical Methods. Cambridge University Press, Cambridge.

Gruber, P., Prieto, C.A., 1976. A collection chamber suitable for recovery of insects from large quantities of host plant material. Environ. Entomol. 5, 343–344.

Gui, H.L., Porter, L.C., Prideax, G.F., 1942. Response of insects to color intensity and distribution of light. Agric. Eng. 23, 51–58.

Gullan, P.J., Cranston, S. (Eds.), 2004. The Insects: An Outline of Entomology. third ed Blackwell Publishing, Oxford [Example of general textbook, including collection, preservation and identification techniques.].

Gulliksen, B., Deras, K.M., 1975. A diver-operated suction sampler for fauna on rocky bottoms. Oikos 26, 246–249.

Gurney, A.B., Kramer, J.P., Steyskal, G.C., 1964. Some techniques for the preparation, study, and storage in microvials of insect genitalia. Ann. Entomol. Soc. Am 57, 240–242.

Gustinich, G., Manfioletti, G., Del Sol, G., Schnieder, C., 1991. A fast method for high-quality genomic DNA extraction from whole human blood. Bio-Techniques 11, 298–301.

Hafraoui, A., Harris, E.J., Chakir, A., 1980. Plastic traps for detection and survey of the Mediterranean fruit fly, *Ceratitis capitata* (Diptera: Tephritidae) in Morocco. Proc. Hawaii Entomol. Soc. 23, 199–203.

Haglund, W.D., Sorg, M. (Eds.), 2002. Advances in Forensic Taphonomy: Method, Theory and Archeological Perspectives. CRC Press, Boca Raton, FL.

Halstead, J.A., Haines, R.D., 1987. Flume collecting: a rediscovered collecting method with notes on insect extracting techniques. Pan-Pac. Entomol. 63, 383–431.

Hammond, H.E., 1960. The preservation of lepidopterous larvae using the inflation and heat-drying technique. J. Lepid. Soc. 14, 67–78.

Hanec, W., Bracken, G.K., 1964. Seasonal and geographical distribution of Tabanidae (Diptera) in Manitoba, based on females captured in traps. Can. Entomol. 96, 1362–1369.

Hansens, E.J., Bosler, E.M., Robinson, J.W., 1971. Use of traps for study and control of saltmarsh greenhead flies. J. Econ. Entomol. 64, 1481–1486.

Hardwick, D.F., 1950. Preparation of slide mounts of lepidopterous genitalia. Can. Entomol. 82, 231–235.

Hardwick, D.F., 1968. A brief review of the principles of light trap design with a description of an efficient trap for collecting noctuid moths. J. Lepid. Soc. 22, 65–75.

Hargrove, J.W., 1977. Some advances in the trapping of tsetse (*Glossina* spp.) and other flies. Ecol. Entomol. 2, 123–137.

Harris, R.H., 1964. Vacuum dehydration and freeze-drying of entire biological specimens. Ann. Mag. Nat. Hist. 7 (13), 65–74.

Harris, T.L., McCafferty, W.P., 1977. Assessing aquatic insect flight behavior with sticky traps. Gr. Lakes Entomol. 10, 233–239.

Harris, E.J., Nakagawa, S., Urago, T., 1971. Sticky traps for detection and survey of three tephritids. J. Econ. Entomol. 64, 62–65.

Hartstack, A.W., Hollingsworth, J.P., Lindquist, D.A., 1968. A technique for measuring trapping efficiency of electric insect traps. J. Econ. Entomol. 61, 546–552.

Harwood, R.F., 1961. A mobile trap for studying the behavior of flying bloodsucking insects. Mosq. News 21, 35–39.

Harwood, J., Areekul, S., 1957. A rearing trap for producing pomace flies for bioassay of insecticides. J. Econ. Entomol. 50, 512–513.

Hathaway, D.O., 1981. Codling Moth: Field Evaluations of Blacklight and Sex Attractant Traps. U.S. Dept. Agric., Agric. Res. Serv., AAT-W-19.

Hazeltine, W.R., 1962. A new insect clearing technique. J. Kans. Entomol. Soc. 35, 165–166.

Heathcote, G.D., 1957a. The comparison of yellow cylindrical, flat, and water traps to Johnson suction traps, for sampling aphids. Ann. Appl. Biol. 45, 133–139.

Heathcote, G.D., 1957b. The optimum size of sticky aphid traps. Plant Pathol. 6, 104–107.

Heathcote, G.D., Palmer, J.M.P., Taylor, L.R., 1969. Sampling for aphids by traps and by crop inspection. Ann. Appl. Biol. 63, 155–166.

Heintzman, P.D., Elias, S.A., Moore, K., Paszkiewicz, K., Barnes, I., 2014. Characterizing DNA preservation in degraded specimens of *Amara alpine* (Carabidae: Coleoptera). Mol. Ecol. Resour. 14, 606–615.

Hendrix III, W.H., Showers, W.B., 1990. Evaluation of differently colored bucket traps for black cutworm and armyworm (Lepidoptera: Noctuidae). J. Econ. Entomol. 83, 596–598.

Hernandez-Triana, L.M., Prosser, S.W., Rodriguez, M.A., Chaverri, L.G., Hebert, P.D.N., Gregory, T.R., 2014. Recovery of DNA barcodes from blackfly museum specimens (Diptera: Simuliidae) using primer sets that target a variety of sequence lengths. Mol. Ecol. Resour. 14, 508–518.

Herting, B., 1969. Tent Window Traps Used for Collecting Tachinids (Dipt.) at Delemont, Switzerland. Commonw. Inst. Biol. Control Tech. Bull.

Hienton, T.E., 1974. Summary of Investigations of Electric Insect Traps. U.S. Dept. Agric. Tech. Bull. 1498.

Hill, M.N., 1971. A bicycle-mounted trap for collecting adult mosquitoes. J. Med. Entomol. 8, 108–109.

Hillis, D.M., Moritz, C., Maple, B.K., 1966. Molecular Systematics, second ed Sinauer Associates, Sunderland, MA.

Hinton, H.E., 1951. Myrmecophilous Lycaenidae and other Lepidoptera—a summary. Proc. S. London Entomol. Nat. Hist. Soc. 1949–1950, 111–175.

Hocking, B., 1953. Plastic embedding of insects—a simplified technique. Can. Entomol. 85, 14–18.

Hocking, B., 1963. The use of attractants and repellents in vector control. Bull. WHO 29 (suppl), 121–126.

Hocking, B., Hudson, G.E., 1974. Insect wind traps: improvements and problems. Queensl. Entomol. 10, 275–284.

Hodgson, C.E., 1940. Collection and laboratory maintenance of Dytiscidae (Coleoptera). Entomol. News 64, 36–37.

Hoffard, W.H., 1980. How to Collect and Prepare Forest Insect and Disease Organisms and Plant Specimens for Identification. U.S. Dept. Agric., Forest Serv., Southeast Area, State and Private Forestry, Gen. Rep. SA-GR13.

Holbrook, J.E.R., 1927. Apparatus and method used to remove pins from insect specimens. J. Econ. Entomol. 20, 642–643.

Holbrook, R.F., Beroza, M., 1960. Gypsy moth (*Porthetria dispar*) detection with the natural female sex lure. J. Econ. Entomol. 53, 751–756.

Hollingsworth, J.P., Hartstack Jr., A.W., 1972. Effect of components on insect light trap performance. Trans. Am. Soc. Agric. Eng. 15, 924–927.

Hollingsworth, J.P., Briggs, C.P., Glick, P.A., Graham, H.M., 1961. Some factors influencing light trap collections. J. Econ. Entomol. 54, 305–308.

Hollingsworth, J.P., Hartsock, J.G., Stanley, J.M., 1963. Electrical Insect Traps for Survey Purposes. U.S. Dept. Agric., Agric. Res. Serv. 42–3–1.

Animal identification: a reference guide. In: Hollis, D. (Ed.), Insects. Brit. Mus. Nat. Hist., vol. 3. John Wiley & Sons, London and New York.

Holtcamp, R.H., Thompson, J.I., 1985. A lightweight, self-contained insect suction sampler. J. Aust. Entomol. Soc. 24, 301–302.

Holzapfel, E.P., Clagg, H.B., Goff, M.L., 1978. Trapping of air-borne insects on ships on the Pacific. Pac. Insects 19, 65–90.

Hood, J.D., 1947. Microscopical Whole-Mounts of Insects, third ed Cornell University, Ithaca, NY [Mimeographed; out of date, but of considerable interest.].

Hopkins, G.H.E., 1949. The host associations of the lice of mammals. Proc. Zool. Soc. (London) 119 (pt. II), 387–604 [Methods on pp. 395–401.].

Hottes, F.C., 1951. A method for taking aphids in flight. Pan-Pac. Entomol. 27, 190.

Houseweart, M.W., Jennings, D.T., Rea, J.C., 1979. Large-capacity pitfall trap. Entomol. News 90, 51–54.

Howell, J.F., 1980. Codling Moth: Blacklight Trapping and Comparisons With Fermenting Molasses Bait and Pheromone Traps. U.S. Dept. Agric., Agric. Res. Serv. ARR-W-22.

Howell Jr., J.R., Cheikh, M., Harris, E.J., 1975. Comparison of the efficiency of three traps for the Mediterranean fruit fly baited with minimum amounts of Trimedlure. J. Econ. Entomol. 68, 277–279.

Hower, R.O., 1979. Freeze-Drying Biological Specimens: A Laboratory Manual. Smithsonian Institution Press, Washington, DC.

Howland, A.F., Debolt, J.W., Wolf, W.W., Toba, H.H., Gibson, T., Kishaba, A.N., 1969. Field and laboratory studies of attraction of the synthetic sex pheromone to male cabbage looper moths. J. Econ. Entomol. 62, 117–122.

Hoy, J.B., 1970. Trapping the stable fly by using CO_2 or CO as attractants. J. Econ. Entomol. 63, 792–795.

Hubbell, T.H., 1956. A new collecting method: the oatmeal trail. Entomol. News 67, 49–51.

Huber, J.T., 1986. Systematics, biology, and hosts of the Mymaridae and Mymarommatidae (Insecta: Hymenoptera): 1758–1984. Entomography 4, 185–243.

Huffacker, C.B., Back, R.C., 1943. A study of methods of sampling mosquito populations. J. Econ. Entomol. 36, 561–569.

Humason, G.L., 1979. Animal Tissue Techniques, fourth ed W. H. Freeman, San Francisco, CA.

Hunter, S.J., Goodall, T.I., Walsh, K.A., Owen, R., Day, J.C., 2008. Nondestructive DNA extraction from blackflies (Diptera: Simuliidae): retraining voucher specimens for DNA barcoding projects. Mole Ecol. Resour. 8, 56–61.

Hurd, P.D., 1954. Myiasis resulting from the use of the aspirator method in the collection of insects. Science (Washington, DC) 119, 814–815.

Inouye, D.W., 1991. Quick and easy insect labels. J. Kans. Entomol. Soc. 64, 242–243.

Jacobson, M., 1972. Insect Sex Pheromones. Academic Press, New York.

Jacobson, M., Beroza, M., 1964. Insect attractants. Sci. Am. 211 (2), 20–27 [Includes excellent figures of Steiner, McPhail, and USDA beetle traps.].

Jacques, H.E., 1947. How to Know the Insects, second ed Wm. C. Brown, Dubuque, IA (See also Bland and Jacques, 1978.).

James, S.H., Nordby, J.J. (Eds.), 2005. Forensic Science: An Introduction to Scientific and Investigative Techniques. second ed Taylor & Francis, New York.

Jenkins, J., 1991. A simple device to clean insect specimens for museums and scanning electron microscopy. Proc. Entomol. Soc. Wash. 93, 204–205.

Jeppson, L.R., Kiefer, H.H., 1975. Mites Injurious to Economic Plants. University of California Press, Berkeley, CA [Includes data on collection and preservation.].

Johnson, C.G., 1950. The comparison of suction trap, sticky trap, and tow-net for the quantitative sampling of small airborne insects. Ann. Appl. Biol. 37, 268–285.

Johnson, C.G., Taylor, L.R., 1955. Development of large suction traps for airborne insects. Ann. Appl. Biol. 43, 51–62.

Johnson, C.G., Southwood, T.R.E., Entwistle, H.M., 1957. A new method of extracting arthropods and molluscs from grassland and herbage with a suction apparatus. Bull. Entomol. Res. 48, 211–218.

Jonasson, P.M., 1954. An improved funnel trap for capturing emerging aquatic insects, with some preliminary results. Oikos 5, 179–188.

Joossee, E.N.G., 1975. Pitfall-trapping as a method for studying surface-dwelling Collembola. Z. Morphol. Oekol. Tiere 55, 587–596.

Junqueira, A.C.M., Lessinger, A.C., Azeredo-Espin, A.M.L., 2002. Methods for the recovery of mitochondrial DNA sequences from museum specimens of myiasis-causing flies. Med. Vet. Entomol. 16, 39–45.

Kasprzak, K., 1993. Methods for fixing and preserving soil animals. In: Gorny, M., Grum, L. (Eds.), Methods on Soil Zoology. Elsevier, Amsterdam, pp. 321–345.

Kempson, D., Lloyd, M., Ghelardi, R., 1962. A new extractor for woodland litter. Pedobiologia 3, 1–21.

Kevan, D.K.Mc.E. (Ed.), 1955. Soil Zoology. Butterworth, London.

Kevan, D.K.Mc.E., 1962. Soil Animals. Witherby, London [Sampling and extraction on pp. 102–125.].

Kieckhefer, R.W., Dickmann, D.A., Miller, E.L., 1976. Color responses of cereal aphids. Ann. Entomol. Soc. Am. 69, 721–724.

Kim, K.C., Pratt, H.D., Stojanovich, C.J., 1986. The Sucking Lice of North America. Pennsylvania State University Press, University Park, PA.

Kimerle, R.A., Anderson, N.H., 1967. Evaluation of aquatic insect emergence traps. J. Econ. Entomol. 60, 1255–1259.

Kissinger, D.G., 1982. Insect label production using a personal computer. Proc. Entomol. Soc. Wash. 84, 855–857.

Klein, M.G., Lawrence, K.O., Ladd Jr., T.L., 1973. Japanese beetles: shielded traps to increase captures. J. Econ. Entomol. 66, 562–563.

Klots, A.B., 1932. Directions for Collecting and Preserving Insects. Ward's Nat. Sci. Estab, Rochester, NY.

Knodel, J.J., Agnello, A.M., 1990. Field comparison of nonsticky and sticky pheromone traps for monitoring fruit pests in western New York. J. Econ. Entomol. 63, 197–204.

Knox, P.C., Hays, K.L., 1972. Attraction of *Tabanus* spp. (Diptera: Tabanidae) to traps baited with carbon dioxide and other chemicals. Environ. Entomol. 1, 323–326.

Knudsen, J.W., 1966. Biological Techniques. Harper & Row, New York.

Knudsen, J.W., 1972. Collecting and Preserving Plants and Animals. Harper & Row, New York [Insects on pp. 128–176.].

Kogan, M., Herzog, D.C. (Eds.), 1980. Sampling Methods in Soybean Entomology. Springer Verlag, New York.

Kosztarab, M., 1966. How to build a herbarium of insect and mite damage. Turtox News 44, 290–294.

Kovrov, B.G., Monchadskii, A.S., 1963. The possibility of using polarized light to attract insects. Entomol. Obozr. 42, 49–55 (In Russian; transl. into English in Entomol. Rev. 42: 25–28.).

Krantz, G.W., 1978. A Manual of Acarology, second ed Oregon State University Book Stores, Corvallis, OR [Collection and preservation on pp. 77–98.].

Kring, J.B., 1970. Red spheres and yellow panels combined to attract apple maggot flies. J. Econ. Entomol. 63, 466–469.

Krogmann, L., Holstein, J., 2010. Preserving and specimen handling: insects and other invertebrates. In: Eymann, J., Degreef, J., VandenSpiegel, C.H. (Eds.), Manual of Field Recording Techniques and Protocols for All Taxa Biodiversity Inventories, vol. 8. ABC Taxa, pp. 464–481.

Krombein, K.V., 1967. Trap-Nesting Wasps and Bees: Life Histories, Nests, and Associates. Smithsonian Institution Press, Washington, DC [Trap-nesting technique described on pp. 8–14.].

Kronblad, W., Lundberg, S., 1978. Bilhåvning-en intressant fangstmetod för skalbaggar och andra insekter. Entomol. Tidskr. 99, 115–118 [Describes net attached to roof of automobile.].

Kühnelt, W., 1976. Soil Biology, With Special Reference to the Animal Kingdom. Michigan State University Press, East Lansing, MI, pp. 385–466, Observation and collecting techniques, pp. 35–65; bibliography.

LaGasa, E.H., Smith, S.D., 1978. Improved Insect Emergence Trap for Stream Community Population Sampling: Environmental Impact of Insecticides. U.S. Dept. Agric. Res. Note PSW, U.S. Pac. Southwest. Forest and Range Exp. Stn. 328.

Laird, N., 1981. Blackflies: The Future Biological Methods in Integrated Control. Academic Press, London [Includes chapters on trapping and sampling methods.].

Lambert, H.L., Franklin, R.T., 1967. Tanglefoot traps for detection of the balsam woolly aphid. J. Econ. Entomol. 60, 1525–1529.

Lammers, R., 1977. Sampling insects with a wetland emergence trap: design and evaluation of the trap with preliminary results. Am. Midl. Nat. 97, 381–389.

Landlin, J., 1976. Methods of sampling aquatic beetles in the transitional habitats at water margin. Freshw. Biol. 6, 81–87.

Lane, J., 1965. The Preservation and Mounting of Insects of Medical Importance. WHO/Vector Control 152.65. Mimeographed.

Lane, R.S., Anderson, J.R., 1976. Extracting larvae of *Chrysops hirsuticallus* (Diptera: Tabanidae) from soil: efficiency of two methods. Ann. Entomol. Soc. Am. 69, 854–856.

Langford, T.E., Daffern, J.R., 1975. The emergence of insects from a British river warmed by power station cooling waterPt. 1. The use and performance of insect emergence traps Hydrobiologia 46, 71–114.

Lawson, D.L., Merritt, R.W., 1979. A modified Ladell apparatus for the extraction of wetland macroinvertebrates. Can. Entomol. 111, 1389–1393.

Lee, W.L., Bell, B.M., Sutton, J.F. (Eds.), 1982. Guidelines for Acquisition and Management of Biological Specimens. Assoc. Syst. Collect., Lawrence, KS.

Leech, H.B., 1955. Cheesecloth flight traps for insects. Can. Entomol. 87, 200.

Lehker, G.E., Deay, H.O., 1969. How to Collect, Preserve, and Identify Insects. Purdue Univ., Coop. Ext. Serv. Ext. Circ. 509.

Lehmann, K., Werner, D., Hoffmann, B., Kampen, H., 2012. PCR identification of culicoid biting midges (Diptera, Ceratopogonidae) of the Obsoletus complex including putative vectors of bluetongue and Schmallenberg viruses. Parasite Vector 5, 213.

LeSage, L., 1991. An improved technique for collecting large samples of arthropods. Entomol. News 102, 97.

LeSage, L., Harrison, A.D., 1979. Improved traps and techniques for the study of emerging aquatic insects. Entomol. News 90, 65–78.

Levin, M.D., 1957. Artificial nesting burrows for *Osmia ligniaria* Say. J. Econ. Entomol. 50, 506–507.

Lewis, G.G., 1965. A new technique for spreading minute moths. J. Lepid. Soc. 19, 115–116.

Lewis, T., Taylor, L.R., 1965. Diurnal periodicity of flight by insects. Trans. R. Entomol. Soc. Lond. 116, 393–469.

Lincoln, R.J., Sheals, J.G., 1979. Invertebrate animals. Collection and preservation. Brit. Mus. (Nat. Hist.). Cambridge University Press, Cambridge.

Lindahl, T., 1993. Instability and decay of the primary structure of DNA. Nature 362 (6422), 709–715.

Lindeberg, B., 1958. A new trap for collecting emerging insects from small rockpools, with some examples of the results obtained. Ann. Entomol. Fenn. 24, 186–191.

Lindroth, C.H., 1957. The best method for killing and preserving beetles. Coleopt. Bull. 11, 95–96.

Lipovsky, L.J., 1951. A washing method of ectoparasite recovery with particular reference to chiggers. J. Kans. Entomol. Soc. 24, 151–156.

Lipovsky, L.J., 1953. Polyvinyl alcohol with lacto-phenol, a mounting and clearing medium for chigger mites. Entomol. News 64, 42–44.

Lis, J.A., Ziaja, D.J., Lis, P., 2011. Recovery of mitochondrial DNA for systematic studies of Pentatomoidea (Hemiptera: Heteroptera): successful PCR on early 20th century dry museum specimens. Zootaxa 2748, 18–28.

Loschiavo, S.R., 1974. The detection of insects by traps in grain-filled boxcars during transit, Proc. Int. Working Conf. Stored-Product Entomol. vol. 1. pp. 639–650.

Lowe, R.B., Putnam, L.G., 1964. Some simple and useful technological improvements in light traps. Can. Entomol. 96, 129.

Lozier, J.D., Cameron, S.A., 2009. Comparative genetic analyses of historical and contemporary collections highlight contrasting demographic histories for the bumble bees *Bombus pennsylvanicus* and *B. impatiens* in Illinois. Mole Ecol. 18, 1875−1886.

Luff, M.L., 1968. Some effects of formalin on the numbers of Coleoptera caught in pitfall traps. Entomol. Mon. Mag. 104, 115−116.

Luff, M.L., 1975. Some factors influencing the efficiency of pitfall traps. Oecologia (Berl.) 19, 345−357.

Lumsden, W.H.R., 1958. A trap for insects biting small vertebrates. Nature (Lond.) 181, 819−820.

Macan, T.T., 1964. Emergence traps and the investigation of stream faunas. Rev. Hydrobiol. 3, 75−82.

MacFadyen, A., 1962. Soil arthropod sampling. Adv. Ecol. Res. 1, 1−34.

Macleod, J., Donnelly, J., 1956. Methods for the study of blowfly populations. I. Bait trapping: significance of limits for comparative sampling. Ann. Appl. Biol. 44, 80−104.

MacSwain, J.W., 1949. A method for collecting male Stylops. Pan-Pac. Entomol. 25, 89−90.

Mandrioli, M., 2008. Insect collections and DNA analyses: how to manage collections? Museum Manage. Curator. 23 (2), 193−199.

Martin, J.E.H., 1977. The Insects and Arachnids of Canada. Pt. 1. Collecting, Preparing, and Preserving Insects, Mites, and Spiders. Can. Dept. Agric., Biosyst. Res. Inst. Publ. 1643.

Masner, L., Gibson, G.A.P., 1979. The separation bag—a new device to aid in collecting insects. Can. Entomol. 111, 1197−1198.

Mason, H.C., 1963. Baited traps for sampling *Drosophila* populations in tomato field plots. J. Econ. Entomol. 56, 897−899.

Mason, W.T., Sublette, J.E., 1971. Collecting Ohio River basin Chironomidae (Diptera) with a floating sticky trap. Can. Entomol. 103, 397−404.

Masteller, E.C., 1977. An aquatic insect emergence trap on a shale stream in western Pennsylvania. Melsheimer Entomol. Ser. 23, 10−15.

Matheson, R., 1951. Entomology for Introductory Courses, second ed Comstock, Ithaca, NY.

Maxwell, C.W., 1965. Tanglefoot traps as indicators of apple maggot activities. Can. Entomol. 97, 110.

McCafferty, W.P., 1981. Aquatic Entomology. Jones & Bartlett, Boston, MA.

McCauley, V.J.E., 1976. Efficiency of a trap for catching and retaining insects emerging from standing water. Oikos 27, 339−346.

McClain, D.C., Rock, G.C., Woolley, J.B., 1990. Influence of trap color and San Jose scale (Homoptera: Diaspidiidae) pheromone on sticky trap catches of 10 aphelinid parasitoids (Hymenoptera). Environ. Entomol. 19, 926−931.

McClung, C.E., Jones McC, R. (Eds.), 1964. Handbook of Microscopical Technique for Workers in Animal and Plant Tissues. third ed Hafner, New York.

McDaniel, B., 1979. How to Know the Mites and Ticks. Wm. C. Brown, Dubuque, IA.

McDonald, J.L., 1970. A simple, inexpensive alcohol light trap for collecting Culicoides. Mosq. News 30, 652−654.

McKillop, W.B., Preston, W.B., 1981. Vapor degreasing: a new technique for degreasing and cleaning entomological specimens. Can. Entomol. 113, 251−253.

McNutt, D.N., 1976. Insect Collecting in the Tropics. Center for Overseas Pest Research, London.

Menzies, D.R., Hagley, E.A.C., 1977. A mechanical trap for sampling populations of small, active insects. Can. Entomol. 109, 1405−1407.

Merrill, W., Skelly, J.M., 1968. A window trap for collection of insects above the forest canopy. J. Econ. Entomol. 61, 1461–1462.

Merritt, R.W., Poorbaugh Jr, J.H., 1975. A laboratory collection and rearing container for insects emerging from cattle droppings. Calif. Vector Views 22, 43–46.

Merritt, R.W., Cummins, K.W., Resh, V.H., 1984. Collecting, sampling, and rearing methods for aquatic insects. In: Merritt, R.W., Cummins, K.W. (Eds.), An Introduction to the Aquatic Insects of North America, second ed Kendall/Hunt, Dubuque, IA, pp. 11–26.

Meyers, E.G., 1959. Mosquito collections by light traps at various heights above the ground. Proc. Calif. Mosq. Control Assoc. 27, 61–63.

Meyerdirk, D.E., Hart, W.G., Burnside, J., 1979. Evaluation of a trap for the citrus blackfly, *Aleurocanthus woglumi* (Homoptera: Aleyrodidae). Can. Entomol. 111, 1127–1129.

Miller, T.A., Stryker, R.G., Wilkinson, R.N., Esah, S., 1970. The influence of moonlight and other environmental factors on the abundance of certain mosquito species in light-trap collections in Thailand. J. Med. Entomol. 7, 555–561.

Miller, R.S., Passoa, S., Waltz, R.D., Mastro, V., 1993. Insect removal from sticky traps using a citrus oil solvent. Entomol. News 104, 209–213.

Minter, D.M., 1961. A modified Lumsden suction trap for biting insects. Bull. Entomol. Res. 52, 233–238.

Mitchell, E.R., Webb, J.C., Baumhover, A.H., Hines, R.W., Stanley, J.W., Endris, R.G., et al., 1972. Evaluation of cylindrical electric grids as pheromone traps for loopers and tobacco hornworms. Environ. Entomol. 1, 365–368.

Mitchell, E.R., Webb, J.C., Benner Jr., J.C., Stanley, J.M., 1973. A portable cylindrical electric grid trap. J. Econ. Entomol. 66, 1232–1233.

Mitchell, E.R., Stanley, J.M., Webb, J.C., Baumhover, A.H., 1974. Cylindrical electric grid traps: the influence of elevation, size and electrode spacing on captures of male cabbage loopers and tobacco hornworms. Environ. Entomol. 3, 49–50.

Mitchell, R.D., Cook, D.R., 1952. The preservation and mounting of water-mites. Turtox News 30, 169–172.

Montgomery, S.L., 1982. Biogeography of the moth genus *Eupithecia* in Oceana and the evolution of ambush predation in Hawaiian caterpillars (Lepidoptera: Geometridae). Entomol. Gen. 8, 549–556.

Montgomery, S.L., 1983. Carnivorous caterpillars: the behavior, biogeography and conservation of *Eupithecia* (Lepidoptera: Geometridae) in the Hawaiian Islands. GeoJournal 7 (6), 549–556.

Moore, R., Clarke, R.T., Creer, S., 1993. An insect sorting device to be used in conjunction with insect suction samplers. Bull. Entomol. Res. 83, 113–120.

Moreau, C.S., Wray, B.D., Czekanski-Moir, J.E., Rubin, B.E.R., 2013. DNA preservation: a test of commonly used preservatives for insects. Invertebr. Syst. 27, 81–86.

Moreland, C.R., 1955. A wind frame for trapping insects in flight. J. Econ. Entomol. 47, 944.

Morgan, C.V.G., Anderson, N.H., 1958. Techniques for biological studies of tetranychid mites. Can. Entomol. 90, 212–215.

Morgan, N.O., Uebel, E.C., 1974. Efficacy of the Assateague insect trap in collecting mosquitoes and biting flies in a Maryland salt marsh. Mosq. News 34, 196–199.

Morgan, N.C., Waddell, A.W., Hall, W.B., 1963. A comparison of the catches of emerging aquatic insects in floating box and submerged funnel traps. J. Anim. Ecol. 32, 203–219.

Morrill, W.L., 1975. Plastic pitfall trap. Environ. Entomol. 4, 596.

Morrill, W.L., Whitcomb, W.H., 1972. A trap for alate imported fire ants. J. Econ. Entomol. 65, 1194–1195.

Morris, K.R.S., 1961. Effectiveness of traps in tsetse surveys in the Liberian rain forest. Am. J. Trop. Med. Hyg. 10, 905–913.

Morris, C.D., DeFoliart, G.R., 1969. A comparison of mosquito catches with miniature light traps and CO_2-baited traps. Mosq. News 29, 424–426.

Mulhern, T.D., 1942. New Jersey mechanical trap for mosquito surveys. N.J. Agric. Exp. Stn. Circ. 421, 1–8.

Mullen, M.A., 1992. Development of a pheromone trap for monitoring *Tribolium castaneum*. J. Stored Prod. Res. 28, 245–249.

Mullen, M.A., Highland, H.A., Taggart, R.E., Lingren, B.W., 1992. Insect Monitoring System. U.S. Dept. Agric. Patent 5,090, 153.

Muma, M.H., 1975. Long-term can trapping for population analyses of ground-surface, arid-land arachnids. Fla. Entomol. 58, 257–270 [Includes bibliography of can and pitfall trapping.].

Mundie, J.H., 1956. Emergence traps for aquatic insects. Mitt. Int. Ver. Theor. Angew. Limnol. 7, 113.

Mundie, J.H., 1964. A sampler for catching emerging insects and drifting materials in streams. Limnol. Oceanogr. 9, 456–459.

Mundie, J.H., 1966. Sampling emerging insects and drifting materials in deep flowing water. Gewaess. Abwaess. 41/42, 159–162.

Mundie, J.H., 1971. Insect emergence traps. In: Edmondson, W.T., Winberg, G.G. (Eds.), Manual for Estimating Secondary Production in Fresh Waters. Blackwell Scientific (IBP Handb. 17), Oxford, pp. 80–93. [References on pp. 80–93.].

Murphy, P.W. (Ed.), 1962. Progress in Soil Zoology. Butterworth, London [Includes reference on sticky traps.

Murphy, W.L., 1985. Procedures for removal of insect specimens from sticky-trap material. Ann. Entomol. Soc. Am. 78, 881.

Murphy, D.H., Gisin, H., 1959. The preservation and microscopic preparation of anopheline eggs in a lacto-glycerol medium. Proc. R. Entomol. Soc. Lond. Ser. A, Gen. Entomol. 34 (10–12), 171–174.

Murray, T.D., Charles, W.N., 1975. A pneumatic grab for obtaining large, undisturbed mud samples: its construction and some applications for measuring the growth of larvae and emergence of adult Chironomidae. Freshwater Biol. 5, 205–210.

Nagy, Z.T., 2010. A hands-on overview of tissue preservation methods for molecular genetic analyses. Org. Diver. Evol. 10, 91–105.

Nakagawa, S., Chambers, D.L., Urago, T., Cunningham, R.T., 1971. Trap-lure combinations for surveys of Mediterranean fruit flies in Hawaii. J. Econ. Entomol. 64, 1211–1213.

Nakagawa, S., Suda, D., Urago, T., Harris, E.J., 1975. Gallon plastic tub, a substitute for the McPhail trap. J. Econ. Entomol. 68, 405–406.

Nantung Institute of Agriculture, 1975. Effects of light traps equipped with two lamps on capture of insects. Acta Entomol. Sin. 18, 289–294 (In Chinese, with English summary.).

Neal, J.W., Jr. (Ed.), 1979. Pheromones of the Sesiidae (Formerly Aegeriidae). U.S. Dept. Agric., Agric. Res. Serv., ARR-NE-6. [Proceedings of a Symposium, Including Trapping and Perspective of Pheromone Research.]

Needham, J.G. (Ed.), 1937. Culture Methods for Invertebrate Animals. Comstock, Ithaca, NY [Includes methods of collecting and rearing some insects.].

Newell, I.M., 1955. An autosegregator for use in collecting soil-inhabiting arthropods. Trans. Am. Microsc. Soc. 74, 389–392.

Newhouse, V.T., Chamberland, R.W., Johnston, J.G., Sudia, W.D., 1966. Use of dry ice to increase mosquito catches of the CDC miniature light trap. Mosq. News 26, 30–35.

Newton Jr., A.F., 1990. Insecta: Coleoptera Staphylinidae adults and larvae. In: Dindal, D.L. (Ed.), Soil Biology Guide. John Wiley & Sons, New York, pp. 1137–1174.

Newton Jr., A.F., Peck, S.B., 1975. Baited pitfall traps for beetles. Coleopt. Bull. 29, 45–46.

Nicholls, C.F., 1960. A portable mechanical insect trap. Can. Entomol. 92, 48–51.

Nicholls, C.F., 1970. Some Entomological Equipment, second ed. Can. Dept. Agric. Inf. Bull. 2.

Nichols, S.W., Schuh, R.T. (Eds.), 1989. The Torre-Bueno Glossary of Entomology. New York Entomol. Soc.

Nielsen, B.O., 1974. Registrering af insektaktivitet på bøgestammer ved hjaelp af fangtragte. Entomol. Meddel. 42, 1–18 [Describes catches of insects in funnel traps on trunks of beech trees.].

Nielsen, E.T., 1960. A note on stationary nets. Ecology 41, 375–376.

Nijholt, W.W., Chapman, J.A., 1968. A flight trap for collecting living insects. Can. Entomol. 100, 1151–1153.

Norris, K.R., 1966. The Collection and Preservation of Insects. Aust. Entomol. Soc. Handb. 1.

Oldroyd, H., 1958. Collecting, preserving and studying insects. Hutchinson, London, Macmillan, New York.

Olmi, M., 1984a. A revision of the Dryinidae (Hymenoptera). Mem. Am. Entomol. Inst. (Gainesville) 37 (1), 1–946.

Olmi, M., 1984b. A revision of the Dryinidae (Hymenoptera). Mem. Am. Entomol. Inst. (Gainesville) 37 (2), 947–1913.

Onsager, J.A., 1976. Influence of Weather on Capture of Adult Southern Potato Wireworm in Blacklight Traps. U.S. Dept. Agric. Tech. Bull. 1527.

Park, D.-S., Maw, E., Hebert, P.D.N., 2011. Barcoding bugs: DNA-based identification of the true bugs (Insecta: Hemiptera: Heteroptera). PLoS One. Available from: https://doi.org/10.13271/journal.pone.0018749.

Parker, S.P. (Ed.), 1982. Synopsis and Classification of Living Organisms, vol. 2. McGraw-Hill, New York.

Parkman, J.P., Frank, J.H., 1993. Use of a sound trap to inoculate *Steinernema scapterisci* (Rhabditida: Steinernematidae) into pest mole cricket populations (Orthoptera: Gryllotalpidae). Florida Entmol. 65, 105–110.

Parman, D.C., 1931. Construction of the Box-Type Trap for Eye Gnats and Blow Flies. U.S. Dept. Agric. Bull. E-299.

Parman, D.C., 1932. A Box-Type Trap to Aid in the Control of Eye Gnats and Blowflies. U.S. Dept. Agric. Circ. 247.

Peacock, J.W., Cuthbert, R.A., 1975. Pheromone-baited traps for detecting the smaller European elm bark beetle. Coop. Econ. Insect Rep. 25, 497–500.

Pearce, M.J., 1990. A new trap for collecting termites and assessing their foraging activity. Trop. Pest Manage. 36, 310–311.

Peck, S.B., 1990. Insecta: Coleoptera Silphidae and the associated families Agrytidae and Leiodidae. In: Dindal, D.L. (Ed.), Soil Biology Guide. John Wiley & Sons, New York, pp. 1113–1136.

Peck, S.B., Davies, A.E., 1980. Collecting small beetles with large-area "window" traps. Coleopt. Bull. 34, 237–239.

Pennak, R.W., 1978. Fresh-Water Invertebrates of the United States, second ed John Wiley & Sons, New York [Apparatus, formulas, etc. on pp. 769–782.].

Pennington, N.E., 1967. Comparison of DDVP and cyanide as killing agents in mosquito light traps. J. Med. Entomol. 4, 518.

Pérez Pérez, R., Hensley, S.D., 1973. A comparison of pheromone and blacklight traps for attracting sugar borer [*Diatraea saccharalis* (F.)] adults from a natural population. P. R. J. Agric. 57, 320–329.

Peterson, A., 1948. Larvae of Insects. Pt. 1. Lepidoptera and Hymenoptera. Pt. 2. Coleoptera, Diptera, Neuroptera, Siphonaptera, Mecoptera. Edwards, Ann Arbor, MI.

Peterson, A., 1964. Entomological Techniques, tenth ed. Edwards, Ann Arbor, MI.

Peterson, B.V., McWade, J.W., Bord, E.F., 1961. A simple method for preparing uniform minuten-pin double mounts. Bull. Brooklyn Entomol. Soc. 56, 19–21.

Phillips, A.J., Simon, C., 1995. Simple, efficient, and nondestructive DNA extraction protocol for arthropods. Ann. Entomol. Soc. Am. 88 (3), 281–283.

Pickens, L.G., Morgan, N.O., Miller, R.W., 1972. Comparison of traps and other methods for surveying density of populations of flies in dairy barns. J. Econ. Entomol. 65, 144–145.

Pieczynski, E., 1961. The trap method of capturing water mites (Hydracarina). Ekol. Pol. B7, 111–115.

Pinniger, D.B., 1975. The use of bait traps for assessment of stored-product insect populations. Coop. Econ. Insect Rep 25, 907909.

Pokulda, P., Cizek, L., Srtribrna, E., Drag, L., Lukes, J., Novotny, V., 2014. A goodbye letter to alcohol: an alternative method for field preservation of arthropod specimens and DNA suitable for mass collecting methods. Eur. J. Entomol. 111 (2), 175–179.

Post, R.J., Flook, P.K., Millest, A.L., 1993. Methods for the preservation of insects for DNA studies. Biochem. Syst. Ecol. 21 (1), 85–92.

Powers, W.J., 1969. A light-trap bag for collecting live insects. J. Econ. Entomol. 62, 735–736.

Pratt, H.D., 1944. Studies on the comparative attractiveness of 25-, 50-, and 100-watt bulbs for Puerto Rican *Anopheles*. Mosq. News 4, 17–18.

Preiss, F.J., Barefoot, H.L., Stryker, R.G., Young, W.W., 1970. Effectiveness of DDVP as a killing agent in mosquito killing jars. Mosq. News 30, 417–419.

Pritchard, M.H., Kruse, G.O.W., 1982. The Collection and Preservation of Animal Parasites. Univ. Nebr. Tech. Bull. 1.

Prokopy, R.J., 1968. Sticky spheres for estimating apple maggot adult abundance. J. Econ. Entomol. 61, 1082–1085.

Prokopy, R.J., 1973. Dark enamel spheres capture as many apple maggot flies as fluorescent spheres. Environ. Entomol. 2, 953–954.

Quicke, L.J., Belshaw, R., Lopez-Vaamonde, C., 1999. Preservation of hymenopteran specimens for subsequent molecular and morphological study. Zool. Scr. 28, 262–267.

Race, S.R., 1960. A comparison of two sampling techniques for lygus bugs and stink bugs on cotton. J. Econ. Entomol. 53, 689–690.

Reeves, R.M., 1980. Use of barriers with pitfall traps. Entomol. News 91, 10–12.

Reeves, W.C., 1951. Field studies on carbon dioxide as possible host stimulant to mosquitoes. Proc. Soc. Exp. Biol. Med. 77, 64–66.

Reeves, W.C., 1953. Quantitative field studies on a carbon dioxide chemotropism of mosquitoes. Am. J. Trop. Med. Hyg. 2, 325–331.

Reierson, D.A., Wagner, R.E., 1975. Trapping yellowjackets with a new standard plastic wet trap. J. Econ. Entomol. 68, 395–398.

Reiss, R., Schwert, D.F., Ashworth, A., 1995. Field preservation of Coleoptera for molecular genetic analyses. Environ. Entomol. 24 (3), 716–719.

Rennison, B.D., Robertson, D.H., 1959. The use of carbon dioxide as an attractant for catching tsetse. Rep. E: Afr. Trypan. Res. Organ. 1958, 26.

Richards, W.R., 1964. A short method for making balsam mounts of aphids and scale insects. Can. Entomol. 96, 963–966.

Richards, O.W., Davies, R.G., 1977. Imm's general textbook of entomology, tenth edStructure, Physiology, and Development; 2, Classification and Biology, vol. 1. London, Chapman & Hall; New York, John Wiley & Sons.

Riley, G.B., 1957. A modified sweep-net for small insects. Bull. Brooklyn Entomol. Soc. 52, 95–96.

Riley, C.V., 1892. Directions for Collecting and Preserving Insects. Bull. U.S. Natl. Mus., Part F. 39.

Roberts, R.H., 1972. Relative attractiveness of CO_2 and a steer to Tabanidae, Culicidae, and *Stomoxys calcitrans* (L.). Mosq. News 32, 208–211.

Roe, R.M., Clifford, C.W., 1976. Freeze-drying of spiders and immature insects using commercial equipment. Ann. Entomol. Soc. Am. 69, 497–499.

Rogers, D.J., Smith, D.T., 1977. A new electric trap for tsetse flies. Bull. Entomol. Res. 67, 153–159.

Rogoff, W.M., 1978. Methods for Collecting Eye Gnats (Diptera: Chloropidae). U.S. Dept. Agric., Agric. Res. Serv., ARM-W-2 (iii). [Includes extensive bibliography.]

Rohlf, F.J., 1957. A new technique in the preserving of soft-bodied insects and spiders. Turtox News 35, 226–229.

Roling, M.P., Kearby, W.H., 1975. Seasonal flight and vertical distribution of Scolytidae attracted to ethanol in an oak-hickory forest in Missouri. Can. Entomol. 107, 1315–1320.

Ross, H.H., Ross, C.A., Ross, J.R.P., 1982. A Textbook of Entomology, fourth ed John Wiley & Sons, New York.

Rowley, D.L., Coddington, J.A., Gates, M.W., Nurrbom, A.L., Ochoa, R.A., Vandenberg, N.J., et al., 2007. Vouchering DNA-barcoded specimens: test of a non-destructive extraction protocol for arthropods. Mol. Ecol. Notes 7 (6), 915–924.

Rudd, W.G., Jensen, R.L., 1977. Sweep net and ground cloth sampling for insects in soybeans. J. Econ. Entomol. 70, 301–304.

Sabrosky, C.W., 1966. Mounting insects from alcohol. Bull. Entomol. Soc. Am. 12, 349.

Sabrosky, C.W., 1971. Packing and shipping of pinned insects. Bull. Entomol. Soc. Am. 17, 6–8.

Saferstein, R., 2004. Criminalistics: An Introduction to Forensic Science. Prentice Hall, Upper Saddle River, NJ.

Salmon, J.T., 1946. A portable apparatus for the extraction from leaf mould of Collembola and other minute organisms. Dom. Mus. Rec. Entomol. (Wellington) 1, 13–18.

Salt, G., Hollick, F.S.J., 1944. Studies of wireworm populations. 1. A census of wireworms in pasture. Ann. Appl. Biol. 31, 53–64.

Sanders, D.P., Dobson, R.C., 1966. The insect complex associated with bovine manure in Indiana. Ann. Entomol. Soc. Am. 59, 955–959.

Sauer, R.J., 1976. Rearing insects in the classroom. Am. Biol. Teach. 1976 (Apr.), 216–221.

Schlee, D., 1966. Präparation und Ermittlung von Messwerton an Chironomiden (Diptera). Gewaess. Abwaess. 41/42, 169–192 [Describes preparation of Chironomidae for study.].

Schmid, J.M., Mitchell, J.C., Schroeder, M.H., 1973. Bark Beetle Emergence Cages Modified for Use as Pitfall Traps. U.S. Dept. Agric., For. Serv. RM-244.

Schmidt, O.A., Hausmann, Cancian de Araujo, B., Peggie, D., Schmidt, S., 2017. Biochem. Data J 5, e20006. Available from: https://doi.org/10.3897//BDJ.5.e200006.

Scott, J.A., 1986. The Butterflies of North America: A Natural History and Field Guide. Stanford University Press, Stanford, CA.

Seber, G.A.F., 1973. The Estimation of Animal Abundance. Griffin, London.

Service, M.W., 1976. Mosquito Ecology: Field Sampling Methods. Applied Science, London [Includes extensive bibliographies.].

Shipley, A.E., 1904. The orders of insects. Zool. Anz. 28, 259−262.

Sholdt, L.L., Neri, P., 1974. Mouth aspirator with holding cage for collecting mosquitoes and other insects. Mosq. News 34, 236.

Shorey, H.H., 1973. Behavioral responses to insect pheromones. Annu. Rev. Entomol. 18, 349−380.

Shorey, H.H., McKelvey Jr., J.J., 1977. Chemical Control of Insect Behavior: Theory and Application. John Wiley & Sons, New York.

Shubeck, P.P., 1976. An alternative to pitfall traps in carrion beetle studies (Coleoptera). Entomol. News 87, 176−178.

Singer, G., 1964. A simple aspirator for collecting small arthropods directly into alcohol. Ann. Entomol. Soc. Am. 57, 796−798.

Singer, G., 1967. A comparison between different mounting techniques commonly employed in acarology. Acarologia 9, 475−484.

Sladeckova, A., 1962. Limnological investigation methods for the periphyton ("aufwuchs") community. Bot. Rev. 28, 286−350.

Smith, B.J., 1976. A new application in the pitfall trapping of insects, to investigate migratory patterns. Trans. Ky. Acad. Sci. 37, 94−97.

Smith, J.S., Jr. 1974a. S-1 Black-Light Insect-Survey Trap Plans and Specifications. U.S. Dept. Agric., Agric. Res. Serv., ARS-S-31.

Smith, K.V.G., 1974b. Rearing the Hymenoptera Parasitica. Leaf. Amat. Entomol. Soc. 35.

Smith, J.G., Pereira, A.C., Correa, B.S., Panizzi, A.R., 1977. Confeção de aparelhos de baix custo para coleta e criaçã de insetos. An. Soc. Entomol. Bras. 6, 132−135 (In Portuguese). [Describes low-cost carrying cages, pitfall traps, and rearing cages.].

Snoddy, E.L., Hays, K.L., 1966. Carbon dioxide trap for Simuliidae. J. Econ. Entomol. 59, 242−243.

Sokolova, Y.Y., Sokolov, I.M., Carlton, C.E., 2010. New microsporidia parasitizing bark lice (Insecta: Psocoptera). J. Invertebr. Pathol. 104, 186−194.

Sommerman, K.M., 1967. Modified car-top insect trap functional to 45 mph. Ann. Entomol. Soc. Am. 50, 857.

Southwood, T.R.E., 1979. Ecological Methods With Particular Reference to the Study of Insect Populations, second ed John Wiley & Sons, New York [Includes extensive bibliographies.].

Sparks, A.N., Raulston, J.R., Lingren, P.D., Carpenter, J.E., Klun, J.A., Mullinix, B.G., 1980. Field response of male *Heliothis virescens* to pheromonal stimuli and traps. Bull. Entomol. Soc. Am. 25, 268−274 [Includes pictures of several types of traps used.].

Stamper, T., Wong, E.S., Timm, A., DeBry, R.W., 2017. Validating sonication as a DNA extraction method for use with carrion flies. Forensic Sci. Int. 275, 171−177.

Stanley, J.M., Dominick, C.B., 1970. Funnel size and lamp wattage influence on light-trap performance. J. Econ. Entomol. 63, 1423−1426.

Stanley, J.M., Webb, J.C., Wolf, W.W., Mitchell, E.R., 1977. Electrocutor grid insect traps for research purposes. Trans. Am. Soc. Agric. Eng. 20, 175−178.

Steck, W., Bailey, B.K., 1978. Pheromone traps for moths: evaluation of cone trap design and design parameters. Environ. Entomol. 7, 419−455.

Stehr, F.W., 1987. Immature Insects, vol. 1. Kendall/Hunt, Dubuque, IA.

Stehr, F.W., 1991. Immature Insects, vol. 2. Kendall/Hunt, Dubuque, IA.

Stein, J.D., 1976. Insects: A Guide to Their Collection, Identification, Preservation and Shipment. U.S. Dept. Agric., For. Serv., RM-311.

Stein, R., Eisenberg, W.V., Brickey Jr., P.M., 1968. Extraneous materials. Staining technique to differentiate insect fragments, bird feathers, and rodent hairs from plant tissue. J. Assoc. Off. Anal. Chem. 51, 513–518.

Steininger, S., Storer, C., Huler, J., Lucky, A., 2015. Alternative preservatives of insect DNA for citizen science and other low-cost applications. Invertebr. Syst. 29, 468–472.

Stewart, J.W., Payne, T.L., 1971. Light trap screening for collecting small soft-bodied insects. Entomol. News 82, 309–311.

Stewart, P.A., Lam, J.J., 1968. Catch of insects at different heights in traps equipped with blacklight lamps. J. Econ. Entomol. 61, 1227–1230.

Steyskal, G.C., 1957. The relative abundance of flies (Diptera) collected at human feces. Z. Angew. Zool. 44, 79–83.

Steyskal, G.C., 1977. History and use of the McPhail trap. Fla. Entomol. 60, 11–16.

Steyskal, G.C., 1981. Bibliography of the Malaise trap. Proc. Entomol. Soc. Wash. 83, 225–229.

Steyskal, G.C., Murphy, W.L., Hoover, E.M., 1986. Insects and Mites: Techniques for Collection and Preservation. U.S. Dept. of Ag. Misc. Pub. No. 1443.

Still, G.W., 1960. An improved trap for deciduous tree fruit flies. J. Econ. Entomol. 53, 967.

Stoecke, B.C., Dworschak, K., Gossner, M.M., Kuehn, R., 2010. Influence of arthropod sampling solutions on insect genotyping reliability. Entomol. Exp. Appl. 135, 217–223.

Stoetzel, M.B., 1985. Pupiform larvae in the Phylloxeridae (Homoptera: Aphidoidea). Proc. Entomol. Soc. Wash. 87, 535–537.

Stoetzel, M.B., 1989. Common Names of Insects and Related Organisms. Entomological Society of America, College Park, MD.

Strandtmann, R.W., Wharton, G.W., 1958. Manual of mesostigmatid mites parasitic on vertebrates. Univ. Md. Inst. Acarol. Contrib. 4, 1–330.

Strange, J.P., Knoblett, J., Griswold, T., 2009. DNA amplification from pin-mounted bumble bees (*Bombu*s) in a museum collection: effects of fragment size and specimen age on successful PCR. Apidologie 40, 134–139.

Strenzke, K., 1966. Empfohlene Methodenzur Aufzucht und Präparation terrestrischer Chironomiden. Gewaess. Abwaess. 41/42, 163–168 [Methods for rearing and preparation of terrestrial Chironomidae (Diptera).].

Stryker, R.G., Young, W.W., 1970. Effectiveness of carbon dioxide and L (+) lactic acid in mosquito light trap with and without light. Mosq. News 30, 388–393.

Stubbs, A., Chandler, P. (Eds.), 1978. A Dipterist's Handbook. Vol. 2, The Amateur Entomologist. J. Amateur Entomol. Soc, Hanworth, England, Collecting and recording.

Stys, P., Kerzhner, I., 1975. The rank and nomenclature of higher taxa in recent Heteroptera. Acta Entomol. Bohemoslov. 72, 65–79.

Summers, C.G.S., Garrett, R.E., Zalom, F.G., 1984. New suction device for sampling arthropod populations. J. Econ. Entomol. 77, 817–823.

Szinwelski, N., Fialho, V.S., Yotoko, K.S.C., Seleme, L.R., Sperber, C.F., 2012. Ethanol fuel improves arthropod capture in pitfall traps and preserves DNA. Zookeys 196, 11–22. Available from: https://doi.org/10.389//zookeys.196.3130.

Tachikawa, T., 1981. Hosts of encyrtid genera in the world (Hymenoptera: Chalcidoidea). Mem. Coll. Agric. Ehime Univ. 25, 85–110.

Tagestad, A.D., 1974. A technique for mounting Microlepidoptera. J. Kans. Entomol. Soc. 47, 26–30.

Tagestad, A.D., 1977. Easily constructed individual mounting blocks for macro- and microlepidoptera. J. Kans. Entomol. Soc. 50, 27–30.

Tarshis, I.B., 1968a. Use of fabrics in streams to collect blackfly larvae. Ann. Entomol. Soc. Am. 61, 260−261.

Tarshis, I.B., 1968b. Collecting and rearing blackflies. Ann. Entomol. Soc. Am. 61, 1072−1083.

Tashiro, H., 1990. Insecta: Coleoptera Sacarabaeidae larvae. In: Dindal, D.L. (Ed.), Soil Biology Guide. John Wiley & Sons, New York, pp. 1191−1209.

Taylor, L.R., 1962a. The absolute efficiency of insect suction traps. Ann. Appl. Biol. 50, 405−421.

Taylor, L.R., 1962b. The efficiency of cylindrical sticky insect traps and suspended nets. Ann. Appl. Biol. 50, 681−865.

Tedders, W.L., Edwards, G.W., 1972. Effects of blacklight trap design and placement on catch of adult hickory shuckworms. J. Econ. Entomol. 65, 1624−1627.

Teskey, H.J., 1962. A method and apparatus for collecting larvae of tabanid (Diptera) and other inhabitants of wetland. Proc. Entomol. Soc. Ont. 92, 204−206.

Thomas, D.B., Sleeper, E.L., 1977. Use of pit-fall traps for estimating the abundance of arthropods, with special reference to the Tenebrionidae (Coleoptera). Ann. Entomol. Soc. Am. 70, 242−248.

Thompson, P.H., 1969. Collecting methods for Tabanidae (Diptera). Ann. Entomol. Soc. Am. 62, 50−57.

Thompson, P.H., Gregg, E.J., 1974. Structural modifications and performance of the modified animal trap and the modified Manitoba trap for collection of Tabanidae (Diptera). Proc. Entomol. Soc. Wash. 76, 119−122.

Thompson, S.R., Brandenburg, R.L., 2004. A modified pool design for collecting adult mole crickets (Orthoptera: Gryllotalpidae). Florida Entomol. 87 (4), 582−584.

Thorsteinson, A.J., Bracken, B.G., Hanec, W., 1965. The orientation behavior of horse flies and deer flies (Tabanidae: Diptera). III. The use of traps in the study of orientation of tabanids in the field. Entomol. Exp. Appl. 8, 189192.

Tindale, N.B., 1962. The chlorocresol method for field collecting. J. Lepid. Soc. 15, 195−197.

Townes, H., 1962. Design for a Malaise trap. Proc. Entomol. Soc. Wash. 83, 225−229.

Townes, H., 1972. A light-weight Malaise trap. Entomol. News 83, 239−247.

Townson, H., Harbach, R., Callan, T., 1999. DNA identification of museum specimens of the *Anopheles gambiae* complex: an evaluation of PCR as a tool for resolving the formal taxonomy of sibling species complexes. Syst. Entomol. 24, 95−100.

Traver, J.R., 1940. Compendium of Entomological Methods, pt. I. Collecting Mayflies (Ephemeroptera). Ward's Nat. Sci. Estab, Rochester, NY.

Tretzel, E., 1955. Technik und Bedeutung des Fallenfanges für ökologische Untersuchungen. Zool. Anz. 155, 276−287 [Technique and significance of the pitfall trap for ecological research.].

Triplehorn, C.A., Johnson, N.F., 2005. Borror and DeLong's Introduction to the Study of Insects., second ed Brooks/Cole Cengage Learning, Belmont, CA.

Turnbull, A.L., Nicholls, C.F., 1966. Quick trap for area sampling of arthropods in grassland communities. J. Econ. Entomol. 59, 1100−1104.

Turnock, W.J., 1957. A trap for insects emerging from the soil. Can. Entomol. 89, 455−456.

U.S. Department of Agriculture, Agricultural Research Service, 1961. Response of Insects to Induced Light. Presentation Papers. U.S. Dept. Agric., Agric. Res. Serv. ARS 20−10. [Includes 11 pp. of bibliography on insects and light.]

U.S. Department of Agriculture, Extension Service, 1970. 4-H Clubs Entomology Publications (Loose Leaf, in 4 pts.): 1, How to Make an Insect Collection; 2, Key to Orders, Rearing Cages, Experimental Activities; 3, Teen and Junior Leader's Guide; 4, Reference Material.

U.S. Department of Agriculture, Plant Pest Control Division, 1966−1970. Survey Methods; Selected References 1942−1970. Pts. 1−35. USDA Coop. Econ. Insect Rep. 17: 97−112, 326−336,

862−868, 977−982, 1071−1076; 18: 70−72, 369−376, 564−568, 746−750, 911−914, 969−974, 990−996, 1054−1058, 1103−1106; 19: 88−96, 335−337, 353−354, 505−506, 785−786, 797−798, 838−842, 865−870; 20: 11−12, 34−36, 43−44, 151−152, 207−208, 239−242, 345−346, 527−528, 727−732, 763−767; 21: 35−40, 87−94; 22: 285−298. [References categorized into population measurement, forecasting, rearing, equipment and technology, traps, attractants, and pictorial keys. The various parts are not in chronological order.]

Urquhart, F.A., 1965. Introducing the Insect. Frederick Warne, London, pp. 1−19 [Elementary directions for making insect collections.

Usinger, R.L. (Ed.), 1956. Aquatic Insects of California With Keys to North American Genera and California Species. University of California Press, Berkeley, CA.

Van Cleave, J., Ross, J.A., 1947. A method of reclaiming dried zoological specimens. Science (Washington, DC) 105, 318.

Vick, K.W., Mankin, R.W., Cogburn, R.R., Mullen, M., Throne, J.E., Wright, V.F., et al., 1990. Review of pheromone-baited sticky traps for detection of stored-product insects. J. Kans. Entomol. Soc. 63, 526−532.

Viggiani, G., 1984. Bionomics of the Aphelinidae. Annu. Rev. Entomol. 29, 257−276.

Viggiani, G., 1990. Hyperparasites. In: Rosen, D. (Ed.), Armored-Scale Insects, Their Biology, Natural Enemies and Control, vol. B. Elsevier Science, Amsterdam.

Vite, J.P., Baader, E., 1990. Present and future use of semiochemicals in pest management of bark beetles. J. Chem. Ecol. 16, 3031−3041.

von Hagen, K.B., Kadereit, J., 2001. The phylogeny of Gentianella (Gentianaceae) and its colonization of the southern hemisphere are revealed by nuclear and chloroplast DNA sequence variation. Organ. Divers. Evol. 1, 61−79.

Wagstaffe, R., Fidler, J.H., 1955. The preservation of natural history specimens, Invertebrates, vol. 1. Philosophical Library, New York.

Walker, T.J., 1982. Sound traps for sampling mole cricket flights (Orthoptera: Gryllotalpidae: Scapteriscus). Florida Entomol. 65 (1), 105−110.

Walker, A.R., Boreham, P.F.L., 1976. Saline, as a collecting medium for culicoides (diptera: ceratopogonidae) in blood-feeding and other studies. Mosq. News 36, 18−20.

Walsh, G.B., 1933. Studies in the British necrophagous Coleoptera. II. The attractive powers of various natural baits. Entomol. Mon. Mag. 69, 28−32.

Waltz, R.D., McCafferty, W.P., 1984. Indication of mounting media information. Entomol. News 95, 31−32.

The insects of Australia. In: Waterhouse, D. (Ed.), CSIRO. Melbourne University Press, Melbourne.

Waters, T.F., 1969. Subsampler for dividing large samples of stream invertebrate drift. Limnol. Oceanogr. 14, 813−815.

Watson, G.E., Amerson, A.B., Jr., 1967. Instructions for Collecting Bird Parasites. Smithson. Inst. Mus. Nat. Hist., Inf. Leafl. 477.

Watson, J.A.L., Smith, G.B., 1991. Archaeognatha (microcoryphia). In: Nauman, I.D. (Ed.), The Insects of Australia: A Textbook for Students and Research Workers. Melbourne University Press, Melbourne, Australia, pp. 272−274.

Watts, P.C., Thompson, D.J., Allen, K.A., Kemp, S.J., 2007. How useful is DNA extracted from the legs of archived insects for microsatellite-based population genetic analyses? J. Insect Conserv. 11, 195−198.

Weatherston, J., 1976. A new insect trap for use with lepidopteran sex pheromones. Can. Dept. Fish. For. Bi-Mon. Res. Notes 32, 9−10.

Weaver III, J.S., White, T.R., 1980. A rapid, steam bath method for relaxing dry insects. Entomol. News 91, 122–124.

Weber, R.G., 1987. An underwater light trap for collecting bottom-dwelling aquatic insects. Entomol. News 98, 246–252.

Welch, P.S., 1948. Limnological Methods. McGraw-Hill, New York.

Wells, J.D., Sperling, F.A.H., 2001. DNA-based identification of forensically important Chrysomyinae (Diptera: Calliphoridae). Forensic Sci. Int. 120, 110–115.

Wells, J.D., Introna, F., Di Vella, G., 2001a. Human and insect mitochondrial DNA analysis from maggots. J. Forensic Sci. 46 (3), 685–687.

Wells, J.D., Pape, T., Sperling, S.H., 2001b. DNA-based identification and molecular systematics of forensically important Sarcophagidae (Diptera). J. Forensic Sci. 46 (5), 1098–1102.

Wellso, S.G., Fischer, R.L., 1971. Insects taken at Japanese beetle traps baited with anethole-eugenol in southern Michigan in 1968. Mich. Entomol. 4, 105–108.

Weseloh, R.M., 1974. Relationships between different sampling procedures for the gypsy moth, *Porthetria dispar* (Lepidoptera, Lymantriidae), and its natural enemies. Can. Entomol. 106, 225–231.

White, E.G., 1964. A design for the effective killing of insects caught in light traps. N. Z. Entomol. 3, 25–27.

Whitsel, R.H., Schoeppner, R.G., 1965. The attractiveness of carbon dioxide to female *Leptoconops torrens* Tns. and *L. kerteszi* Kieff. Mosq. News 25, 403–410.

Whittaker, R.H., 1952. A study of summer foliage insect communities in the Great Smoky Mountains. Ecol. Monogr. 22 (1), 1–44.

Wiens, J.E., Burgess, L., 1972. An aspirator for collecting insects from dusty habitats. Can. Entomol. 104, 1557–1558.

Wild, A.L., Madison, D.R., 2008. Evaluating nuclear protein-coding genes for phylogenetic utility in beetles. Mol. Phylogenet. Evol. 48, 877–891.

Wilkinson, R.S., 1969. A blacklight box trap for nocturnal insects. Newsl. Mich. Entomol. Soc. 14 (3/4), 3.

Willey, R.L., 1971. Microtechniques: A Laboratory Guide. Macmillan, New York.

Williams, C.B., 1948. The Rothamsted light trap. Proc. R. Entomol. Soc. Lond., Ser. A, Gen. Entomol. 23, 80–85.

Williams, C.B., 1951. Comparing the efficiency of insect traps. Bull. Entomol. Res. 42, 513–517.

Williams, R.W., 1954. A study of the filth flies in New York City—1953. J. Econ. Entomol. 47, 556–562.

Williams, D.F., 1973. Sticky traps for sampling populations of *Stomoxys calcitrans*. J. Econ. Entomol. 66, 1279–1280.

Williams, C.B., Milne, P.S., 1935. A mechanical insect trap. Bull. Entomol. Res. 26, 543–551.

Williams III, L., O'Keeffe, L.E., 1990. An ultrasonic procedure for cleaning sticky-trapped arthropods. J. Kans. Entomol. Soc. 63 (3), 461–462.

Williamson, K., 1954. The Fair Isle apparatus for collecting bird ectoparasites. Br. Birds 47, 234–235.

Willows-Munro, S., Schoeman, M.C., 2015. Influence of killing method on Lepidoptera DNA barcode recovery. Mol. Ecol. Resour. 15, 613–618.

Wilson, L.F., 1969. The window-pane insect trap. Newsl. Mich. Entomol. Soc. 14 (3/4), 1–3.

Wilson, B.H., Tugwell, N.O., Burns, E.C., 1966. Attraction of tabanids to traps baited with dry ice under field conditions in Louisiana. J. Med. Entomol. 3, 148–149.

Wilson, J.G., Kinzer, D.R., Sauer, J.R., Hair, J.A., 1972. Chemo-attraction in the lone star tick (Acarina: Ixodidae): I. Response of different developmental stages to carbon dioxide administered via traps. J. Med. Entomol. 9, 245–252 [A modification of this method is described in Eads et al., 1982.].

Wilson, S.W., Smith, J.L., Purcell III, A.H., 1993. An inexpensive vacuum collector for insect sampling. Entomol. News 104, 203–208.

Wilton, D.P., 1963. Dog excrement as a factor in community fly problems. Proc. Hawaii. Entomol. Soc. 118, 311–317.

Wirth, W.W., Marston, N., 1968. A method for mounting small insects on microscope slides in Canada balsam. Ann. Entomol. Soc. Am. 61, 783–784.

Woke, P.A., 1955. Aspirator-cage combinations for delicate and infected arthropods. Ann. Entomol. Soc. Am. 48, 485–488.

Wood, D.M., Davies, D.M., 1966. Some methods of rearing and collecting black flies. Proc. Entomol. Soc. Ont. 96, 81–90.

Wood, D.M., Danf, P.T., Ellis, R.A., 1979. The Insects and Arachnids of Canada. Pt. 6. The Mosquitoes of Canada: Diptera: Culicidae. Can. Dept. Agric., Biosyst. Res. Inst., Publ. 1686 [Methods of collecting, preservation, and rearing on pp. 45–53.]

Woodring, J.P., 1968. Automatic collecting device for tyroglyphid (Acaridae) mites. Ann. Entomol. Soc. Am. 61, 1030–1031.

Woodring, J.P., Blum, M.S., 1963. Freeze-drying of spiders and immature insects. Ann. Entomol. Soc. Am. 56, 138–141.

Woolley, T.A., 1988. Acarology: Mites and Human Welfare. John Wiley & Sons, New York.

Yates, M., 1974. An emergence trap for sampling adult tree-hole mosquitoes. Entomol. Mon. Mag. 109, 99101.

Zimmerman, E.C., 1978. Microlepidoptera. Insects Hawaii 9, 1–1903.

Zolnerowich, G., Heraty, J.M., Wooley, J.B., 1990. Separation of insect and plant material from screen-sweep samples. Entomol. News 101, 301–306.

Zycherman, L.A., Schrock, J.R. (Eds.), 1988. A guide to museum pest control. Association of Systematic Collections. [Excellent summary of pests, their biology and identification, and control methods.]

Index

Note: Page numbers followed by "*f*" refer to figures.

A

Abbreviations, 112
Acari, 135—136, 139—142
Acaridae, 240
Acariformes, 140
Acarina, 23—24
Acetone, 15, 64, 71, 95
Aculeate, 233—234
Acyrthosiphon pisum, 64
Adhesive, 17, 52, 79, 81, 89, 120—121
Adult mosquito, 226*f*
Aerial net, 9—13
AGA (alcohol—glycerin—acetic acid), 72, 209
Alcohol, 6, 13—14, 23, 25, 27—28, 37, 71—72, 79, 82, 95—96, 98—99, 101, 105—106, 108, 115, 119
Alcohol-glycerin-acetic acid solution, 237
Alderflies, 218
Aleyrodidae, 211
Amara alpina, 67
Amblycera, 150, 167
Amblypygi, 138
Ambrosia beetles, 63
Ambush bug, 211*f*
Ambylcera, 208
Ametabolous, 148—149
Ammonia water, 15
Anamorphic, 187
ANDE, 68
Angelwing, 221*f*
Animal kingdom, 130—131
Anisoptera, 192
Anoplura, 149—150, 167—168, 207—208
Antennules, 142—143
Anthophoridae, 149
Antlions, 151, 218—219, 218*f*
Ants, 148, 151, 187—188, 195, 200, 210, 212—213, 218—219, 222, 229, 231*f*, 232—233
Aphididae, 64
Aphids, 28, 64, 196—197, 200, 205, 209—212, 237, 239
APHIS quarantine inspectors, 122
Apidae, 149
Apocrita, 229
Apotele, 141
Apterygota, 145
Aquarium, 46
Aquatic canvas net, 41*f*
Aquatic dip net, 41*f*
Aquatic insects, 2, 40—42

Aquatic net, 9, 12
Arachnida, 133, 135, 139—140
Araneae, 23—24
Araneida, 136
Archaeognatha, 145, 147, 189—190
Arctiid, 221*f*
Armored scales, 211—212
Arthropods, 23—24
Artificial diets, 50
Artificial refuge traps, 36
Aspirators, 7, 18—20
Assassin bug, 211*f*
Association, 45, 112—113
Astigmata, 140
Atta spp., 63
Attractants, 36—39
Auchenorrhyncha, 150, 154—155, 163, 183, 211—212
Augochlorella striata, 64

B

Backswimmer, 211*f*
Baetidae, 190—191
Bags, 8
Baits, 36—39
Barber's fluid, 237
Bark beetles, 151, 214*f*
Barklice, 205—206
Barriers, 25
Basonym, 131
Basswood, 88—89
Bat flies, 224
Battery, 32—33
Beating sheets, 20
Bed bugs, 168, 210
Bees, 151, 183, 216, 229—234
Beetle mites, 23—24
Beetles, 23—24, 151, 212—215, 214*f*, 223—224
Behavior, 2—4
Berlese funnel, 22—23
Binomen, 130—131
Biological supply houses, 7, 9—11, 88, 99, 103, 110, 114, 121—122
Birge—Ekman dredge, 41*f*
Bite, 3
Bittacidae, 216—217
Black fly, 226*f*, 227
Blattaria, 195—196
Blattodea, 148—149, 157, 178, 195—196

Bleach, 99, 107
Bleaching, 99
Blue traps, 28
Body lice, 208
Bombidae, 149
Booklice, 150, 205–206
Boreidae, 217
Bostrichidae, 149
Botanist's vasculum, 8
Bottle flies, 224
Brachycera, 224–226
Bristletails, 147, 189–190
Brush, 8, 20–21, 51, 82, 107, 118
Buchidae, 149
Buchnera sp., 64
Bulb cutter, 28
Bumble bees, 67
Butterflies, 148–151, 220–224

C

Caddisflies, 151, 219–220, 219*f*
Cages, 29–30
Calcium cyanide, 16
Calliphoridae, 68
Camel's hair, 8
Campodeidae, 187–188
Canada balsam, 237, 239–240
Carabidae, 66–67, 149
Carbon dioxide, 40
Carbon tetrachloride, 15, 76, 78, 118
Card point, 83–85, 113–114
Care of the collection, 115–118
Carnoy's solution, 62, 64
Carrion beetle, 214*f*
Case-making larvae, 219–220
Castes, 201
Cat flea, 228*f*
Caterpillars, 222
Cellosolve, 35, 79
Cellulose, 201
Centipedes, 134, 142
Cerci, 153–154, 157, 160, 162, 166, 172–173,
    177–180, 187–188, 224–226
Chalcidae, 62
Chelisochidae, 199–200
Chelonethida, 138
Chewing lice, 150, 207–208
Chilopoda, 134, 142
Chloroform, 15, 76, 118
Chrysomelidae, 149
Chrysopidae, 217–218
Cicadas, 150, 209–212
Cicindelidae, 149

Cimicidae, 210
Classification, 129, 218
Click beetle, 214*f*
CO_2 Trap, 40*f*
Cockroaches, 149, 195–196, 198–199
Coenagrionidae, 67
Coleoptera, 23–24, 63, 66–67, 148–149, 151, 154,
    170, 172, 178, 181, 183–184, 212–215
Collecting nets, 9–13
Collection, 5, 9, 19–20, 26–27, 31–32, 40–42, 50,
    72–75, 79, 87–88, 93, 100–101, 109, 112, 122
Collection data, 73–75, 79, 93, 111, 114–115
Collection information, 42–43, 93, 114
Collector, 112
Collectors' bag, 5
Collembola, 23–24, 145, 147, 172, 188–189
Color traps, 34–35
Combination traps for student studies, 36
Common name, 131, 189–190, 198–200, 202, 205,
    208–209
Common scorpionflies, 217
Communications & Taxonomic Services Unit (CTSU),
    266
Composite trap, 36
Computer generated labels, 110–111
Coniopterygidae, 164
Containers, 6, 13–17, 45–48, 77, 118
Corpse, 3
Corrodentia, 150, 205
Corydalidae, 218
Coverslips, 102
Crab louse, 208
Crepuscular, 193
Crickets, 148–149, 193–195, 198–199
Criminal investigations, 3
Critical point drying (CPD), 62, 64, 95, 108
Crochets, 222
Crustacea, 133
Cryofreezing, 63
Cryptostigmata, 140
Cryptozoic, 198–199
Ctenidia, 228
Culicidae, 227
Curculionidae, 63
Curing, 104
Curved forceps, 5
Cyanide, 52
Cytochrome c oxidase (COI), 52

D

Damselflies, 148, 192–193
Danaidae, 149
Date, 112

DDVP, 16
Deep freeze, 78
Deer flies, 151
Degreasing, 78
Dehydration, 98–99
Dermaptera, 148–150, 153, 176, 199–200
Detergent, 65
Diagnosis, 129
Diapause, 20–21, 49
Dichlorvos, 16, 118
Dichotomy, 145
Dictyoptera, 148–149
Diethyl ether, 64
Dimethyl sulfoxide (DMSO), 63, 65
Diplopoda, 134, 142
Diplura, 145, 147, 173, 187–188
Diplurans, 147
Diptera, 64, 67–68, 131, 148–149, 151, 158, 168, 172,
    181–183, 185, 215–216, 224–228, 232
Direct pinning, 80–83, 86–87, 113–114
Diseases, 3
Dish trap, 27–29
Diurnal, 193, 203, 220, 222
Diving beetle, 214*f*
DNA, 62–65, 67
DNase activity, 62
Dobsonflies, 151, 218
Dog flea, 228
Dogface, 221*f*
Dormancy, 49
Dorsal, 139
Double mounts, 79–80, 83–87, 113–114
Drag cloth, 20
Dragonflies, 148, 192–193
Drawing pens, 110–111
Dredge, 40–42
Drosophila simulans, 52, 62
Dry preservation, 73–74
Drying, 95–96, 122–123
Drying fresh specimens, 64

E

Earwigs, 150, 199–200
Echinophthiridae, 208
Ectognatha, 145, 147–148, 189–190
Ectoparasites, 42
Electrical grid traps, 36
Embalming fluid, 237
Embedding, 96
Embiidina, 148–150, 165, 180, 202–203
Embioptera, 150, 202–203
Emergence traps, 29–30
Endopterygota, 148, 150–151

Entognatha, 143–144
Entomobryidae, 188–189
Envelopes, 7, 73
Environmental disturbance, 2
Ephemeridae, 190–191
Ephemeroptera, 144, 148–149, 158, 161, 170, 190–192
Ephydrid, 227
Eradication, 2–3
Eriophyidae, 240–241
Essig's fluid, 239
Ethanol, 19, 63–64
Ethanol fuel, 65
Ether, 15, 76, 78
Ethyl acetate, 15, 18, 35, 52–61, 64, 76, 118
Ethyl alcohol, 71
Ethylenediaminetetraacetic acid (eDTA), 64, 66
Euparal, 102, 106
Euplexoptera, 150
Evidence, 3
Exopterygota, 143–144, 148–150
Exoskeleton, 144
Extension service, 243–262
Extractors, 21–24
Exuviae, 45

F

Families, 131, 159, 168, 187–191, 193–194, 199–200,
    207, 212–213, 217–219, 224, 226–228
Feces, 38
Firebrats, 147, 190
Fishflies, 218
Flag, 20
Flagellate protozoans, 201
Flashlight, 9, 22, 38–39
Flat headed wood borer, 214*f*
Fleas, 148, 151, 188–189, 228–229
Flesh fly, 226*f*
Flowerpot cage, 47–48, 47*f*
Flushing agents, 3–4
Food, 43–44, 46–47, 50, 77, 109, 122–123
Footspinners, 202–203
Forceps, 5
Forensic, 44–45, 109–110, 113, 127
Forensic scientists, 44–45
Forficulidae, 199–200
Form, ARS-748, 264, 267
Formaldehyde, 66
Formalin, 66, 72
Formicidae, 63
Free-living caddisfly larva, 219–220
Freeze-drying, 72, 95, 108
Freezer, 13, 52, 73, 78, 95
Freezing, 62–63

Fringe-winged insects, 208—209
Fumigation, 118
Fungivorous, 206—207, 209
Furcula, 188

G

Gall wasps, 233
Gardener's trowel, 7—8, 8*f*
Genitalia, 100, 104—106
Genus, 129, 193, 232—233
Giant water bug, 211*f*
Glaciers, 198—199
Glassine envelopes, 7, 74—75, 90
Glue, 44, 83, 85, 90
Glue traps, 44
Glycerine, 72
Glyptol, 100—101, 103
Gnats, 151, 224
Gonopore, 187—188
Gonostyli, 187—188
Grasshoppers, 148—149, 193—195
Ground beetles, 67, 151, 214*f*
Ground litter, 20—21
Gryllidae, 68
Grylloblattaria, 198—199
Grylloblattodea, 179, 198—199
Gut bacteria, 64

H

Habitat, 2—4, 23, 44, 46, 112—113
Haematopinidae, 208
Hairstreak, 221*f*
Halictidae, 64
Haller's organ, 140
Halteres, 151, 158, 224—226
Hand lens, 8
Hand sanitizer, 63
Hangingflies, 151, 216—217
Head lice, 208
Headlamp, 9, 38—39
Heliconid line, 221*f*
Hellgrammites, 151
Hemimetabolous, 148—149
Hemiptera, 64, 148—150, 154—155, 158, 163, 168,
    182—183, 209—212, 215—216
Hemoerobiidae, 218
Heptageniidae, 190—191
Herbarium sheets, 85, 110
Herkol, 16
Heteroptera, 150, 154—155, 163, 182, 209—210
Hexamethyldisilazane (HMDS), 62
Hexapoda (Insecta), 134, 143—145, 151—185, 187
Hister beetle, 214*f*
Historical baseline, 2

Hobby, 127
Holometabolous insects, 150—151, 224
Holothyrina, 141
Homoptera, 149—150, 210—212
Honey, 130—131, 196, 232
Honeybee, 231*f*
Honeydew, 211—212
Horse flies, 224, 226*f*
Hospitals, 3
Host, 3—4
Host plant, 47, 50, 112—113, 122—123
House flies, 151, 224—227, 226*f*
Housing the collection, 115—118
Hoyer's medium, 99—100, 116—117, 238, 240—241
Hymenoptera, 52, 63—64, 148—149, 151, 164,
    175—176, 183—184, 210—212, 215—216, 229—234
Hymenopterans, 52
Hypermetamorphic, 215—216, 231—232
Hypodermic syringe, 13—14
Hypostome, 140—141

I

Ichneumon wasp, 231*f*
Ichneumondidae, 62
Ichneumonidae, 52
Ichneumons, 151
Identification, 3, 122, 126—128, 263—270
Identification labels, 115
Imago, 144
Incandescent lamp, 23
Indicators, 2
Inflation of larvae, 94
Injection, 13—14, 71, 106—107
Ink, 110, 114
Insect audits, 2
Insect drawer, 117
Insect killing agents, 52
Insect pin, 80, 82, 90, 93, 106—107, 117
Insecta, 143—145, 147—185, 187, 206—207
Inspections, 3
Instar, 143—144, 202, 208—209, 215—216, 231—232
Interception nets, 26
International Code, 130—131
Invasion, 2—3
Ischnocera, 150
Ishnocera, 208
Isopoda, 142—143
Isopropyl alcohol, 71
Isoptera, 148—150, 156, 165, 179—180, 200—202
Ixodida, 141

J

Japygidae, 187—188

K

Katydids, 148−149, 193−195
Kerosene−acetic acid solution, 238
Kerosene, alcohol, acetic acid, and dioxane mixture
    (KAAD), 238
Keys, 129
Kick screen, 40−42
Killing agents, 52−62
Killing container, 13−17
Killing jar, 6, 11−12, 14, 22, 26, 30−31, 38−39,
    71−72, 77−78
King, 201
Knife, 7−8, 83−84, 93

L

Labeling, 3−4, 52, 104, 109, 114
Labels, 110−111, 114, 122−123
Labiduridae, 199−200
Labiidae, 199−200
Labium, 171, 192−193, 208−209
Lacewings, 151, 218−219
Lanyard, 8
Larva, 143−144, 150−151, 218−220, 228−229,
    231−232
Leaf beetles, 151
Leaf insects, 149, 197−198
Leaf-cutter bee, 231f
Leafhoppers, 211−212
Lepidoptera, 148−149, 151, 159, 163, 171, 174,
    180−181, 183, 185, 217, 220−224, 221f, 231−232
Lepidopterans, 52
Lepismatidae, 190
Life stages, 127
Light, 26, 30−33, 49−50
Light sheets, 33−34
Light traps, 30−33
Liquid killing agents, 15−16
Liquid-preserved specimens, 79−80
Litigation, 3
Loading cartons, 121
Lobster trap, 30
Locality, 111
Locusts, 148−149
Longhorned beetle, 214f
Lures, 36−39
Lycaenidae, 222
Lyctidae, 149
Lygaeidae, 149

M

Maceration, 97−98, 240
Machilidae, 189−190
Maggot, 226f
Malacostraca, 142−143

Malaise traps, 26−27, 36, 66
Mallophaga, 145, 149−150, 207
Manitoba trap, 34−35, 34f, 40
Mantidflies, 151
Mantids, 149, 193−194, 196−197, 212
Mantispid, 218−219
Mantodea, 148−149, 157, 178, 196−197, 215−216
Mayflies, 148, 158, 190−192
Mealybugs, 211−212
Mecoptera, 148−149, 151−185, 216−217
Medico-criminal investigations, 44−45
Megaloptera, 161, 170, 218
Meinertellidae, 189−190
Mengeidae, 215
Mercury vapor lamp, 32
Mesostigmata, 141
Metallic wasp, 231f
Metamorphosis, 143−144, 148−151, 188−200, 203,
    207−208, 211, 216−217
Methanol, 62, 64
Microcoryphia, 145, 147, 189−190
Microlepidoptera, 89, 92−93, 98
Microscope slides, 114−115, 118, 121−122
Microsporidia, 66
Microthelyphonida, 137
Microvial, 100, 105−106, 109
Midges, 151, 224
Millipedes, 134, 142
Minuten, 80, 86−87, 90, 92−93, 100, 105−107
Mites, 129, 133, 136, 139−140, 142
Moisture, 48
Mold, 118
Molecular research, 52−68
Molts, 143, 187, 215−216
Monarch butterfly, 221f
Monarch caterpillar, 221f
Monitoring, 3
Monocondylic, 189−190
Mosquito, 151, 183, 224, 227−228
    larva (wiggler), 226f
    pupa (tumbler), 226f
Moths, 151, 159, 220, 222−224
Mound-building ants, 233
Mounting, 75−109, 239
    boards, 88−96
    larvae, 108−109
    medium, 96−109, 114−115
    mites, 239−241
    small specimens, 75−109
    soft body specimens, 239
    thrips, 240
    wings, 107
Multivoltine, 209
Museum pests, 120, 122
Museum specimens, 66

Mycetocyte symbiont, 64
Myiasis, 226–227
Myriapoda, 142
Myrmecophilous, 232–233
Myrmeleontidae, 218
Myrmeleontids, 218–219

N

Naiads, 143–144, 203–204
Names, 130–131, 139–140, 147, 150, 202, 205, 217, 224
Necrophilous, 206–207
Nematocera, 224–226
Nemobius fasciatus, 68
Neoptera, 148, 228
Neotenics, 201
Neuroptera, 148–149, 151–185, 217–219
Neuroptra, 165
New Jersey trap, 31
Nocturnal, 138, 189–190, 193–194, 199–200, 220, 222–223
Nogos, 16
Nonbeetle mites, 23–24
Nondestructive methods, 68
No-pest strips, 16
No-see-ums, 224
Notebook, 7
Notoptera, 149
Nuptial chamber, 201
Nuvan, 16, 118
Nymphalidae, 149
Nymphs, 143–144, 148–149, 190–200, 202–203, 206–208

O

Oatmeal, 39
Odonata, 67, 148–149, 162, 171, 192–193
Ootheca, 195–197
Opilioacarida, 141
Opiliones, 139
Orders, 187
Ordinal classification, 145
Oribatida, 23–24
Orthoptera, 65, 68, 148–149, 156, 177, 193–195, 199, 215–216
Owlflies, 217–219

P

Packaging and shipping specimens, 119–123
Packing materials, 119
Paleoptera, 148, 190–191
Palpigradi, 137

Panorpidae, 217
Paper, 6, 14, 16, 22, 31, 38, 45–46, 73–77, 79–80, 83, 90, 92–93, 110, 121
Papering, 74–75
Paradichlorobenzene (PDB), 118
Paralyzing a butterfly, 11–12
Parasitic wasps, 28
Parasitiformes, 140–141
Parthenogenesis, 188–189, 209, 232
Parthenogenetic, 195–196
Passinae, 67
Paurometabolous, 148–149
Pauropoda, 135, 142
Pediculidae, 208
Pediculus, 208
Pedipalpida, 137
Pencil, 52, 110, 114
Permanent mounting media, 101–103
Pest control industry, 127
Pest insects and mites, 127
Pest management audits, 43–44
Petiole, 164, 176, 229, 232
Phalangida, 139
Phasmatodea, 149, 156–157, 197–198, 197f
Phasmida, 149, 197–198
Pheromones, 39–40
Philopteridae, 208
Photographers, 127
Photographs, 3–4
Phragmotic, 201
Phthiraptera, 148–150, 167–168, 207–208
Phylloxeridae, 211
Phylogeneticists, 145
Physogastry, 201
Picture-wing fly, 226f
Pinned damselfly, 67
Pinned specimens, 66–68, 119–120
Pinning block, 87–88, 113–114
Pitfall trap, 27–29
Placing labels, 113–114
Plaited-winged insects, 203–205
Planipennia, 164–165, 181
Plant bugs, 210–211
Plant clippers, 7–8
Plant press, 9
Plasticine, 96–97
Plecoptera, 148–150, 160, 171, 203–205
Poduridae, 188–189
Point punch, 83–84
Poison, 15–16
Pollination, 232
Pollinators, 28, 226–227, 232
Pollution, 2
Polymerase chain reaction (PCR), 64, 66

Pooter, 18
Ports of entry, 2–3
Potassium cyanide, 16
Praying mantids, 196
Prehensile, 192–193
Preservatives in traps, 65–66
Primary reproductives, 201
Primitive, 147
Printed labels, 111
Procampodeidae, 187–188
Professional entomologists, 85, 127
Projapygidae, 187–188
2-Propanol, 64
Propano-2-ol, 64
Proper pin positions, 82f
Propodeum, 229
Propylene glycol, 66
Prostigmata, 140
Protecting specimens, 118
Protocols, 2–3
Protura, 145, 147, 172, 187
Proturans, 147, 187
Pseudoscorpionida, 138
Psocids, 150
Psocoptera, 66, 148–150, 166, 177, 205–206
Psyllids, 205, 209–212
Pterygota, 143–144, 147–151
Pthiraptera, 167
Punkies, 224
Pupa, 144, 150–151, 208–209, 216–217, 222, 226, 228
Puparia, 45
Pycnogonida, 132

Q

Queen, 201

R

Radical, 148–149
Raphidiodea, 164, 181
Rapidograph, 110
Rearing, 3–4, 29–30, 45–51
Reference collections, 93–94, 127
Refrigeration, 73
Regulated insects, 42–43
Regulating, 2–3
Relaxing chamber, 77
Retinaculum, 188
Rhaphidioidea, 218
Rhynchophthirina, 208
Ricinulei, 139
Riker mounts, 93–94, 118
Ringing, 103
RNA, 52

RNALater, 63
Robber flies, 224, 226f
Robinson trap, 32
Rock crawlers, 149, 198–199
Rock roller, 219f

S

Sample vials, 6
Sarcophagidae, 67–68
Satyridae, 149
Sawdust piles, 23
Sawflies, 151, 217, 224, 229, 231–232, 231f
Scale insects, 158, 209–212, 224–226, 239
Scales, 150
Schizomida, 138
Scientific name, 130–131, 188, 203, 206–207, 232–233
    italicized, 130–131
Sciomyzidae, 227
Scorpionflies, 151, 216–217, 224
Scorpionida, 137
Secondary reproductives, 201
Sejugal furrow, 140
SEL, 264
Sensitive situations, 2
Sheep ked, 226f
Shipping live specimens, 122–123
Shore flies, 227
Sialidae, 218
Sifters, 22–23
Silica gel, 64
Silica gel-dried specimens, 64
Silverfish, 147, 190
Simuliidae, 64, 67–68, 227
Simulium posticatum, 68
Siphonaptera, 148–149, 151, 167, 183, 228–229
Size of a label, 111
Skin beetle, 214f
Skippers, 220, 221f
Sminthuridae, 188–189
Snakeflies, 151, 218
Snap traps, 35–36
Snipe flies, 224
Snow scorpionflies, 217
Social system, 201
Sodium cyanide, 16
Sodium dodecyl sulfate (SDS), 64, 66
Soft forceps, 6f
Soft scales, 211–212
Soil insects, 42
Soldiers, 148, 200–201
Solid killing agents, 16–17
Solpugida, 138

Sonication, 68
Soothsayers, 196
Sounds, 40
Species, 104–106, 126–131, 134–135, 142, 149, 151, 187
Specimens in vials, 120–121
Specimens preservation, 62–65
Spermatophore, 187–189
Spider wasp, 231*f*
Spiders, 23–24, 132, 135
Spittlebugs, 150
Spot plate, 105–106
Spreading blocks, 92–93
Spreading boards, 88–96
Springtails, 23–24, 147, 188–189
Stadium, 143–144
Staining, 99, 105–107
Standard insect pins, 80
Staphylinidae, 66
Sternorrhyncha, 150, 154–155, 158, 163, 183, 211–212
Sticky traps, 35
Sting, 3
Stinging insects, 12
Stoneflies, 150, 203–205
Storage, 73–75, 100, 104–107, 115–116
Storage containers, 116–117
Strepsiptera, 148–149, 151, 156, 181, 183, 215–216
Student, 127
Styli, 173, 187–188, 190
Stylopidae, 216
Subfamilies, 131
Subimago, 144, 190–192
Submittal, 263–270
Submitting specimens, 263–270
Succession, 3
Suchenorrhyncha, 212
Sucking lice, 150, 207–208
Sugaring, 37–38
Swallowtail, 221*f*
Sweeping net, 9, 11
Symbiotic bacteria, 201
Symphyla, 135, 142
Symphyta, 229, 231–232
Syrphid fly, 226*f*
Systematic Entomology Laboratory (SEL), 239, 263–270

T

Tabanidae, 34–35
Tachnid fly, 226*f*
Taxonomic hierarchy, 130
Taxonomic information, 114–115
Temperature, 48–49
Temporary mounting, 100–101

Tenebrionidae, 63
Teneral, 196
Teneral adults, 193
Tephiid wasp, 231*f*
Termitaria, 201
Termites, 148–150, 187–188, 200–202, 206–207, 212–213
Tetranychidae, 240
Thelyphonida, 137
Thread-waisted wasp, 231*f*
Thrips, 150, 208–209
Thysanoptera, 150, 159, 182, 208–209, 208*f*
Thysanura, 145, 147, 173, 188–190, 215–216
Ticks, 20, 40
Tiger beetle, 214*f*
Tooth, 192
Tow trap, 40–42
Tracing paper, 90, 93
Trap, 24–36, 38, 40, 48, 71–72, 112
Traps, preservatives in, 65–66
Treehoppers, 150
Triassomachilidae, 189–190
Trichodectidae, 208
Trichoptera, 148–149, 151–185, 219–220
Triungulinid, 215–216
Trochanter, 188–189, 215–216
True bugs, 150, 209–212
True flies, 151
True lice, 207–208
Twisted-winged parasites, 151, 215–216

U

Ultraviolet light, 32
Uropygi, 137
U.S. Department of Agriculture, 122

V

Vapona, 16, 118
Vapor degreasing, 78
Vehicle-mounted net, 13*f*
Vertical illuminator, 100–101
Vials, 114
Viviparity, 209
Viviparous, 195–196
Voucher specimens, 9, 52
Vouchuring specimens, 69

W

Walking sticks, 149, 193–194, 197–198
Washing, 98–99
Wasps, 151, 183, 212, 216, 229, 232–233
Watchmaker forceps, 6*f*
Water and detergent, 65
Water boatman, 211*f*

Water strider, 211*f*
Waterproofing spray, 110
Web spinners, 150, 202—203
Weevils, 151, 214*f*
Whiteflies, 150, 209—212, 239
Wilkinson trap, 31
Windowpane trap, 25—26
Wing-flexion, 148
Wingless, 147—152
Wings, 134, 144, 147—154, 156, 158—166, 184, 187,
        190—202, 204—209, 212—213, 215—219,
        222—224, 227—229, 232—233
Wings spread, 82, 88, 113—114
Wolbachia, 64
Wood boring beetles, 151
Wrigglers, 183
Written accounts, 3—4

X
Xiphosura, 132
Xylene, 22, 35, 79—80, 83, 95
Xylosandrus compactus, 63

Y
Yellow jacket, 231*f*
Yellow traps, 28, 36

Z
Zoraptera, 148—150, 166, 177, 206—207
Zorapterans, 150, 206—207
Zygoptera, 192

Printed in the United States
by Baker & Taylor

Printed in the United States
By Bookmasters